Advances in Intelligent Systems and Computing

Volume 852

The series "Advances in Intelligent Systems and Computing" contains publications on theory, applications, and design methods of Intelligent Systems and Intelligent Computing. Virtually all disciplines such as engineering, natural sciences, computer and information science, ICT, economics, business, e-commerce, environment, healthcare, life science are covered. The list of topics spans all the areas of modern intelligent systems and computing such as: computational intelligence, soft computing including neural networks, fuzzy systems, evolutionary computing and the fusion of these paradigms, social intelligence, ambient intelligence, computational neuroscience, artificial life, virtual worlds and society, cognitive science and systems, Perception and Vision, DNA and immune based systems, self-organizing and adaptive systems, e-Learning and teaching, human-centered and human-centric computing, recommender systems, intelligent control, robotics and mechatronics including human-machine teaming, knowledge-based paradigms, learning paradigms, machine ethics, intelligent data analysis, knowledge management, intelligent agents, intelligent decision making and support, intelligent network security, trust management, interactive entertainment, Web intelligence and multimedia.

The publications within "Advances in Intelligent Systems and Computing" are primarily proceedings of important conferences, symposia and congresses. They cover significant recent developments in the field, both of a foundational and applicable character. An important characteristic feature of the series is the short publication time and world-wide distribution. This permits a rapid and broad dissemination of research results.

More information about this series at http://www.springer.com/series/11156

Leszek Borzemski · Jerzy Świątek
Zofia Wilimowska
Editors

Information Systems Architecture and Technology: Proceedings of 39th International Conference on Information Systems Architecture and Technology – ISAT 2018

Part I

Editors
Leszek Borzemski
Faculty of Computer Science
and Management
Wrocław University of Science
and Technology
Wrocław, Poland

Zofia Wilimowska
University of Applied Sciences
in Nysa
Nysa, Poland

Jerzy Świątek
Faculty of Computer Science
and Management
Wrocław University of Science
and Technology
Wrocław, Poland

ISSN 2194-5357 ISSN 2194-5365 (electronic)
Advances in Intelligent Systems and Computing
ISBN 978-3-319-99980-7 ISBN 978-3-319-99981-4 (eBook)
https://doi.org/10.1007/978-3-319-99981-4

Library of Congress Control Number: 2018952643

This Springer imprint is published by the registered company Springer Nature Switzerland AG
The registered company address is: Gewerbestrasse 11, 6330 Cham, Switzerland

Preface

Variability of the environment increases the risk of the business activity. Dynamic development of the IT technologies creates the possibility of using them in the dynamic management process modeling and decision-making processes supporting. In today's information-driven economy, companies uncover the most opportunities. Contemporary organizations seem to be knowledge-based organizations, and in connection with that information becomes the most critical resource. Knowledge management is the process through which organizations generate value from their intellectual and knowledge-based assets. It consists of the scope of strategies and practices used in corporations to explore, represent, and distribute knowledge. It is a management *philosophy,* which combines good practice in purposeful information management with a culture of organizational learning, to improve business performance. An improvement of the decision-making process is possible to be assured by the analytical process supporting. Applying some analytical techniques such as computer simulation, expert systems, genetic algorithms can improve the quality of managerial information. Combining analytical techniques and building computer hybrids give synergic effects—additional functionality—which makes managerial decision process better. Different technologies can help in accomplishing the managerial decision process, but no one is in favor of information technologies, which offer differentiable advantages. Information technologies take place a significant role in this area. A computer is a useful machine in making managers' work more comfortable. However, we have to remember that the computer can become a tool only, but it cannot make the decisions. You can not build computers that replace the human mind. Computers can collect, select information, process it and create statistics, but decisions must be made by managers based on their experience and taking into account computer use. Different technologies can help in accomplishing the managerial decision process, but no one like information technologies, which offer differentiable advantages.

Computer science and computer systems, on the one hand, develop in advance of current applications, and on the other hand, keep up with new areas of application. In today's all-encompassing cyber world, nobody knows who motivates. Hence, there is a need to deal with the world of computers from both points of view.

In our conference, we try to maintain a balance between both ways of development. In particular, we are trying to get a new added value that can flow from the connection of the problems of two worlds: the world of computers and the world of management. Hence, there are two paths in the conference, namely computer science and management science.

This three-volume set of books includes the proceedings of the 2018 39th International Conference Information Systems Architecture and Technology (ISAT), or ISAT 2018 for short, held on September 16–18, 2018, in Nysa, Poland. The conference was organized by the Department of Computer Science and Department of Management Systems, Faculty of Computer Science and Management, Wrocław University of Science and Technology, Poland, and University of Applied Sciences in Nysa, Poland.

The International Conference Information Systems Architecture has been organized by the Wrocław University of Science and Technology from the seventies of the last century. The purpose of the ISAT is to discuss a state-of-the-art of information systems concepts and applications as well as architectures and technologies supporting contemporary information systems. The aim is also to consider an impact of knowledge, information, computing, and communication technologies on managing of the organization scope of functionality as well as on enterprise information systems design, implementation, and maintenance processes taking into account various methodological, technological, and technical aspects. It is also devoted to information systems concepts and applications supporting the exchange of goods and services by using different business models and exploiting opportunities offered by Internet-based electronic business and commerce solutions.

ISAT is a forum for specific disciplinary research, as well as on multi-disciplinary studies to present original contributions and to discuss different subjects of today's information systems planning, designing, development, and implementation.

The event is addressed to the scientific community, people involved in a variety of topics related to information, management, computer and communication systems, and people involved in the development of business information systems and business computer applications. ISAT is also devoted as a forum for the presentation of scientific contributions prepared by MSc. and Ph.D. students. Business, Commercial, and Industry participants are welcome.

This year, we received 213 papers from 34 countries. The papers included in the three proceedings volumes have been subject to a thoroughgoing review process by highly qualified peer reviewers. The final acceptance rate was 49%. Program Chairs selected 105 best papers for oral presentation and publication in the 39th International Conference Information Systems Architecture and Technology 2018 proceedings.

The papers have been grouped into three volumes:

Part I—discoursing about essential topics of information technology including, but not limited to, computer systems security, computer network architectures, distributed computer systems, quality of service, cloud computing and high-performance computing, human–computer interface, multimedia systems, big data, knowledge discovery and data mining, software engineering, e-business systems, web design, optimization and performance, Internet of things, mobile systems and applications.

Part II—addressing topics including, but not limited to, model-based project and decision support, pattern recognition and image processing algorithms, production planning and management systems, big data analysis, knowledge discovery and knowledge-based decision support and artificial intelligence methods and algorithms.

Part III—is gain to address very hot topics in the field of today's various computer-based applications—is devoted to information systems concepts and applications supporting the managerial decisions by using different business models and exploiting opportunities offered by IT systems. It is dealing with topics including, but not limited to, knowledge-based management, modeling of financial and investment decisions, modeling of managerial decisions, organization and management, project management, risk management, small business management, software tools for production, theories, and models of innovation.

We would like to thank the program committee and external reviewers, essential for reviewing the papers to ensure a high standard of the ISAT 2018 conference and the proceedings. We thank the authors, presenters, and participants of ISAT 2018, without them the conference could not have taken place. Finally, we thank the organizing team for the efforts this and previous years in bringing the conference to a successful conclusion.

September 2018

Leszek Borzemski
Jerzy Świątek
Zofia Wilimowska

ISAT 2018 Conference Organization

General Chair

Zofia Wilimowska, Poland

Program Co-chairs

Leszek Borzemski, Poland
Jerzy Świątek, Poland
Zofia Wilimowska, Poland

Local Organizing Committee

Zofia Wilimowska (Chair)
Leszek Borzemski (Co-chair)
Jerzy Świątek (Co-chair)
Mariusz Fraś (Conference Secretary, Website Support)
Arkadiusz Górski (Technical Editor)
Anna Kamińska (Technical Secretary)
Ziemowit Nowak (Technical Support)
Kamil Nowak (Website Coordinator)
Danuta Seretna-Sałamaj (Technical Secretary)

International Program Committee

Zofia Wilimowska (Chair), Poland
Jerzy Świątek (Co-chair), Poland
Leszek Borzemski (Co-chair), Poland
Witold Abramowicz, Poland
Dhiya Al-Jumeily, UK
Iosif Androulidakis, Greece

Patricia Anthony, New Zealand
Zbigniew Banaszak, Poland
Elena N. Benderskaya, Russia
Janos Botzheim, Japan
Djallel E. Boubiche, Algeria
Patrice Boursier, France
Anna Burduk, Poland
Andrii Buriachenko, Ukraine
Udo Buscher, Germany
Wojciech Cellary, Poland
Haruna Chiroma, Malaysia
Edward Chlebus, Poland
Gloria Cerasela Crisan, Romania
Marilia Curado, Portugal
Czesław Daniłowicz, Poland
Zhaohong Deng, China
Małgorzata Dolińska, Poland
Ewa Dudek-Dyduch, Poland
Milan Edl, Czech Republic
El-Sayed M. El-Alfy, Saudi Arabia
Peter Frankovsky, Slovakia
Mariusz Fraś, Poland
Naoki Fukuta, Japan
Bogdan Gabryś, UK
Piotr Gawkowski, Poland
Arkadiusz Górski, Poland
Manuel Graña, Spain
Wiesław M. Grudewski, Poland
Katsuhiro Honda, Japan
Marian Hopej, Poland
Zbigniew Huzar, Poland
Natthakan Iam-On, Thailand
Biju Issac, UK
Arun Iyengar, USA
Jürgen Jasperneite, Germany
Janusz Kacprzyk, Poland
Henryk Kaproń, Poland
Yury Y. Korolev, Belarus
Yannis L. Karnavas, Greece
Ryszard Knosala, Poland
Zdzisław Kowalczuk, Poland
Lumír Kulhanek, Czech Republic
Binod Kumar, India
Jan Kwiatkowski, Poland
Antonio Latorre, Spain
Radim Lenort, Czech Republic
Gang Li, Australia
José M. Merigó Lindahl, Chile
Jose M. Luna, Spain
Emilio Luque, Spain
Sofian Maabout, France
Lech Madeyski, Poland
Zbigniew Malara, Poland
Zygmunt Mazur, Poland
Elżbieta Mączyńska, Poland
Pedro Medeiros, Portugal
Toshiro Minami, Japan
Marian Molasy, Poland
Zbigniew Nahorski, Poland
Kazumi Nakamatsu, Japan
Peter Nielsen, Denmark
Tadashi Nomoto, Japan
Cezary Orłowski, Poland
Sandeep Pachpande, India
Michele Pagano, Italy
George A. Papakostas, Greece
Zdzisław Papir, Poland
Marek Pawlak, Poland
Jan Platoš, Czech Republic
Tomasz Popławski, Poland
Edward Radosinski, Poland
Wolfgang Renz, Germany
Dolores I. Rexachs, Spain
José S. Reyes, Spain
Małgorzata Rutkowska, Poland
Leszek Rutkowski, Poland
Abdel-Badeeh M. Salem, Egypt
Sebastian Saniuk, Poland
Joanna Santiago, Portugal
Habib Shah, Malaysia
J. N. Shah, India
Jeng Shyang, Taiwan
Anna Sikora, Spain
Marcin Sikorski, Poland
Małgorzata Sterna, Poland
Janusz Stokłosa, Poland
Remo Suppi, Spain
Edward Szczerbicki, Australia

Eugeniusz Toczyłowski, Poland
Elpida Tzafestas, Greece
José R. Villar, Spain
Bay Vo, Vietnam
Hongzhi Wang, China
Leon S. I. Wang, Taiwan
Junzo Watada, Japan
Eduardo A. Durazo Watanabe, India
Jan Werewka, Poland
Thomas Wielicki, USA
Bernd Wolfinger, Germany
Józef Woźniak, Poland
Roman Wyrzykowski, Poland
Yue Xiao-Guang, Hong Kong
Jaroslav Zendulka, Czech Republic
Bernard Ženko, Slovenia

ISAT 2018 Reviewers

Hamid Al-Asadi, Iraq
Patricia Anthony, New Zealand
S. Balakrishnan, India
Zbigniew Antoni Banaszak, Poland
Piotr Bernat, Poland
Agnieszka Bieńkowska, Poland
Krzysztof Billewicz, Poland
Grzegorz Bocewicz, Poland
Leszek Borzemski, Poland
Janos Botzheim, Hungary
Piotr Bródka, Poland
Krzysztof Brzostkowski, Poland
Anna Burduk, Poland
Udo Buscher, Germany
Wojciech Cellary, Poland
Haruna Chiroma, Malaysia
Witold Chmielarz, Poland
Grzegorz Chodak, Poland
Andrzej Chuchmała, Poland
Piotr Chwastyk, Poland
Anela Čolak, Bosnia and Herzegovina
Gloria Cerasela Crisan, Romania
Anna Czarnecka, Poland
Mariusz Czekała, Poland
Y. Daradkeh, Saudi Arabia
Grzegorz Debita, Poland
Anna Dobrowolska, Poland
Maciej Drwal, Poland
Ewa Dudek-Dyduch, Poland
Jarosław Drapała, Poland
Tadeusz Dudycz, Poland
Grzegorz Filcek, Poland
Mariusz Fraś, Poland
Naoki Fukuta, Japan
Piotr Gawkowski, Poland
Dariusz Gąsior, Poland
Arkadiusz Górski, Poland
Jerzy Grobelny, Poland
Krzysztof Grochla, Poland
Bogumila Hnatkowska, Poland
Katsuhiro Honda, Japan
Zbigniew Huzar, Poland
Biju Issac, UK
Jerzy Józefczyk, Poland
Ireneusz Jóźwiak, Poland
Krzysztof Juszczyszyn, Poland
Tetiana Viktorivna Kalashnikova, Ukraine
Anna Kamińska-Chuchmała, Poland
Yannis Karnavas, Greece
Adam Kasperski, Poland
Jerzy Klamka, Poland
Agata Klaus-Rosińska, Poland
Piotr Kosiuczenko, Poland
Zdzisław Kowalczyk, Poland
Grzegorz Kołaczek, Poland
Mariusz Kołosowski, Poland
Kamil Krot, Poland
Dorota Kuchta, Poland
Binod Kumar, India
Jan Kwiatkowski, Poland
Antonio LaTorre, Spain
Arkadiusz Liber, Poland
Marek Lubicz, Poland

Emilio Luque, Spain
Sofian Maabout, France
Lech Madeyski, Poland
Jan Magott, Poland
Zbigniew Malara, Poland
Pedro Medeiros, Portugal
Vojtěch Merunka, Czech Republic
Rafał Michalski, Poland
Bożena Mielczarek, Poland
Vishnu N. Mishra, India
Jolanta Mizera-Pietraszko, Poland
Zbigniew Nahorski, Poland
Binh P. Nguyen, Singapore
Peter Nielsen, Denmark
Cezary Orłowski, Poland
Donat Orski, Poland
Michele Pagano, Italy
Zdzisław Papir, Poland
B. D. Parameshachari, India
Agnieszka Parkitna, Poland
Marek Pawlak, Poland
Jan Platoš, Czech Republic
Dolores Rexachs, Spain
Paweł Rola, Poland
Stefano Rovetta, Italy
Jacek, Piotr Rudnicki, Poland
Małgorzata Rutkowska, Poland
Joanna Santiago, Portugal
José Santos, Spain
Danuta Seretna-Sałamaj, Poland
Anna Sikora, Spain
Marcin Sikorski, Poland
Małgorzata Sterna, Poland
Janusz Stokłosa, Poland
Grażyna Suchacka, Poland
Remo Suppi, Spain
Edward Szczerbicki, Australia
Joanna Szczepańska, Poland
Jerzy Świątek, Poland
Paweł Świątek, Poland
Sebastian Tomczak, Poland
Wojciech Turek, Poland
Elpida Tzafestas, Greece
Kamila Urbańska, Poland
José R. Villar, Spain
Bay Vo, Vietnam
Hongzhi Wang, China
Shyue-Liang Wang, Taiwan, China
Krzysztof Waśko, Poland
Jan Werewka, Poland
Łukasz Wiechetek, Poland
Zofia Wilimowska, Poland
Marek Wilimowski, Poland
Bernd Wolfinger, Germany
Józef Woźniak, Poland
Maciej Artur Zaręba, Poland
Krzysztof Zatwarnicki, Poland
Jaroslav Zendulka, Czech Republic
Bernard Ženko, Slovenia
Andrzej Żołnierek, Poland

ISAT 2018 Keynote Speaker

Professor Dr. Abdel-Badeh Mohamed Salem, Faculty of Science, Ain Shams University, Cairo, Egypt
Topic: Artificial Intelligence Technology in Intelligent Health Informatics

Contents

Keynote Speech

Artificial Intelligence Technology in Intelligent Health Informatics

Abdel-Badeeh M. Salem(✉)

Head of Artificial Intelligence and Knowledge Engineering Research Labs
Faculty of Computer and Information Sciences, Ain Shams University,
Abbassia Cairo, Egypt
absalem@cis.asu.edu.eg, abmsalem@yahoo.com
http://staff.asu.edu.eg/Badeeh-Salem
http://aiasulab.000webhostapp.com/

Abstract. Artificial Intelligence (AI) is devoted to creating computer software and hardware that imitates the human mind. The primary goal of AI technology is to make computers smarter by creating software that will allow a computer to mimic some of the functions of the human brain in selected applications. Applications of AI technology include; general problem solving, expert systems, natural language processing, computer vision, robotics, and education. All of these applications employ knowledge base and inferencing techniques to solve problems or help make decisions in specific domains.

In the last years various AI paradigms and computational intelligence (CI) techniques have been proposed by the researchers in order to develop efficient and smart systems in the areas of health informatics and health monitoring systems.AI and CI offers robust, intelligent algorithms and smart methods that can help to solve problems in a variety of health and life sciences areas.

This talk discusses the potential role of the AI and CI approaches, techniques, and theories which are used in developing the intelligent health informatics and health monitoring systems. The talk presents the following techniques: (a) case-based reasoning; (b) intelligent data mining; (c) rough sets; (d) genetic algorithms; and (e) ontological engineering. Strengths and weaknesses of these approaches were enunciated. Also, the talk presents the challenges as well as the current research directions in the areas of health informatics. Examples of the research performed by the author and his associates for developing knowledge-based systems for cancer, heart, brain tumor, and thrombosis diseases are discussed.

L. Borzemski et al. (Eds.): ISAT 2018, AISC 852, p. 3, 2019.
https://doi.org/10.1007/978-3-319-99981-4_1

Computer Systems Security

Concurrent, Coherent Design of Hardware and Software Embedded Systems with Higher Degree of Reliability and Fault Tolerant

Mieczysław Drabowski(✉)

Faculty of Electrical and Computer Engineering, Cracow University of Technology, 24 Warszawska Street, 31-155 Kraków, Poland
drabowski@pk.edu.pl

Abstract. The paper includes a proposal of a new model of synthesis multi-processors systems with higher degree of reliability. Optimal task scheduling and optimal partition at resources are basic problems in high-level synthesis of computer systems. Concept of reliability is a system idea that integrates hardware and software. Reliability is a feature of the system, which reflects the degree of user's dependency to the system and is reflected in the continuity of actions of the equipment and installed programs. Achieving a higher degree of reliability of the system is implemented during the system operation, which limits damage caused by failures. This implementation is manifested by introducing internal control into the system, diagnostic of damaged components and using redundant modules, which allows tolerance of damage, achievement of the ability to soft fall and survival of the system. Concurrent, coherent and computer aided synthesis may have a practical application in developing tools for rapid prototyping of such systems.

Keywords: Synthesis · Reliability · Fault tolerance · Embedded systems

1 Introduction

The goal of high-level synthesis of computer systems is to find an optimum solution satisfying the requirements and constraints enforced by the given specification of the system. The following criteria of optimality are usually considered: costs of system implementation, its operating speed [1] and, in this paper presented, reliability, availability and fault tolerant.

A specification describing a computer system may be provided as a set of interactive tasks. In any computer system certain tasks are implemented by hardware. The basic problem of system synthesis is partitioning system functions due to their hardware and software implementation. The goal of the resources assignment is to specify what hardware and software resources are needed for the implementation and to assign them to specific tasks of the system, even before designing execution details. In the synthesis methods used so far, software and hardware parts are developed separately and then

L. Borzemski et al. (Eds.): ISAT 2018, AISC 852, pp. 7–18, 2019.
https://doi.org/10.1007/978-3-319-99981-4_2

composed, what results in cost increasing and decreasing quality and reliability of the final product.

Task scheduling is one of the most important issues occurring in [2] the synthesis of operating systems responsible for controlling allocation of tasks and resources in computer systems.

Another important issue that occurs in designing computer systems is assuring their fault-free operation. Such synthesis concentrates on developing fault-tolerant architectures and constructing dedicated operating systems for them. In this system an appropriate strategy of self-testing during regular exploitation must be provided. In general, fault tolerant architectures of computer systems are multiprocessor ones. The objective of operating systems in multiprocessor systems is scheduling tasks and their allocation to system resources. For fault tolerant operating system, this means scheduling usable and also testing tasks, that should detect errors of executive modules, in particular processors.

Modeling fault tolerant systems consists of resource identification and task scheduling. These problems (exactly, their decision versions) are NP-complete [3, 8]. Algorithms for solving such problems are usually based on heuristic approaches [4]. The objective of this paper is to present the concept of combined approach to the problem of fault tolerant system synthesis, i.e. a coherent solution to task scheduling and resource assignment problems. The solution includes also the system self-testing strategies.

2 Model for the Problem of Designing Reliable Computer System

Synthesis of computer systems is a multi-criteria optimization problem. The starting point for constructing our approach to the issues of hardware and software synthesis is the deterministic theory of task scheduling [1, 5]. This theory may serve as a methodological basis for fault tolerant multiprocessor systems synthesis. Accordingly, decomposition of the general task scheduling model is suggested, adequate to the problems of fault tolerant system synthesis.

From the control point of view such a model should take into account the tasks, which may be either preemptable or nonpreemptable. These characteristics are defined according to the scheduling theory. Tasks are preemptable when each task can be interrupted and restarted later without incurring additional costs. In such a case the schedules are called to be preemptive. Otherwise, tasks are nonpreemptable and schedules nonpreemptive. Preemptability of tasks in our approach cannot be a feature of the searched schedule – as in the task scheduling model so far. The schedule contains all assigned tasks with individual attributes: preemptive, nonpreemptive. From the point of view of the system synthesis, the implementation of certain tasks from the given set must be nonpreemptible, for the other may be preemptible (what, in turn, influences significantly selection of an appropriate scheduling algorithm) [6]. The above approach allows for inclusion into the discussed set of tasks also the system functions that are operating in real time. They must be realized by the hardware elements, often specialized ones. In such cases, the model is relevant to the synthesis of multiprocessors computer system. The set of functions not only models software units realized by a versatile hardware (processors and memories), but may represent all functions of the system, including those which must be realized by specialized components (e.g. hardware testing tasks). Moreover, we wish to specify the model of task scheduling in a way suitable for finding optimum control methods (in terms of certain criteria) as well as optimum assignment of tasks to universal and specialized hardware components.

Accordingly, we shall discuss the system [7]:

$$\sum = \{\mathbf{R}, \mathbf{T}, \mathbf{C}\} \tag{1}$$

where: R – is the set of resources (hardware and software), T – is the set of the tasks (operations), C – is the set of optimization criteria.

Resources. We assume that processor set (parallel and dedicated) $P = \{P_1, P_2, \ldots, P_m\}$ consists of *m* elements and the set additional resources $A = \{A_1, A_2, \ldots, A_p\}$ consist of *p* elements.

Tasks. We consider a set of *n* tasks to be processed with a set of resources. The set of tasks is divided into 2 subsets: $T^1 = \{T_1^1, T_2^1, \ldots, T_{n1}^1\}, \{T_1^2, T_2^2, \ldots, T_{n2}^2\}$, where $n = n1 + n2$. Each task T_i^1 $(i = 1, 2, \ldots, n1)$ requires one arbitrary processor for its processing and its processing time is equal to t_i^1. Similarly, each task T_i^2 $(i = 1, 2, \ldots, n2)$, requires 2 arbitrary processors simultaneously for its processing during a period of time whose length is equal to t_i^2. We will call this task as a two-processor task [2, 8]. A schedule is called feasible if, besides the usual conditions, each task T_i^1 is processed by one processor and each task T_i^2 is processed by 2 processors at a time. A feasible schedule is optimal, if it is to minimized length. Each task is defined by a set of parameters:

- Resource requirements. The task may additionally require *j* units of resource A_i.
- Execution time.
- Ready time and Deadline.
- Attribute - preemptable or nonpreemptable.

The tasks set may contain defined precedence constraints represented by a digraph with nodes representing tasks, and directed edges representing precedence constraints. If there is at least one precedence constraint in a task set, we shall refer it to as a set of dependent tasks; otherwise they are a set of independent tasks.

Let us assume that fault tolerant system resources include universal parallel processors and specialized processors. As for tasks, we assume one-processor tasks used for modeling usable preemptable/nonpreemptable and dependent tasks, and two-processor tasks, for which we assume time and resource non-preemptable. Two-processor tasks model the system testing tasks (e.g. one processor checks the other). Testing tasks may be dependent on the defined time moments of readiness to perform and to complete assigned tasks. Two-processor tasks (multiprocessor) may realize a defined strategy of testing a fault tolerant system [9].

Optimality criteria. As for the optimality criteria for the system being designed, we shall assume its minimum cost and maximum operating speed. In addition, we will consider the degree of system reliability and its fault tolerant.

3 Concurrent Process of Reliable Computer System Synthesis

Modeling the joint search for the optimum task schedule and resource partition of the designed system into hardware and software parts is fully justified. Simultaneous consideration of these problems may be useful in implementing optimum solutions, e.g. the cheapest hardware structures or the fastest schedule. With such approach, the optimum task distribution is possible on the universal and specialized hardware and defining resources with maximum efficiency. We propose the following schematic diagram of a coherent process of fault tolerant systems synthesis which is shown in Fig. 1.

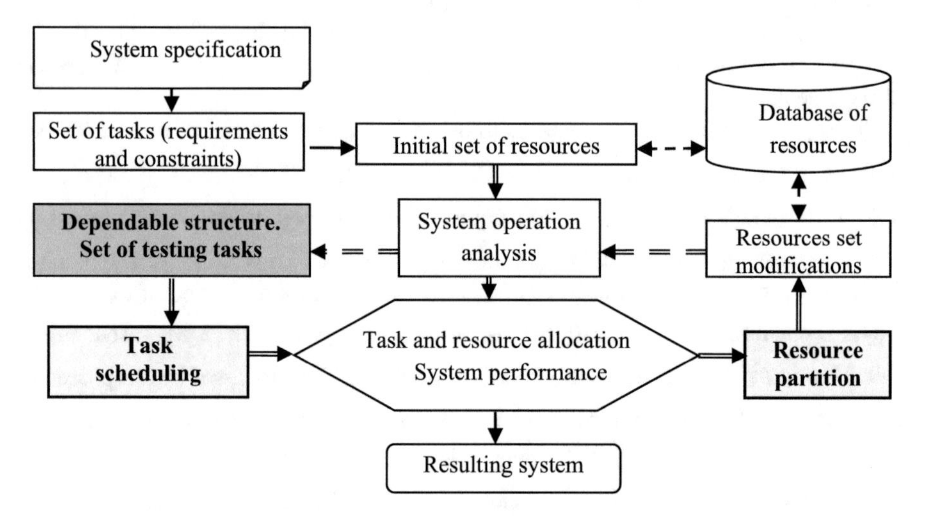

Fig. 1. The process coherent synthesis of dependable computer system

The suggested coherent analysis consists of the following steps:

1. specification of requirements for the system,
2. specification of tasks,
3. assuming the initial values of resource set,
4. defining testing tasks and the structure of system and his redundancy, testing strategy selection,
5. task scheduling,
6. evaluating the operating speed and system cost, multi-criteria optimization,
7. the evaluation should be followed by a modification of the resource set, a new system partitioning into hardware and software parts and an update of test tasks and new test structure (step 4).

In this approach a combined search for optimal resources partition and optimal tasks scheduling occur. Iterative calculations are executed till satisfactory design results are obtained – i.e. optimal system structure, level of self-testing and fault tolerant and optimal fast schedule.

4 Example of Higher Degree of Reliability System Synthesis

We will consider the synthesis of the computer specialized system which is testing computer board type packages. A system will be optimized as regards cost (cost minimization) and execution speed (schedule length minimization). We will consider also optimization self-testing of system. Tested system will be implemented in the common structure – first without fault tolerant - and in the fault-tolerant structure.

4.1 System Specification

Testing system should realize tasks, which can be presented by a digraph (Fig. 2). Testing of the package should be realized in the following order: T_0 test of power supply circuits, T_1 test of processor, T_2 test of buses, T_3 test of interrupt controller, T_4 test of graphics controller, T_5 test of memory, T_6 test of disc controller, T_7 test of network controller, T_8 test of audio-video controller, T_9 test of program executing. Precedence constraints of the functions, which are to be realized by the system, ensue from package test procedures.

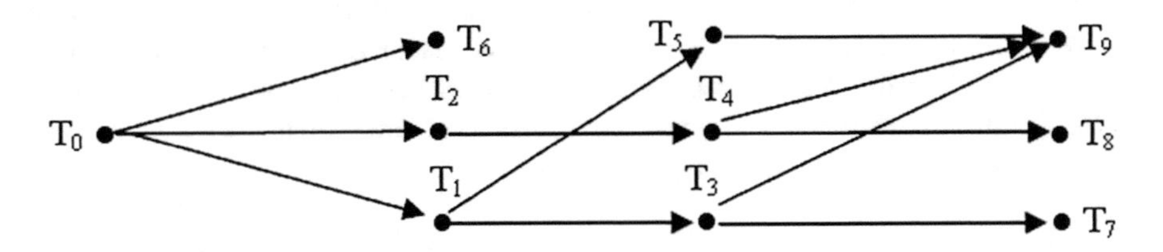

Fig. 2. The digraph of tasks showing test executing sequence

Test of power supply circuits is the root of the digraph, it is hard core [10] of board for his testing. Next in order, there are tests of processor and test of system buses. Interrupt controller can be tested after checking the processor. Test of network controller requires that interrupt controller is in working order and therefore T_7 task follows T_3 task. Next package test procedure realizes memory test (T_5) with processor (T_1) in working order (because previously checked), and tests of controllers of disc and graphics. Graphics controller checked and in working order makes it possible to realize the following test of audio-video controller. In the end of the testing procedure, test of checking cooperation between processor and memory is being executed (T_9). We will assume executing times of tasks the following: $t_0 = 3$, $t_1 = 2$, $t_2 = 2$, $t_3 = 1$, $t_4 = 3$, $t_5 = 3$, $t_6 = 4$, $t_7 = 3$, $t_8 = 1$, $t_9 = 1$. As an optimal criterion for designed system we will assume minimization of executing time for all testing tasks – minimization of testing tasks schedule length. The second considered optimal criterion is system cost which takes into consideration cost of all hardware resources. Cost of system is:

$$C_S = C_P + i * C_M \tag{2}$$

where: C_P – processor cost, C_M – memory cost, i – number of memory modules.

Additional requirements of designing system, we will assume as the following:

RI There is necessary a deadline of all tasks without delays executing which equals 13 time units.

RII There is necessary execute task T_6 (because disc controller usually operating in real-time) as a nonpreemptable task.

RIII There is necessary a critical line for T_6 task which equals 9 time units.

RIV There is desirable a deadline, executing all processes without delay, equal 9 time units.

We will consider synthesis as parallel multiprocessors embedded system.

4.2 Project of Structure and Schedule Without Providing Fault-Tolerant

If we consider two identical and parallel processors, then this optimal schedule fulfills requirement RI that can be implemented by Muntz-Coffman algorithm [7] (Fig. 3). Cost of this system – with assumption that every processor needs one memory unit, equals $C_S = 2 * C_P + 10 * C_M$. Taking into consideration requirement RII it is necessary to correct tasks schedule (Fig. 3). System cost doesn't change. For requirement RIII realization there is necessary further correction of tasks schedule (Fig. 3). System cost doesn't change, too. At last it turns out; that all requirements (include RIV) can't be realized on two processors. We will apply specialized resource, which can execute tasks: T_0, T_4, T_5, T_6, T_7 with a triple speed (as compared to the standard processor) and its cost is C_{ASIC}. Resources structure and processor schedule we are showing in Fig. 3. System cost equals $C_S = C_P + 6 * C_M + C_{ASIC}$.

<table>
<tr><th>Requirments</th><th>time</th><th>1</th><th>2</th><th>3</th><th>4</th><th>5</th><th>6</th><th>7</th><th>8</th><th>9</th><th>10</th><th>11</th><th>12</th><th>13</th></tr>
<tr><td rowspan="2">RI</td><td>P1</td><td colspan="3">T0</td><td colspan="2">T1</td><td colspan="3">T4</td><td colspan="2">T6</td><td>T7</td><td>T6</td><td>T7</td></tr>
<tr><td>P2</td><td colspan="3"></td><td colspan="2">T2</td><td>T3</td><td>T5</td><td>T6</td><td>T7</td><td colspan="2">T5</td><td>T8</td><td>T9</td></tr>
<tr><td rowspan="2">RII</td><td>P1</td><td colspan="3">T0</td><td colspan="2">T1</td><td colspan="2">T4</td><td colspan="4">T6</td><td colspan="2">T7</td></tr>
<tr><td>P2</td><td colspan="3"></td><td colspan="2">T2</td><td>T3</td><td>T5</td><td>T4</td><td colspan="2">T5</td><td>T7</td><td>T8</td><td>T9</td></tr>
<tr><td rowspan="2">RIII</td><td>P1</td><td colspan="3">T0</td><td colspan="2">T1</td><td colspan="4">T6</td><td>T3</td><td>T5</td><td colspan="2">T7</td></tr>
<tr><td>P2</td><td colspan="3"></td><td colspan="2">T2</td><td>T5</td><td colspan="3">T4</td><td>T5</td><td>T7</td><td>T8</td><td>T9</td></tr>
<tr><td rowspan="2">RIV</td><td>P1</td><td></td><td colspan="2">T2</td><td>T6</td><td colspan="2">T1</td><td>T3</td><td>T8</td><td>T9</td><td colspan="4"></td></tr>
<tr><td>ASIC</td><td>T0</td><td colspan="2"></td><td>T4</td><td>T6</td><td></td><td>T5</td><td>T7</td><td></td><td colspan="4">← idle time</td></tr>
</table>

Fig. 3. Realized requirements RI, RII, RIII, RIV

4.3 Project of Structure and Schedule Providing Fault-Tolerant

The cost of the system should be as low as possible, and the architecture conformant with the fault tolerant system model is required, with two-processor testing tasks. We shall assume the following labeling for processor testing tasks - T_{gh}, where P_g processor is testing (checking) P_h processor.

Implementing the system satisfying the requirement RI, the architecture of a fault tolerant system [11] was shown in Fig. 4 – Structure 1. For such architecture, the optimum tasks schedule, guaranteeing the requirement RI, has been shown in Fig. 5.

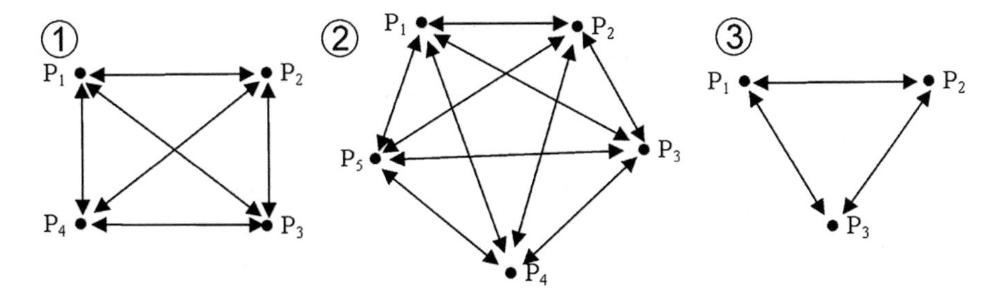

Fig. 4. 1 - Single-processors and two-processor testing tasks in a fault tolerant four-processor structure: T_{12}, T_{13}, T_{14}, T_{23}, T_{24}, T_{21}, T_{34}, T_{31}, T_{32}, T_{41}, T_{42}, T_{43} - Structure 1. 2 - Processors and two-processor testing tasks in a fault tolerant five-processor structure - T_{12}, T_{13}, T_{14}, T_{15}, T_{21}, T_{23}, T_{24}, T_{25}, T_{31}, T_{32}, T_{34}, T_{35}, T_{41}, T_{42}, T_{43}, T_{45}, T_{51}, T_{52}, T_{53}, T_{54} – Structure 2. 3 - Processors and two-processor tasks in a fault tolerant three-processor system with a specialized ASIC processor - T_{12}, T_{13}, T_{23}, T_{21}, T_{31}, T_{32} – Structure 3.

Req.	*time*	*1*	*2*	*3*	*4*	*5*	*6*	*7*	*8*	*9*	*10*	*11*	*12*	*13*
RI	**P1**	T_{12}	T_0	T_0	T_{13}	T_1	T_{31}	T_{14}	T_{21}	T_7	T_{41}	T_7	T_6	T_{12}
	P2	T_{12}	T_{23}		T_1	T_{24}	T_4	T_4	T_{21}	T_{32}	T_6	T_{42}	T_8	T_{12}
	P3	T_0	T_{23}	T_{34}	T_{13}	T_2	T_{31}	T_5	T_4	T_{32}	T_5	T_5	T_{43}	T_7
	P4			T_{34}	T_2	T_{24}	T_3	T_{14}	T_6	T_6	T_{41}	T_{42}	T_{43}	T_9
RII	**P1**	T_{12}	T_0	T_0	T_{13}	T_1	T_{31}	T_{14}	T_{21}	T_7	T_{41}	T_6	T_7	T_{12}
	P2	T_{12}	T_{23}		T_1	T_{24}	T_4	T_4	T_{21}	T_{32}	T_6	T_{42}	T_8	T_{12}
	P3	T_0	T_{23}	T_{34}	T_{13}	T_2	T_{31}	T_5	T_4	T_{32}	T_5	T_5	T_{43}	T_7
	P4			T_{34}	T_2	T_{24}	T_3	T_{14}	T_6	T_6	T_{41}	T_{42}	T_{43}	T_9
RIII	**P1**	T_{12}	T_0	T_0	T_{13}	T_1	T_{31}	T_{14}	T_{21}	T_7	T_{41}	T_4	T_7	T_{12}
	P2	T_{12}	T_{23}		T_1	T_{24}	T_6	T_6	T_{21}	T_{32}	T_4	T_{42}	T_8	T_{12}
	P3	T_0	T_{23}	T_{34}	T_{13}	T_2	T_{31}	T_5	T_4	T_{32}	T_5	T_5	T_{43}	T_7
	P4			T_{34}	T_2	T_{24}	T_3	T_{14}	T_6	T_6	T_{41}	T_{42}	T_{43}	T_9

Fig. 5. Tasks schedule satisfying the requirement RI in a four-processor system (Req. RI), the requirements RI and RII in a four-processor system (Req. RII), the requirements RI, RII, and RIII in a four-processor system (Req. RIII) – Structure 1.

Taking into account the requirement RII, the following correction is done to the task schedule. Thus, we obtain the schedule shown in requirement RII (Fig. 5). The system architecture and costs remain unchanged.

The next requirement RIII is reflected in a corrected schedule presented in requirement RIII (Fig. 5).

Considering the RIV requirement, the system structure change is necessary. Two variants of the structure shall be proposed. The first structure consists of five identical parallel processors, with two-processor testing tasks Fig. 4 (Structure 2). Task schedule in such structure is depicted in Fig. 6. In the second variant, a specialized module (ASIC) was applied, that may perform the tasks: T_0, T_4, T_5, T_6 and T_7 with a triple speed (as compared with the standard universal processor). The system structure and schedule are shown in Figs. 4 (Structure 3) and 6, respectively. The universal processor completes processing of usable tasks in 9 time units, while ASIC processor completes performing its function in 8 time units. Accordingly, the required deadline was reached in 9 units. Set of utility tasks (Fig. 2) with precedence constraints and with execution time: t0 = 3, t1 = 2, t2 = 2, t3 = 1, t4 = 3, t5 = 3, t6 = 4, t7 = 3, t8 = 1, t9 = 1. All tasks are preemptive.

St.	*time*	*1*	*2*	*3*	*4*	*5*	*6*	*7*	*8*	*9*	*10*	*11*	*12*	*13*
2	**P1**	T_{12}				T_{13}	T_3	T_4	T_{14}	T_9	T_{15}	T_{21}		
	P2	T_{12}	T_{23}		T_2	T_2	T_{24}	T_5	T_5	T_{25}	T_8	T_{21}	T_{32}	
	P3		T_{23}	T_{34}	T_6	T_{13}	T_5	T_{35}	T_4	T_4	T_9		T_{32}	T_{43}
	P4	T_0	T_0	T_{34}	T_{45}	T_1	T_{24}	T_6	T_{14}	T_7	T_7			T_{43}
	P5			T_0	T_{45}	T_6	T_6	T_{35}	T_7	T_{25}	T_{15}			
3	**P1**	T_{12}	T_2	T_{13}	T_{21}	T_1	T_{31}	T_{12}	T_8	T_{13}				
	P2	T_{12}	T_{23}	T_2	T_{21}	T_{32}	T_1	T_{12}	T_{23}	T_{13}				
	P3		T_{23}	T_{13}	T_6	T_{32}	T_{31}	T_3	T_{23}	T_9				
	ASIC	T_0			T_4	T_6		T_5	T_7					

Fig. 6. Tasks schedule satisfying the requirements RI, RII, RIII and RIV: in the five-processor system – Structure 2 and in the three-processor system with the specialized processor – Structure 3.

The cost of the developed system shall be estimated as follows. If we assume that each usable task performed by a universal processor needs one memory unit dedicated to such task, and task assigned to ASIC processor do not need dedicated memory units the system cost is:

$$C_S = m * C_P + n_u * C_M + p * C_{ASIC} \tag{3}$$

where: m – the number of identical parallel processors, n_u – the number of tasks assigned to universal processors, p – the number of specialized ASIC processors devoted for processing remaining $(n - n_u)$ tasks.

For the requirements: A, A + B, A + B + C: m = 4, $n_u = 10$, p = 0. For the requirements A + B + C + D, in the first variant, m = 5, $n_u = 10$, p = 0. In the second variant, where one ASIC processor is applied, m = 3, $n_u = 6$ and p = 1.

4.4 Project of Self-testing a Behavior Fault Tolerant Multiprocessor Embedded System and with Attribute of Soft Fall

In the example occurs:

- the redundancy on the level of processors,
- is used the strategy MRM (Multiple Modular Redundancy) [9] of monitoring of behavior all processors,
- dynamic reallocation of processors through the migration of processes from damaged processors,
- the isolation damaged processors,
- soft fall and survival.

The system consists of parallel and identical five processors connected in a self-testing structure – Fig. 4 - Structure 2 with tasks testing processors in the system.

The Gantt chart of the self-testing fault-tolerant system, which is potentially ready to perform utility tasks, shown Fig. 7.

time	*1*	2	3	*4*	5	6	*7*	*8*	*9*	*10*	*11*	*12*	*13*	14
P1	T_{12}				T_{13}			T_{14}		T_{15}	T_{21}			
P2	T_{12}	T_{23}				T_{24}			T_{25}		T_{21}	T_{32}		
P3		T_{23}	T_{34}		T_{13}		T_{35}					T_{32}	T_{43}	
P4			T_{34}	T_{45}		T_{24}		T_{14}					T_{43}	T_{54}
P5				T_{45}			T_{35}		T_{25}	T_{15}				T_{54}

CPU work time, which available for utility tasks

Fig. 7. Self-testing five-processors system with two-processors task (CPU work time, which available for utility tasks)

The allocation of the utility task T0 to the system is shown in Fig. 8 and the allocation of utility tasks T0, T1, T2, T6 to the quantum of time t = 4 is shown in Fig. 9.

time	*1*	2	3	*4*	5	6	*7*	*8*	*9*	*10*	*11*	*12*	*13*	14
P1	T_{12}				T_{13}			T_{14}		T_{15}	T_{21}			
P2	T_{12}	T_{23}				T_{24}			T_{25}		T_{21}	T_{32}		
P3	T_0	T_{23}	T_{34}		T_{13}		T_{35}					T_{32}	T_{43}	
P4		T_0	T_{34}	T_{45}		T_{24}		T_{14}					T_{43}	T_{54}
P5			T_0	T_{45}			T_{35}		T_{25}	T_{15}				T_{54}

Fig. 8. The allocation of task T0 to the self-testing system

The allocation of other utility tasks is shown in Fig. 10. The system will perform utility tasks in 10 quanta of time. The system performed all utility tasks and at the same time carried out tasks to check the operation of the system. If the processor fails, it is possible to automatically reconfigure the system and continue its operation. For example, suppose that at time t = 6, the T_{24} test detected an abnormality; it is shown in Fig. 11. The system from now on does not perform any of the utility tasks by P2 and P4 processors: (1) At time t = 7, the T_{12} test is performed, the P4 processor is "suspended", (2) At time t = 8, the T_{14} test is performed; the P2 processor is "suspended".

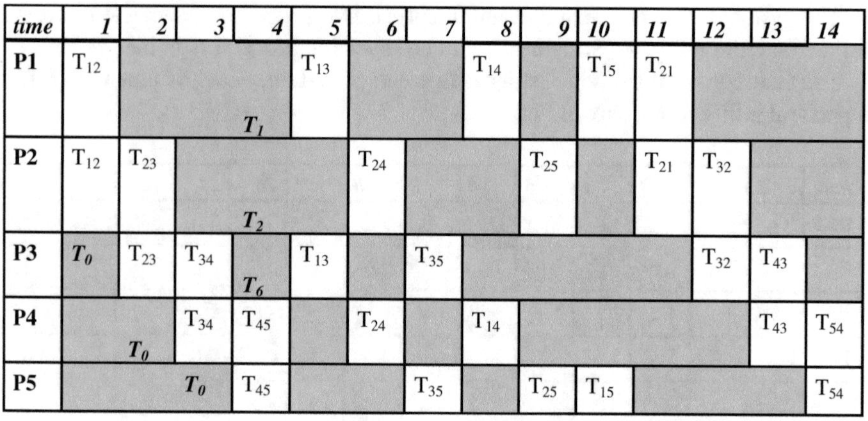

time	1	2	3	4	5	6	7	8	9	10	11	12	13	14
P1	T_{12}			T_1	T_{13}			T_{14}		T_{15}	T_{21}			
P2	T_{12}	T_{23}		T_2		T_{24}			T_{25}		T_{21}	T_{32}		
P3	T_0	T_{23}	T_{34}	T_6	T_{13}		T_{35}					T_{32}	T_{43}	
P4		T_0	T_{34}	T_{45}		T_{24}		T_{14}					T_{43}	T_{54}
P5			T_0	T_{45}			T_{35}		T_{25}	T_{15}				T_{54}

Fig. 9. The allocation of utility tasks T0, T1, T2, T6.

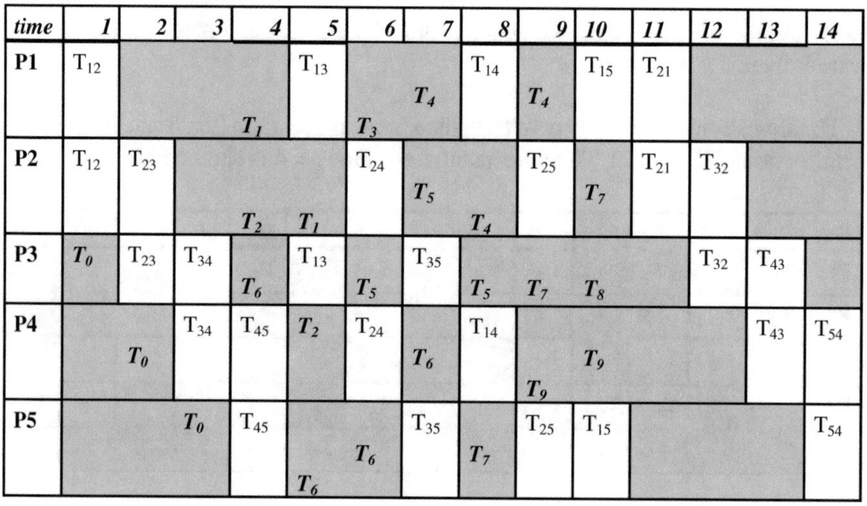

time	1	2	3	4	5	6	7	8	9	10	11	12	13	14
P1	T_{12}			T_1	T_{13}	T_3	T_4	T_{14}	T_4	T_{15}	T_{21}			
P2	T_{12}	T_{23}		T_2	T_1	T_{24}	T_5	T_4	T_{25}	T_7	T_{21}	T_{32}		
P3	T_0	T_{23}	T_{34}	T_6	T_{13}	T_5	T_{35}	T_5	T_7	T_8		T_{32}	T_{43}	
P4		T_0	T_{34}	T_{45}	T_2	T_{24}	T_6	T_{14}	T_9	T_9			T_{43}	T_{54}
P5			T_0	T_{45}	T_6	T_6	T_{35}	T_7	T_{25}	T_{15}				T_{54}

Fig. 10. The allocation of all utility tasks.

Additional T_{12} and T_{14} tests enable indication of a damaged processor: for example, let it be a P4 processor, this processor will be isolated in the system; it is shown in Fig. 12. The processor P4 goes into a high impedance state and is galvanic ally disconnected from all system buses from 9 quanta of time. The system performed all utility tasks – in 12 quanta of time - and at the same time carried out tasks to check the operation of the system. Failure of the P4 processor did not cause the whole system to be immobilized. The system continues to operate, albeit with less possibilities. The system is characterized by a soft fall and survival.

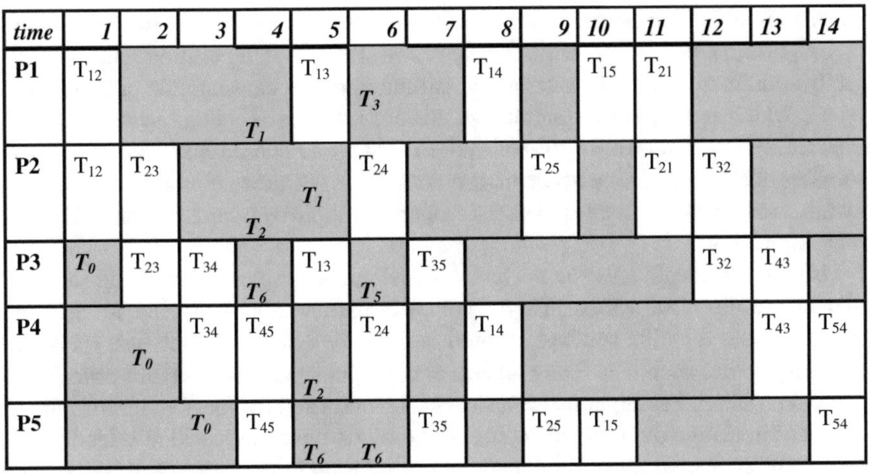

time	1	2	3	4	5	6	7	8	9	10	11	12	13	14
P1	T_{12}			T_1	T_{13}	T_3		T_{14}		T_{15}	T_{21}			
P2	T_{12}	T_{23}		T_2	T_1	T_{24}			T_{25}		T_{21}	T_{32}		
P3	T_0	T_{23}	T_{34}	T_6	T_{13}	T_5	T_{35}					T_{32}	T_{43}	
P4		T_0	T_{34}	T_{45}	T_2	T_{24}		T_{14}					T_{43}	T_{54}
P5			T_0	T_{45}	T_6	T_6	T_{35}		T_{25}	T_{15}				T_{54}

Fig. 11. The T_{24} test detected an abnormality in behavior of system.

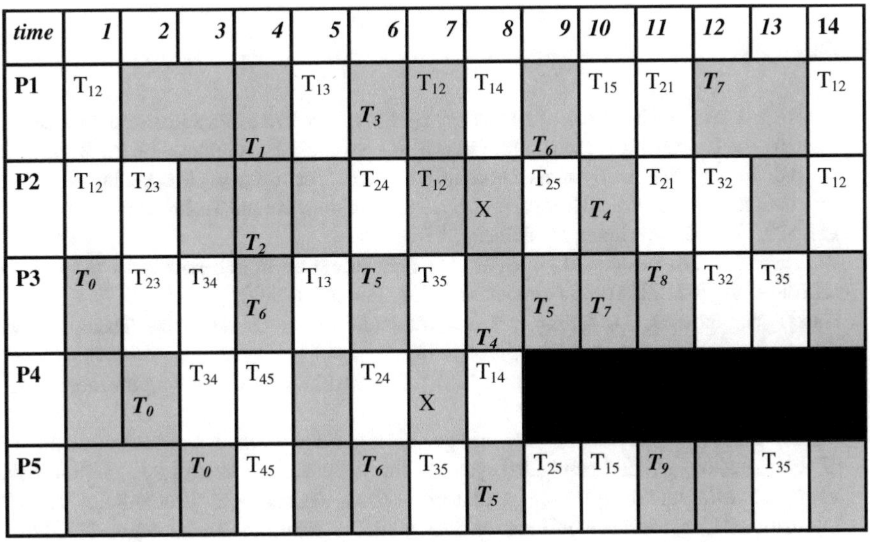

time	1	2	3	4	5	6	7	8	9	10	11	12	13	14
P1	T_{12}			T_1	T_{13}	T_3	T_{12}	T_{14}	T_6	T_{15}	T_{21}	T_7		T_{12}
P2	T_{12}	T_{23}		T_2		T_{24}	T_{12}	X	T_{25}	T_4	T_{21}	T_{32}		T_{12}
P3	T_0	T_{23}	T_{34}	T_6	T_{13}	T_5	T_{35}	T_4	T_5	T_7	T_8	T_{32}	T_{35}	
P4		T_0	T_{34}	T_{45}		T_{24}	X	T_{14}						
P5			T_0	T_{45}		T_6	T_{35}	T_5	T_{25}	T_{15}	T_9		T_{35}	

Fig. 12. The damaged processor (T4) is isolated by the system.

5 Conclusions

The model and examples the synthesis presented in this paper are an attempt of comprehensive and coherent approach to high-level system synthesis. This synthesis is directed to systems with tasks self-testing the system main resources, namely the processors. Such approach in designing fault tolerant systems accounts for indivisible testing tasks. The following system optimization criteria are accepted: operating time minimization and cost minimization. Synergic solution is a result of cooperation between the scheduling algorithms and the algorithms responsible for resource partition. One may also specify additional optimality criteria, e.g. minimum power consumption of the designed system, which is particularly significant for embedded and mobile systems. For the proposed system's relevance to real systems [11], one should take into account the processes of communication between resources and tasks, preventing resource conflicts, as well as extend the available resources sets, for example by programmable and configurable structures.

This work presents only the universal model and examples of providing self-testability of the designed system. Fault tolerant is particularly significant for real time systems, that is why the synthesis should include the criterion of the task scheduling optimality before deadlines. The problem of coherent synthesis is a multi-criteria optimization problem. Taking into account several criteria and the fact that optimization of one criterion results often in worsening of the second one, indicated at selecting optimization in the Pareto sense [7]. The optimum solution in such case shall be the whole expanse of solutions. The above issues and also methods for implementing various instances of problems generated by the model described here are now studied.

References

1. Drabowski, M.: Modification of concurrent design of hardware and software for embedded systems – a synergistic approach. In: Grzech, A., Świątek, J., Wilimowska, Z., Borzemski, L. (eds.) Information Systems Architecture and Technology: Proceedings of 37th International Conference on Information Systems Architecture and Technology – ISAT 2016, vol. 522, pp. 3–13. Springer, Heidelberg (2016)
2. Błażewicz, J., Drabowski, M., Węglarz, J.: Scheduling multiprocessor tasks to minimize schedule length. IEEE Trans. Comput. **C-35**(5), 389–393 (1986)
3. Garey, M., Johnson, D.: Computers and Intractability: A Guide to the Theory of NP-Completeness. Freeman, San Francisco (1979)
4. Dorigo, M., Di Caro, G., Gambardella, L.M.: An algorithms for discrete optimization. Artif. Life **5**(2), 137–172 (1999)
5. Drabowski, M., Kiełkowicz, K.: A hybrid genetic algorithm for hardware–software synthesis of heterogeneous parallel embedded systems. In: Świątek, J., Borzemski, L., Wilimowska, Z. (eds.) Information Systems Architecture and Technology: Proceedings of 38th International Conference on Information Systems Architecture and Technology – ISAT 2017, vol. 656, pp. 331–343. Springer, Heidelberg (2017)
6. Drabowski, M.: Parallel synthesis of computer systems. Monographs no. 225. AGH Press, Kraków (2011)
7. Błażewicz, J., Ecker, K., Pesch, E., Schmidt, G., Węglarz, J.: Handbook on Scheduling. Springer, Heidelberg (2007)
8. Drozdowski, M.: Scheduling multiprocessor tasks – an overview. Eur. J. Oper. Res. **94**, 215–230 (1996)
9. Drozdowski, M.: Scheduling for Parallel Processing. Springer, London (2009)
10. Pricopi, M., Mitra, T.: Task scheduling on adaptive multi-core. IEEE Trans. Comput. **C-59**, 167–173 (2014)
11. Agraval, T.K., Sahu, A., Ghose, M., Sharma, R.: Scheduling chained multiprocessor tasks onto large multiprocessor system. Computing **99**(10), 1007–1028 (2017)

Network's Delays in Timed Analysis of Security Protocols

Sabina Szymoniak[1(✉)], Olga Siedlecka-Lamch[1], and Mirosław Kurkowski[2]

[1] Institute of Computer and Information Sciences, Czestochowa University of Technology, Dabrowskiego 69/73, 42-200 Czestochowa, Poland
{sabina.szymoniak, olga.siedlecka}@icis.pcz.pl
[2] Institute of Computer Sciences, Cardinal Stefan Wyszynski University, Woycickiego 1/3, 01-938 Warsaw, Poland
m.kurkowski@uksw.edu.pl

Abstract. For several years, the analysis of security protocols time properties has been very important in the area of computer networks security. Up to now, however, it has been primarily used for timestamps analysis, without the other time related parameters being taken into account. As we can see in literature using formal, mathematical structures many problems can be considered and solved. In order to present the assets and liabilities of the tested protocol, depending on the known time parameters, we have proposed a mathematical model.

For our research on security protocols time properties we use synchronized networks of timed automata – a specifically designed discrete mathematical model. This model allows to express and investigate the mentioned properties. In our work we also use encoding of structures and properties into Boolean propositional formulas that can be solved using SAT techniques. The investigation based on our model proved that even protocols which are potentially weak can be used with proper time constraints. In this paper we consider one of the popular security protocols: the WooLamPi protocol and its variations. By strengthening the crucial points, a way to improve protocol safety may also be found. Part of the work was to implement a tool which not only helps in the above mentioned activity, but also allows to display experimental results.

Keywords: Security protocols · Modeling and verification · Time analysis

1 Introduction

Security protocols - specially developed algorithms - are used in modern servers, terminals as well as other network communication devices to achieve key security objectives. Many of these protocols use time stamps without detailed analysis of protocol parameters and network time. The increasing number of elements guaranteeing the confidentiality of data, such as encryption keys, time stamps, the number of protocol steps, makes implementation of the protocol itself can be problematic. Administrators are burdened with additional responsibility - proper key management.

L. Borzemski et al. (Eds.): ISAT 2018, AISC 852, pp. 19–29, 2019.
https://doi.org/10.1007/978-3-319-99981-4_3

To ensure secure communication in a moderate time, it is necessary to skillfully select several parameters, such as the network security level, as well as the role of the users'.

A variety of methods, inter alia practical tests of real and virtual systems, or more effective mathematical methods have been used to research and check security protocols properties. In the mathematical modelling, not only can we formally define the properties being investigated as a formula, but we can also formally prove whether the desired, significant security properties are satisfiable or not. Whether or not the Intruder can execute an attack on an honest user or the entire network has been the main focus of security protocols analysis up until now. To prove if the examined protocol is correct and attack resistant, a variety of verification methods based on formal mathematical modeling was used, i.e.: inductive [12], deductive [3] and discrete model checking [6]. Among others, numerous notable projects such as SCYTHER [4], ProVerif [2], Uppal [5], Avispa [1], or native VerICS [9] are connected with security protocols mathematical model checking.

Nonetheless, time, which is a crucial parameter, is commonly overlooked in the analysis carried out using the mentioned methods and tools. One can conclude that the protocol is unsafe if assuming that we have a simple three step protocol, it is discovered that an attack can be executed in ten steps. However, it is possible for the protocol to be secure using a time limit of communication for only three correct steps. Adding timestamps to designed protocols is done intuitively by many protocol designers, but maybe it would be enough for the administrators to know exactly how long the protocol will be protected.

A new formal, discrete, mathematical method of protocol executions has been introduced in work [11]. Due to this, the correctness of a time-dependent security protocol can be proven. This model was used to study the authentication processes. Network delays were taken into account in Penczek and Jakubowska's papers [7, 8], and their method was associated with proper communication session time calculation. The analyzed time constraints allowed to reveal the time influence on the protocol security. Penczek and Jakubowska's studies have not been extended beyond one session.

Our previous studies which used synchronized networks of automata and Boolean SAT techniques, have been broadened by the time parameter and the temporal aspect, what is presented in this work. The developed model has not only shown the strengths and weaknesses of the tested protocol, but also the fact that even potentially weak protocols can be used with appropriate time constraints. A way to make the protocol safer by strengthening the crucial points can also be found. For experimental research, a tool implemented by us was used.

The rest of the paper is organized as follows. In the next section we present a formal language and mathematical, computational structure. Next, we present research assumptions and experimental results for timed versions of a few versions of the WooLamPi (WLP) protocol. The last section presents our conclusions.

2 WooLamPi Protocols

The WLP protocol described in Woo and Lam's work [17] is a very interesting security protocol. This protocol uses symmetric cryptography. The WooLamPi protocol is a one-way authentication using a trusted server. WLP is the basis for the next version of this protocol (WooLamPi1, WooLamPi2 and WooLamPi3). In original WooLamPi protocol are used nonces (pseudorandom numbers). In order to execute carry out our research, we exchange nonces by timestamps. The timed version of this protocol in Common Language is as follows:

$$\begin{array}{lll} \alpha_1 & A \rightarrow B: & I_A, \\ \alpha_2 & B \rightarrow A: & T_B, \\ \alpha_3 & A \rightarrow B: & <T_B>_{K_{AS}}, \\ \alpha_4 & B \rightarrow S: & <I_B, <T_B>_{K_{AS}}>_{K_{BS}}, \\ \alpha_5 & S \rightarrow B: & <T_B>_{K_{BS}}. \end{array} \quad (1)$$

The WLP protocol consists of five steps. In the first step, user A sends to user B its identifier, informing it of the wish to initiate a new session. In response, B generates its timestamp T_B and transmits it to A. Then A composes the message in which timestamp T_B is received. User B forwards this message to the server, labeled S, adding I_A to it. The whole message is encrypted with a symmetric key shared between B and the server. In the last step of this protocol, the server sends to B its timestamp in a message encrypted with a symmetric key shared between the server and B.

Each version of WLP has the same number of steps. In addition, in the first and second step the same operations are performed. From the third step, some modifications to ensure the security of the protocol are introduced. All versions of WLP protocols are described in [17].

For the WooLamPi1 protocol, modified steps are as follows:

$$\begin{array}{lll} \alpha_3 & A \rightarrow B: & <I_A, I_B, T_B>_{K_{AS}}, \\ \alpha_4 & B \rightarrow S: & <I_A, I_B, <T_B>_{K_{AS}}>_{K_{BS}}, \\ \alpha_5 & S \rightarrow B: & <I_A, I_B T_B>_{K_{BS}}. \end{array} \quad (2)$$

The message from the third step includes the generated timestamp T_B and the identifiers of both users. The next two messages also contain the identifiers of both users.

For the WooLamPi2 protocol, proper modified steps are as follows:

$$\begin{array}{lll} \alpha_3 & A \rightarrow B: & <I_B, T_B>_{K_{AS}}, \\ \alpha_4 & B \rightarrow S: & <I_A, <T_B>_{K_{AS}}>_{K_{BS}}, \\ \alpha_5 & S \rightarrow B: & <I_A, T_B>_{K_{BS}}. \end{array} \quad (3)$$

For all messages, from step 3, the initiator ID (user I) has been added.

For the WooLamPi3 protocol, modified steps are as follows:

$$\begin{array}{lll} \alpha_4 & B \rightarrow S: & <I_A, <T_B>_{K_{AS}}>_{K_{BS}}, \\ \alpha_5 & S \rightarrow B: & <I_A, I_B T_B>_{K_{BS}}. \end{array} \quad (4)$$

The message wasn't modified from the third step. In the fourth and fifth messages, the initiator's ID was added.

No currently known attack was detected on any of the WooLamPi protocols. Only the man-in-the-middle attack is possible.

3 Formal Language and Computational Structure

The basic elements of a formal language on which the calculation structure will be built are focused on in this section. All of formal model and computational structure was presented in [14]. Initially, the definition of protocol steps is required, and formally the step of the timed protocol, which takes network delays into account, is defined by two tuples $(\alpha_1, \alpha_2,)$, where:

$$\begin{array}{lll} \alpha^1 & = & (S_{\rightarrow}, R_{\leftarrow}, L), \\ \alpha^2 & = & (\tau, D, X, G, tc). \end{array} \quad (5)$$

In this notation, L is the message being sent in a given step, whereas $S_{\rightarrow}$ and $R_{\leftarrow}$ denote the sender and the receiver respectively. From the security protocols temporal properties examining point of view, the second tuple is very important. τ denotes the time of sending the message, D denotes the network delay, X is the collection of letters necessary to construct message L. G is a set of letters that must be generated by the sender in order to create message L. tc is a set of timed conditions that should be fulfilled to allow the execution of the protocol.

As an example, the formal definition of the WLP Protocol is presented:

$$\begin{gathered} \alpha_1 = \left(\alpha_1^1, \alpha_1^2\right), \alpha_1^1 = (A, B, I_A), \\ \alpha_1^2 = (\tau_1, D_1, \{I_A\}, \{\emptyset\}, \tau_1 + D_1 \leq L_F). \\ \alpha_2 = \left(\alpha_2^1, \alpha_2^2\right), \alpha_2^1 = (B, A, \tau_B), \\ \alpha_2^2 = (\tau_2, D_2, \{\tau_B\}, \{\tau_B\}, \tau_2 + D_2 - \tau_B \leq L_F). \\ \boldsymbol{\alpha_3 = \left(\alpha_3^1, \alpha_3^2\right), \alpha_3^1 = (A, B, <\tau_B>_{K_{AS}}),} \\ \boldsymbol{\alpha_3^2 = (\tau_3, D_3, \{\tau_B, K_{AS}\}, \{\emptyset\}, \tau_3 + D_3 - \tau_B \leq L_F).} \\ \alpha_4 = \left(\alpha_4^1, \alpha_4^2\right), \alpha_4^1 = (B, S, <I_A, <\tau_B>_{K_{AS}}>_{K_{BS}}), \\ \alpha_4^2 = (\tau_4, D_4, \{I_A, <\tau_B>_{K_{AS}}, K_{BS}\}, \{\emptyset\}, \tau_4 + D_4 - \tau_B \leq L_F) \\ \alpha_5 = \left(\alpha_5^1, \alpha_5^2\right), \alpha_5^1 = (S, B, <\tau_B>_{K_{BS}}), \\ \alpha_5^2 = (\tau_5, D_5, \{\tau_B, K_{BS}\}, \{\emptyset\}, \tau_5 + D_5 - \tau_B \leq L_F) \end{gathered} \quad (6)$$

Due to the structure of WLP protocol, we will discuss in details the bolded third step. In this step, user A sent to user B a message. In order to compose this message, the following set of objects is needed: $\{\tau_B, K_{AS}\}$. The set of generated objects is empty because we don't need to generate any objects for this step. The time of sending this message was increased by the delay on the network and reduced by the value of the timestamp. This time must be less or equal to the assumed lifetime. The previous and next steps in WLP protocol should be considered similarly.

Executions of security protocols can be modeled as specially designed discrete, mathematical, transition structures, for example using a synchronized network of timed automata. There are many types of automata networks synchronization [10, 13]. Our network works according to formal definition of synchronized timed automata's network presented in [10, 13]. The global state of the network is the tuple that consists of exactly one state from each automaton. The initial state of the network is a global state that consists of all initial states of all automata. The global state labeled by α can be changed to the next one when all transitions in all automata labeled α enabled.

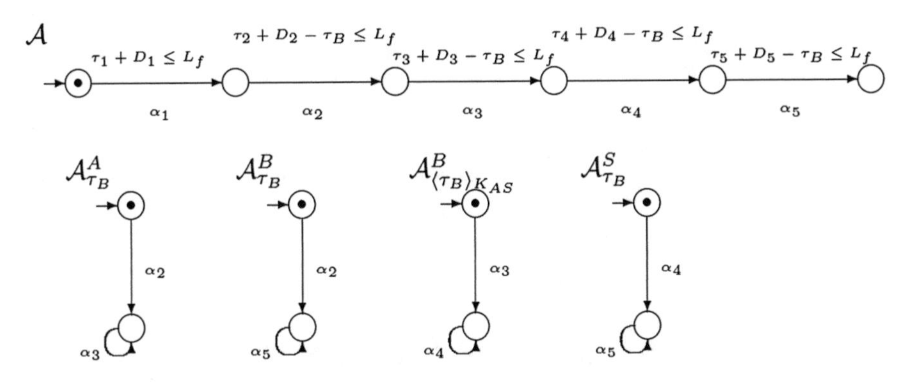

Fig. 1. Network of synchronized timed automata for NSPK protocol.

In our work we consider two types of timed automata which are included in the network. The first ones model protocols executions. The second type are automata that model users knowledge. Execution automata model executions of individual protocol steps while preserving the imposed time conditions (including delays in the network). Knowledge automata model the process of gaining knowledge by users. Knowledge and execution automata are synchronized by labels that allow to model the need to acquire specific knowledge by users, so that it can execute the next protocol step.

The network of synchronized timed automata that models one execution of the WLP protocol (including delays in the network) is presented on Fig. 1. In this picture, the global initial state, which is a tuple of local initial states, of the network is marked. Automaton **A** models the execution of all individual protocol steps and time conditions. The next automata model changes the knowledge of users. For example, automaton $\boldsymbol{A}^A_{\tau_B}$ models gaining knowledge about the timestamp τ_B by user A.

This global state can be changed to the next one what is shown on Fig. 2. Observe that according to protocol execution, in the first step the knowledge of the users is not changed.

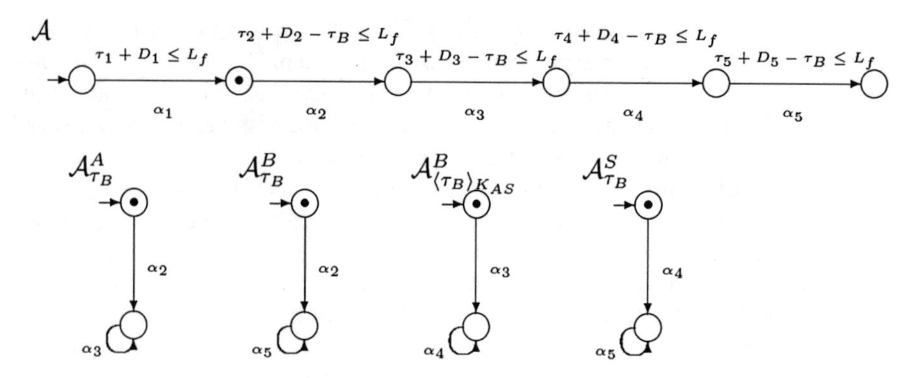

Fig. 2. The first step of WooLamPi protocol.

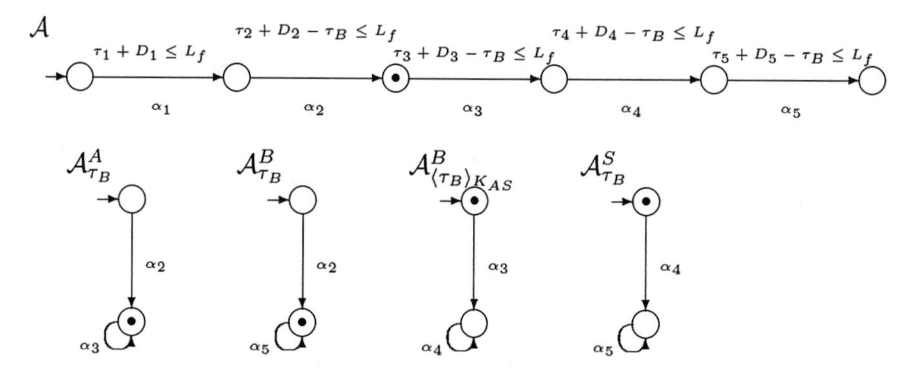

Fig. 3. The second step of WooLamPi protocol.

The second step of the protocol can be now executed (Fig. 3). We can see that the second transition in the automaton **A** is synchronized with the first transitions in automata $\mathbf{A}^A_{\tau_B}$ and $\mathbf{A}^B_{\tau_B}$. This is because the execution of the second step of the WLP protocol requires changing of user *B*'s knowledge about the timestamp τ_B (generating) and user *A*'s knowledge about this ticket (possessing).

Now the third step of the protocol can be performed. In Fig. 4 we can see that the third transition in automaton *A* is synchronized with the loop in automaton $\mathbf{A}^A_{\tau_B}$ because *A* needs this knowledge for the third step execution.

Considering Fig. 5 we can see that in the fourth step user *B* needs $<\tau_B>_{K_{AS}}$ for executing this step. Additionally, the server *S* possesses a ticket τ_B.

In the last step (Fig. 6) the server uses knowledge about a ticked τ_B and the user *B* gets its own ticket again.

The method of automatic generation of all automata that models executions in the investigated protocol and considered space of the users is given in [9].

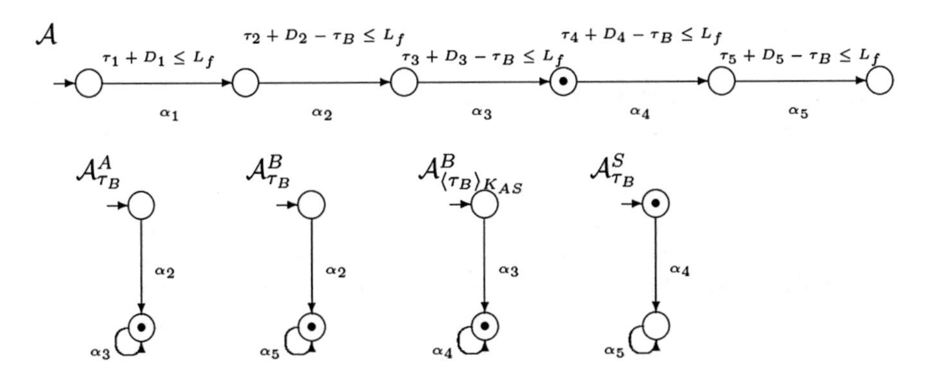

Fig. 4. The third step of WooLamPi protocol.

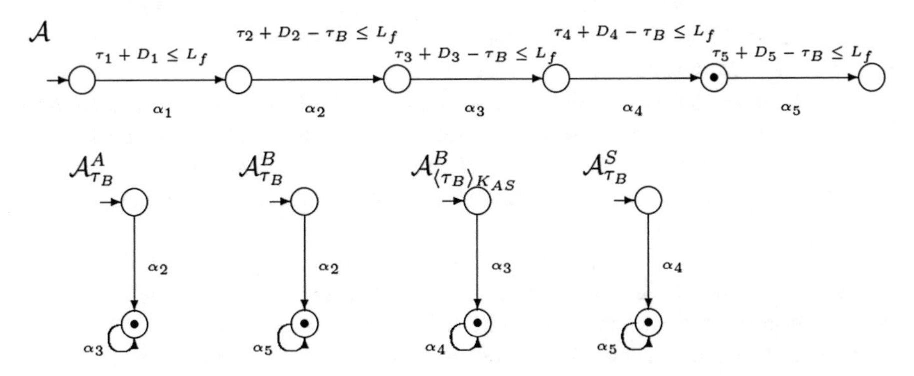

Fig. 5. The fourth step of WooLamPi protocol.

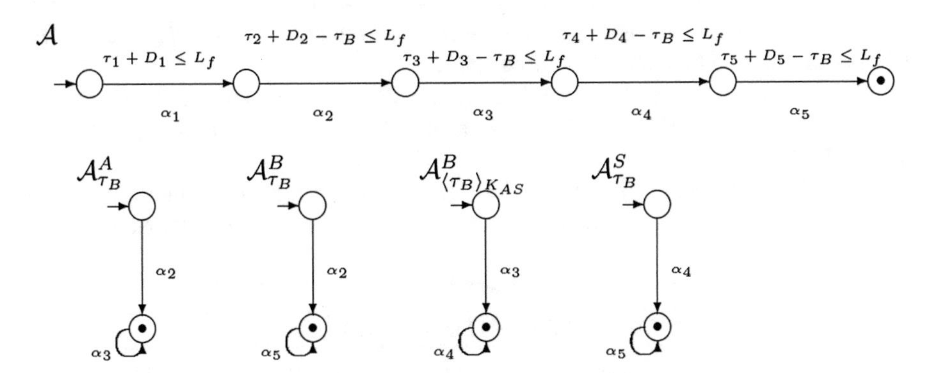

Fig. 6. The fiveth step of WooLamPi protocol

4 SAT Testing

Using our tool for all security protocols (described in [14, 16]), the protocol executions and networks of synchronized timed automata that model them were generated. The tests were performed using a computer unit with the Linux Ubuntu operating system, processor Intel Corei7 and 16 GB RAM.

Subsequently, the network of automata is encoded into a Boolean, propositional, logical formula, into which the tested security property is also encoded. The problem of fulfilling the property by the encoded system (set of protocol executions) is thereby extruded in a way that is mutually unambiguous to the problem of the satisfiability of the conjunction of the received logic of formulas (SAT problem). These formulas are large from a computational point of view, they have at least a few hundred variables, and their conjunction normal form consists of thousands (hundreds of thousands) of clauses.

The formulas acquired in the manner mentioned above can virtually be tested by specialized tools called SAT-solvers, yet it is well-known that the problem of the formulas satisfiability is an NP-complete problem. What is notable is that with the unambiguous encoding method it is known that once there is a validation satisfying the given formula, the property under test is met by the set of executions. An attack upon the protocol exists if we examine the suitability of a protocol set for an attack.

The length of the path, the number of variables and clauses used, the amount of memory used by the processor, and the calculation time are the central parameters of this verification method.

To begin the study, the appropriate SAT-solver was selected. The following five SAT-solvers were tested: Glucose, Lingeling, Treengeling, Clasp and MiniSAT. During research on the selection of SAT-solver, all protocols from the WLP family were used. The MiniSAT turned out to be the most effective. A further part of the study was carried out using MiniSAT.

The synchronized network of timed automata for this protocol consisted of twenty-two automata of executions and twenty-three automata of knowledge (for honest users and the Intruder). These studies dealt with occurrences run, in a given network that terminated in one of the states modeling attacking execution. For a path of length 10, the man in the middle attack was discovered.

The first phase of the WooLamPi study was designed to demonstrate its vulnerability to attack with a delay in the network of 0,07 time unit ([tu]) and a lifetime of 3 [tu] (Table 1).

Table 2 shows test results for the WLP protocol. The table takes into account the relevant values of variables, clauses, memory and time obtained on each path length, as

Table 1. Experimental results for WooLamPi protocol.

Length of path	Variable	Clauses	Memory [MB]	Time [ms]	Result
4	23022	56898	3.04	64	UNSAT
6	36493	88817	3.77	124	UNSAT
8	50028	120832	4.28	272	UNSAT
10	63670	153413	5.03	620	SAT

well as information whether a Boolean formula was satisfied or not at a given level. UNSAT annotation means that the formula has not been satisfied, and the SAT annotation means that the formula is satisfied, so there is an attack upon this protocol.

In the next phase of the WooLamPi protocol, study involved an indication of the range of network delay values for which the protocol remains secure. This means that for a set time value, a SAT-solver returns a value of UNSAT, hence there is no attack upon the protocol.

The range of network delay values for which the protocol remains secure is then presented. We conducted these studies through 100 test series for various network delays and lifetime values. It was concluded, thanks to the selected values and acquired results that the Intruder would not be able to perform a man-in-the-middle attack once the following dependency was satisfied in all WooLamPi protocol steps:

$$D_1 \leq L_f,\ D_2 \leq \frac{L_f}{4},\ D_3 \leq \frac{L_f}{7},\ D_4 \leq \frac{L_f}{8},\ D_5 \leq \frac{L_f}{9}. \quad (7)$$

Adjusting time conditions in particular steps of the protocol was made possible as a result of such defined dependencies. Thanks to accurately selected dependencies, the honest user will have enough time to complete the step, and the Intruder will not be able to acquire additional knowledge to carry out an attack.

Then, tests were conducted considering the value of the time parameters according to the designated dependencies. In the following steps, protocol lifetime was 3 [tu] and delay values were 3,1 [tu], 0,76 [tu], 0,43 [tu], 0,38 [tu] and 0,34 [tu] for the WooLamPi.

Table 2. Experimental results for selected SAT-solver and protocols.

Length of path	Variable	Clauses	Memory [MB]	Time [ms]	Result
4	23276	57540	3.05	44	UNSAT
6	36921	89877	3.80	67	UNSAT
8	52916	127829	4.57	339	UNSAT
10	64341	155382	4.96	652	**UNSAT**

Test results for the WooLamPi protocol are shown in Table 3. It is worth noticing that the values of clauses and variables have increased. Furthermore, the tested formula was not satisfied for the selected time parameters, hence an attack upon the WooLamPi protocol for this value of clocks does not exist.

5 Conclusions

The results of formal, mathematical modeling, as well as the WLP protocols' verification, including delays in the computer network, have been presented in this paper. To achieve this, discrete automata based structures, and propositional Boolean encoding have been used.

We have carried out our computational results using our tool and the SAT-solver MiniSAT. These tests were constructed in order to check the validity of the test formula, and also the attacks on the two well-known protocols, considering delays in the network. The extension of the study range was also enabled by the emergence of delays in the formal model. Thus, security protocols could be considered in a broader sense, the study of the network's transmission time effects on an Intruder's capabilities particularly. Moreover, verifying whether generated executions can actually occur has been made possible thanks to the research. The conducted research has proven that there is a validation satisfying the test formula for each protocol. Hence, an attack was found for every tested protocol.

The results proved that the methods and restrictions used effectively prevent the Intruder from carrying out an attack upon the tested protocols. Accurately selected lifetime values make the protocol safer.

References

1. Armando, A., Basin, D., Boichut, Y., Chevalier, Y., Compagna, L., Cuellar, J., et al.: The AVISPA tool for the automated validation of internet security protocols and applications. In: Proceedings of 17th International Conference on Computer Aided Verification (CAV 2005), vol. 3576. LNCS, pp. 281–285. Springer (2005)
2. Blanchet, B.: Modeling and verifying security protocols with the applied pi calculus and ProVerif. Found. Trends Priv. Secur. **1**(1–2), 1–135 (2016)
3. Burrows, M., Abadi, M., Needham, R.: A logic of authentication. Proc. R. Soc. London A **426**, 233–271 (1989)
4. Cremers, C., Mauw, S.: Operational Semantics and Verification of Security Protocols. Information Security and Cryptography Series. Springer, Heidelberg (2012)
5. David, A., Larsen, K.G., et al.: Uppaal SMC tutorial. Int. J. Softw. Tools Technol. Transfer (STTT) **17**(4), 397–415 (2015)
6. Dolev, D., Yao, A.: On the security of public key protocols. IEEE Trans. Inf. Theory **29**(2), 198–207 (1983)
7. Jakubowska, G., Penczek, W.: Modeling and checking timed authentication security protocols. In: Proceedings of the International Workshop on Concurrency, Specification and Programming (CS&P 2006), Informatik-Berichte, vol. 206, no. 2, str. pp. 280–291. Humboldt University (2006)
8. Jakubowska, G., Penczek, W.: Is your security protocol on time? In: Proceedings of FSEN'07. LNCS, vol. 4767, pp. 65–80. Springer (2007)
9. Kacprzak, M., Nabiałek, W., Niewiadomski, A., Penczek, W., Półrola, A., Szreter, M., et al.: Verics 2007 - a model checker for knowledge and real-time. Fundamenta Informatiace **85**, 313–328 (2008)
10. Kurkowski, M., Penczek, W.: Applying timed automata to model checking of security protocols. In: Wang, J. (ed.) Handbook of Finite State Based Models and Applications, pp. 223–254. CRC Press, Boca Raton (2012)
11. Kurkowski, M.: Formalne metody weryfikacji wlasnosci protokolow zabezpieczajacych w sieciach komputerowych, wyd. Exit, Warszawa (2013)
12. Paulson, L.: Inductive analysis of the internet protocol TLS. ACM Trans. Inf. Syst. Secur. (TISSEC) **2**(3), 332–351 (1999)

13. Penczek, W., Półrola, A.: Advances in Verification of Time Petri Nets and Timed Automata: Temporal Logic Approach. Studies in Computational Intelligence, vol. 20. Springer (2006)
14. Szymoniak, S.: Modeling and verification of security protocols including delays in the network. Ph.D. thesis, Czestochowa University of Technology (2017)
15. Siedlecka-Lamch, O., Kurkowski, M., Piatkowski, J.: Probabilistic model checking of security protocols without perfect cryptography assumption. In: Proceedings of 23rd International Conference, Computer Networks 2016, Brunow, Poland, 14–17 June 2016. Communications in Computer and Information Science, vol. 608, pp. 107–117. Springer (2016)
16. Szymoniak, S., Siedlecka-Lamch, O., Kurkowski, M.: Timed analysis of security protocols. In: Proceedings of 37th International Conference ISAT 2016, Karpacz, Poland, 18–20 September 2017. Advances in Intelligent Systems and Computing, vol. 522, pp. 53–63. Springer (2017)
17. Woo, T.Y.C., Lam, S.S.: A lesson on authentication protocol design. SIGOPS Oper. Syst. Rev. **28**(3), 24 (1994)

Evaluation of an Impact of the DoS Attacks on the Selected Virtualization Platforms

Daniel Kosowski, Grzegorz Kołaczek([✉]), and Krzysztof Juszczyszyn

Wroclaw University of Science and Technology, Wroclaw, Poland
Grzegorz.Kolaczek@pwr.edu.pl

Abstract. The work includes research results characterizing the impact of distributed denial of service attacks (DDoS) on the resources of host operating systems and virtual machines performance. Analysis of existing types of virtualization has been made. It was necessary to discuss the architectures of virtualization systems such as Xen Project, VMware ESXi, KVM and Oracle Virtualbox in order to outline differences in their structure, which will result in often very different results in the performance test. It was also necessary to present the concept of denial of service attack, its history and evolution over the years from simple DoS attacks in layer 3 and 4 of the OSI model by distributed DoS (DDoS) attacks most often implemented in layer 7 to their currently used forms which are multiplied DDoS attacks with reflection exploiting the weaknesses of the DNS protocol. The most attention was paid to the study of the impact of various types of DDoS attacks on the real performance of the virtualized system that based on Intel VT-x technology. To ensure the comprehensiveness of research results, the attacks were carried out in three layers, dividing each of them into several stages characterized by different intensity of network traffic.

Keywords: Virtualization · Network attacks · Network security

1 Introduction

1.1 Virtualization

The use of the possibilities of operating system virtualization is still increasing its popularity both in business applications (mainly corporate clients) and private users. In recent times, possible applications of virtualization have been considered even for embedded systems. Thanks to the separation of the virtual and physical parts of the computing platform through a virtual machine system, virtualization has become a very important and profitable branch of IT market generating billions of profits. The spectrum of possibilities of application of this technology is very wide and includes [1, 9, 10]:

- resource isolation - each device is isolated from all other operating devices on the same place. In the case of accident or hacking resources outside infected machines will be secure,
- load balancing of the computational clusters. The new task will be created on the least loaded host. Sometimes, it is required also to migrate some tasks from one system to the other one.

L. Borzemski et al. (Eds.): ISAT 2018, AISC 852, pp. 30–40, 2019.
https://doi.org/10.1007/978-3-319-99981-4_4

- energy management - virtual machines working on lightly loaded machines can be moved to other hosts by allowing the first one to be disabled.
- ability to work with many different operation systems season under control of one physical machine It helps to simultaneously work with several applications dedicated to one specific OS.
- sandboxing - a virtual machine can be used as a testing ground for example for new or insecure software.

Between operating systems installed on a virtual machine and the native system, there will be differences in performance that result from the architecture of virtual machine. The fact that many virtual systems simultaneously share physical resources is crucial in the context of considerations about the vulnerability of the entire system to one of the currently greatest threats in the global network - DDoS attacks.

1.2 DoS Attacks

The Internet has become an inherent element of everyday life for most of today's people. The consequence of the development of the Internet next to the growing data transfer speeds (development of the network infrastructure) is the increasing requirements for the quality and reliability of the services provided through its services (Quality of Service). This is particularly important from the point of view of large commercial entities, such as banks or stock exchanges in the case of which any failure resulting in the inability to provide or access to the service translates into a multimillion losses. This type of threat is a denial of service (DoS) attack [3]. Its purpose is to prevent or significantly impede access to the service of an ordinary user. A single DoS attack it is easy to defend by adding the Internet Protocol (IP) number of the attacker to the blacklist. However, significantly increasing the number of attacking computers (and their dispersal) usually by infecting subsequent victims with malicious software such as a worm makes the defense much harder. Currently, taking advantage of the growing popularity of IoT (Internet of Things) concept, more and more often computers and devices connected to home networks such as smart TVs or light bulbs can be used to perform DDoS attack [7]. A distributed denial of service attack quickly became the most common and by far the most effective attack on services and services on the network. The main problem to be faced in this case is the distinction of traffic generated by legal users from the traffic generated by the botnet network. The growing problem is reflected in the Kaspersky report for the second quarter of 2016, which presents the current statistics on DDoS attacks [2]:

- DDoS attacks targeted resources in 70 countries,
- over 77% of the targets were in China,
- the longest DDoS attack lasted 291 h, or more than 12 days,
- SYN DDoS, TCP DDoS and HTTP DDoS remain the most common methods the attack,
- over 70% of detected attacks came from Linux botnets,
- the IoT devices and other devices play a major role in the botnet,
- built-in cameras such as CCTV cameras - in one of the discovered botnets they were almost half of the devices.

1.3 The Research Goals

The problem of DDoS attacks with each subsequent year does not diminish and grows stronger. There are many reasons why this happens. First of all, its characteristics make it extremely difficult to prevent it and probably never a defense mechanism will be created, which will be in one hundred percent effective. The methods used to carry out attacks and acquire larger numbers of hosts are evolving. In addition to building more botnets - mainly using poorly secured IoT devices, the weaknesses of the DNS (Domain Name System) protocol and UDP (User Datagram Protocol) faults are also widely used. On the other side, in the last few years, there has been a real revolution in the context of virtual resources being leased. The continuous development of virtualization tools results in the possibility of better and more appropriate allocation of physical resources for each user. Individual virtualization systems differ from each other - from more important issues such as the type of supervisor used (Hypervisor) to less important such as implementation details. From the point of view of the research goals, it is important to give an answer to the question which virtualization system could withstand the most of the different variants of the mentioned threat. The aim of the work is to analyze architecture and test the most popular free solutions available on the market that use hardware virtualizations for the impact of DDoS attacks on the availability, quality, and stability of services provided through them.

2 Tests Environment

2.1 Hardware Configuration

A two personal computers have been used for tests. The brief characteristic of the hardware configuration used is given in Tables 1 and 2.

Table 1. Specification of the attacking machine (generating traffic) - HP Pavilion dv6.

CPU	Intel Core i7-2670QM 2.20 GHz
RAM	8 GB DDR3 1333 MHz SODIMM
HDD	SATA III WD Scorpio Black 7200
Network	Realtek RTL Gigabit Ethernet
Chipset	Intel HM65

Table 2. Host machine specifications (desktop).

CPU	Intel Core i7-860 2.20 GHz
RAM	14 GB DDR3 1333 MHz DIMM
HDD	SATA II SAMSUNG HD103SJ 7200
Network	Realtek RTL Gigabit Ethernet
Chipset	Intel P55

The processors in both machines support Intel VT-x I and II generation technologies. Additionally the i7 860 processor has support for Intel VT-d. Initially, tests were planned for both technologies to check the difference in performance, but the Intel P55 chipset does not supports Intel VT-d technology because the idea was abandoned [4, 8].

Regardless of the virtualization system being tested, 4 vCPUs were assigned to each virtual machine and 7168 MB of operating memory leaving the second part of the resources at the disposal of the host system.

2.2 Network Configuration

To be able to answer the research question correctly it is necessary to provide the high precision of tests. All generated malicious traffic should load the host machine and not other network components, therefore it was decided to connect the attacking machine and the host machine directly. The machine from which the attack is carried out also performs the role of the DHCP server. In order for the virtual machine to be visible to the attacker, it was necessary to replace the default configuration of VM access to the network as it is NAT for bridging the virtual Ethernet interface to its physical counterpart.

2.3 Operating System

Performing tests requires choosing the right operating system and its specific versions from both the host and guest sides. The system installed on the host machine must support several different solutions of virtualization and a specific VMM must offer support for the chosen one system as a guest system. The system should also have a rich repository to enable installations and compilations from the basic software sources necessary for testing. This chosen system was Debian 9.3 64 bit with a *stretch* code name that has been installed on all physical and virtual ones machines used in the test.

2.4 Software Layer Configuration

Target Application for the Attacks. In order to perform tests with some network service using the popular PHP language and MySQL database were used XVWA web applications (Xtreme VulnerableWeb Application). The XVWA project was created to enable the user in legal, secure and easy way to learn about the most popular vulnerabilities encountered in web applications, including: SQL Injection - Error Based, SQL Injection – Blind, XPATH Injection, etc. During the tests, the application has been used for simulation of load generated by a small, simple but using PHP and database web site.

Botnet Simulation. The construction of even a large laboratory network in order to carry out the distributed DoS attacks are not able to reflect the attacks carried out by botnets them is composed of a few or many thousands of hosts. Taking into account that there is only a need to generate a large number of queries from a random pool of IP addresses, it is worth to use ready solutions dedicated to performance tests such as: solutions dedicated to performance tests such as:

- BoNeSi - enables execution of DDoS attacks in layer 3 and 4 of the OSI model such as ICMP and UDP flood. It also allows the implementation of TCP (HTTP) flood layer 7 attack generating queries allows IP spoofing using a database of 50,000 randomly generated addresses and tries to avoid easily recognizable patterns.
- hping3 - is a network tool that lets you send custom packages TCP/IP. It is used, among others, for testing firewalls, advanced scanning ports, network performance tests using different protocols and many more. The key from the point of view of the tests is the possibility of easy and above all efficient generation of TCP SYN packets with a randomly generated address source.

Performance Monitoring. The SysBench program was used to measure processor performance, memory and check the speed of hard disk operations. Measurement of performance performed for different components is carried out in the following way:

- processor efficiency - the test is limited to the determination of prime numbers until the threshold value specified in the calling parameter is reached. The value returned by the program is the time needed to calculate all prime numbers smaller than the set threshold.
- memory performance - the segments of the global or local log memory are allocated and then the read and write operations are performed on them. The most important value is the memory bandwidth given in MB/s.
- I/O operations speed - for the purpose of the test, files are created on which randomly write and read data operations are performed in the next order. The result is the average amount of all data read and written in each second given in MB/s.

The Iperf program was used to measure network performance, which uses the TCP or UDP protocol in 10 s to measure the maximum available bandwidth (Mb/s) between two network hosts. An important difference in the case of the native system and virtualized systems will be the number of context changes and inaccuracies in the cache miss memory, which will be measured using the Iperf program. The last tool used is AB (Apache Benchmark), which is a simple HTTP client that allows sending repeated HTTP GET queries to the server that meet a set of criteria. The result of the test is a report from which the most important value for the conducted tests is the average number of queries serviced at each second of the test duration [5].

System Resources Monitoring. The NMON and NMONVisualizer are tools intended for administrators to monitor and visualize changes in the values of various parameters describing the state of the system's resources. As one of the few tools, it has the ability to collect data only and exclusively within the range of the given time interval, for example from now for exactly 400 next seconds. The program was created as an IBM internal project but after many years in July 2009, it was decided to make it available under the GPL license. It is worth noting that the program allows data analysis in two ways. Statistics can be viewed in real time in the console window as raw data or in the form of charts generated using the basic ASCII table signs. The most useful, however, is the use of the possibility to save the collected data to csv format and then use one of several available programs such as nmonchart, nmon2rrd or NMONVisualizer, which generate accurate charts on their basis. To visualize the test results, the NMONVisualizer has been used, which is a multifunctional tool written in Java and having an extensive GUI. It was created in the same way as NMON as an internal IBM project in 2011 as part of support for performance tests of the IBM SmartCloud service. Unfortunately, not in all tested scenarios, the program was able to collect reliable data on the current consumption of host machine resources. In a case of the virtualization using Xen to collect data on the current level of processor speed by dom0 and virtual machine, the xentop program was used which was part of the xen-tools package. The program works similarly to the standard Linux top program, however, its task is to monitor only domain 0 and currently working virtual machines. It also provides basic information about them, such as the number of vCPUs permanently assigned to them and operating memory.

3 Test Scenarios

The key issues during the selection of virtualization systems to be tested were:

- popularity
- compatibility with the hardware platform
- HVM/PVHVM virtualization - a prerequisite for being used by the system virtualization of hardware support such as Intel VT-x
- compatibility with GNU Linux - the Linux system provides a rich package of serving tools to measure system performance as well as the level of resource consumption.

After considering the above-mentioned issues, it was decided to run tests for Oracle [6], VirtualBox, KVM, Xen and for the native Linux system so that in addition to the differences in performance between different virtualization systems, compare how much overhead generates to the system installed directly on the hardware. The choice of the types of attacks to be performed as part of the tests was made on the basis of a report on the popularity of particular types of DDoS on the Internet made by Kaspersky Lab [2]. It was decided to conduct UDP flood attacks, ICMP (ping) flood, TCP SYN flood and HTTP flood. Analyzing the test results, there is no doubt about the effectiveness of the chosen attacks. Both the host and the virtual machine are Linux systems. They were installed and used for AS IS tests, which means that no additional software apart from the necessary was installed on them and system configuration changes were not made except the interfaces configuration file in order to bridge the host network interface with the interface. Attacks in layers 3 and 4 have been known for many years and there are a lot of mechanisms in the Linux system like SYN cache and SYN cookies, whose task in this case is not to allocate any buffers for half-open connections and overflows in memory queue [2]. Taking these into account, these types of attacks serve more as attacks that aim to use the available network bandwidth of the server than to consume its resources such as operational memory. At the present moment the only DDoS attacks capable of completely using the resources such as the processor's time are attacks in layer 7.

In an experimental way, appropriate parameters for the programs measuring the efficiency were selected.

- processor - the time needed to determine all prime numbers from up to 125,000.
- memory - operating memory bandwidth based on the time needed for a single thread to perform a write to global memory operation or read from it 7 GB of data using a 1 MB buffer for write operations and 1 KB for read operations.
- disk - performing random write and random read operations on previously generated files by the sysbench program. The maximum time of this test is 90 s and the value is returned by the average speed of read/write.
- network - iperf program running in client mode checks based on the TCP protocol maximum bandwidth connecting with the program acting as a server.

The perf program based on information provided by PMU (Performance Monitoring Unit) enables monitoring of more detailed information about operations occurring in the lower layers, among others, context switch or the number of cancellations to LLC (Last-Level) cache). A single session during which data is collected lasts 10 s and is repeated 5 times to obtain averaged results. The AB (Apache Benchmark) program is used to check the current performance of the web server, which is configured in such a way that for 60 s using 2 threads to generate mass HTTP GET queries and then to route them to the virtual machine address - http://192.168.1.3:/xvwa/vulnerabilities/xpath/. The value used is returned by the amount of the average number serviced in each query.

Each of the selected virtualization systems and the native system were subjected to exactly the same test scenario:

1. Test of the processor, operating memory, disk and network performance performed by sysbench, iperf programs at the level of the virtual machine in idle state.
2. UDP flood type attack with a packet size of 1024 bytes: The total time during which the first attack on a virtual machine is carried out is 10 min and is divided into 10 equal time intervals (stages). At the same time, the level of resource utilization is monitored from the level of the host machine. Next, an attack is carried out on the virtual machine, but this time with a fixed number of packages 60,000/s during which the performance test is repeated. Another steady-state attack during which the perf program collects data provided by the PMU.
3. An ICMP flood type attack with a packet size of 512 bytes: First attack carried out on the same principle as described in point 2. The other three attacks were carried out in the same way as in point the only difference is the packet value of 150,000/s
4. TCP SYN flood attack with a packet size of 256 bytes: A variable-sized attack performed analogous to the point. 2 and 3. The remaining attacks were carried out with the intensity of 77,000/s packages.
5. HTTP GET flood attack: Attacks were carried out with the intensity of 4,000/s packages.

4 Results

Analyzing the results of variable-load attacks in terms of CPU time consumption, one can determine that the results obtained except those from the native system are similar. In the case of VirtualBox and Xen, the CPU usage progressively increases with the attack. It looks a bit different in KVM where the characteristic is more linear. In the last phase of the test characterized by the highest level of network traffic, CPU usage in the case of Virtual-Box and KVM is comparable, however, in Xen, the total CPU utilization by dom0 and domU turned out to be a few percentage points higher in each of the attacks. It could be also noticed that there are differences in the way of load distribution between individual processor cores (also virtual). The Oracle product distributes the resulting load uniformly between all available processor cores while KVM tends to load a single one, and in general, as shown by the first three tests to frequently load only 2 or at most 3 of the same cores leaving the rest relatively free. The detailed results of the performed tests have been presented on the Figs. 1, 2, 3, 4, 5 and 6 below.

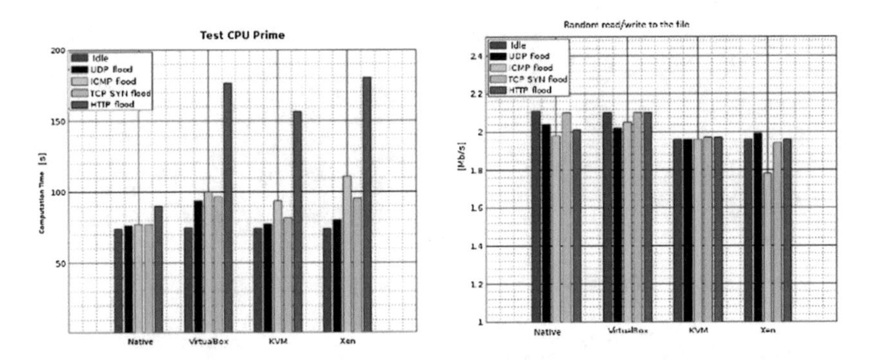

Fig. 1. CPU Prime (left) and random r/w to the file tests results

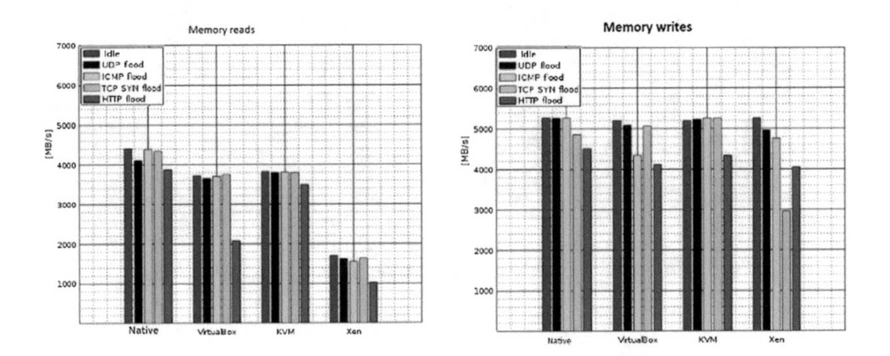

Fig. 2. Memory Read/Write tests results

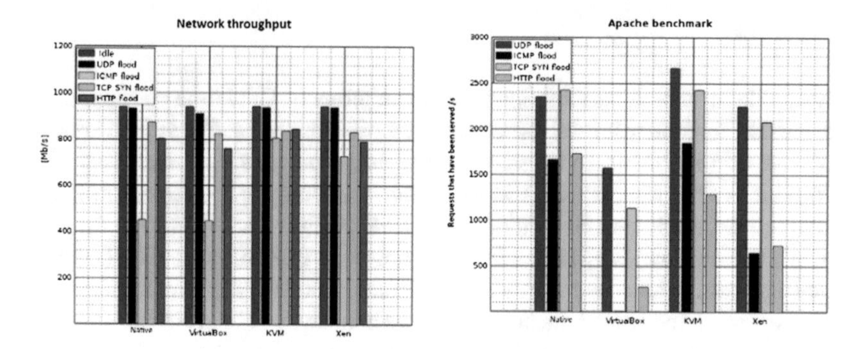

Fig. 3. Network throughput (left) and Apache benchmark (right) tests results

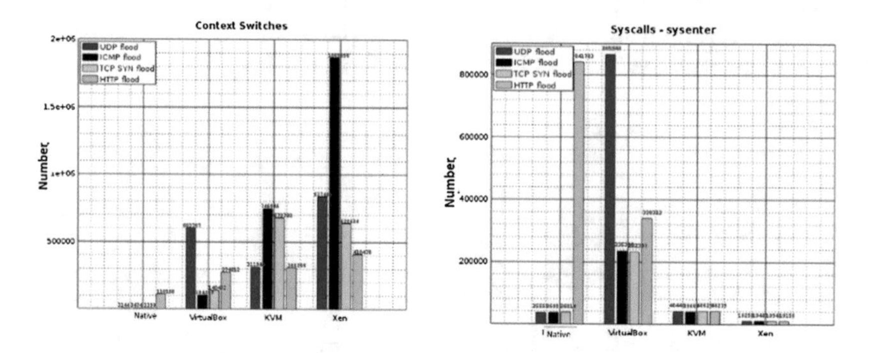

Fig. 4. Context switches (left) and syscalls (right) tests results.

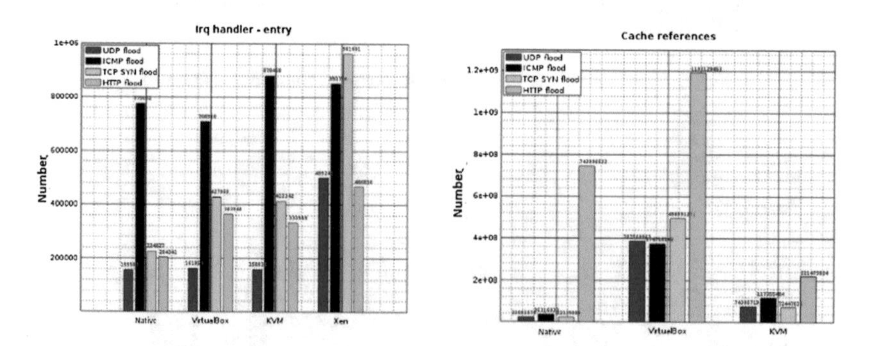

Fig. 5. Irq handler (right) and cache references (right) tests results.

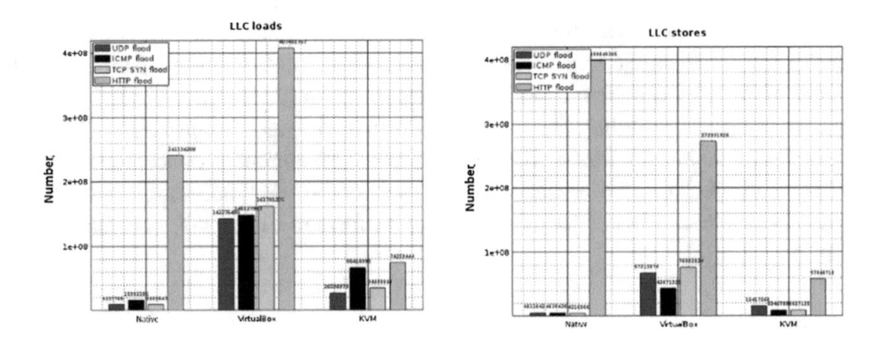

Fig. 6. LLC loads (left) and LLC stores (right) tests results.

5 Summary and Conclusions

Virtualization has become a technology that has changed the way of the IT resources management. There is no longer an obligation to buy a service which is the whole dedicated server. Without any problems, one can buy a virtual machine with a number of resources defined by the customer, which allows to better adjust the costs and scope of the service to individual requirements. Virtualization, however, is not a perfect solution free of problems. Solutions based on hypervisors are known for generating considerable overheads related to the operation of I/O devices. The most important research goal of this work was to find the answer to the question of what will be an impact of the biggest threats of the modern Internet, such as DDoS attacks on the selected virtualized systems. The results confirmed that as one could expect the most resistance was demonstrated by the native system. The worst results were obtained in the tests by Oracle VirtualBox. This could have been also expected considering that it was the only host in the tests which is the hosted hypervisor. This type of hypervisor as proven by the tests is the most vulnerable to DDoS attacks and definitely should not be used to support virtual machines that are web servers. Its susceptibility and low efficiency results from the additional layer that is between hypervisor and the physical components of the machine. A much more efficient and extensive solution is virtualization of the type 1 (bare-metal) hypervisor which has direct access to the machine's resources. The most popular solutions on the market today are based on Linux KVM and Xen as well as the VMware - ESXi product. Of the three products, the first two were selected for testing. The results obtained in the performance test of the WWW server in both cases are much better than the VirtualBox discussed earlier. Additionally, confronting the results of all tests with each other, it is easy to notice that KVM is the solution that is much more efficient and resistant to DDoS attacks. However, it is necessary to mention the virtualization mode which was used to carry out the tests - HVM with additional paravirtualized I/O controllers. This mode is not native to Xen and has been added to it relatively recently, which could have caused weaker results. In order to take advantage of the maximum possibilities offered by Xen, one would have to run the guest system in paravirtualization mode (with the modified kernel). Further tests should be dedicated to tests in a similar way the VMware product, which is the ESXi hypervisor that is part of the VMware vSphere product. It is widely used by private users and commercial entities who appreciate stability and extensive support from the manufacturer. It is also necessary to find new tools to monitor the resources of the host system and read the values from the PMU (if there is such a possibility). An interesting solution that would be worth testing as it is characterized by high efficiency [5] is OpenVZ. It is a virtualization based on container concepts - without a hypervisor. In addition, the testing concept itself could have been subject to several changes. Future tests should focus on attacks performed in layer 7 - not necessarily dispersed.

References

1. Heiser, G.: The role of virtualization in embedded systems. In: Proceedings of the 1st Workshop on Isolation and Integration in Embedded Systems, IIES 2008, pp. 11–16 (2008). https://doi.org/10.1145/1435458.1435461
2. Kasperksy Lab. Kaspersky DDoS Intelligence Report for Q2 2016 (2016). https://securelist.com/analysis/quarterly-malware-reports/75513/kaspersky-ddos-intelligence-report-for-q2-2016/
3. Shea, R., Liu, J.: Understanding the impact of denial of service attacks on virtual machines. In: 2012 IEEE 20th International Workshop on Quality of Service, pp. 1–9 (2012). https://doi.org/10.1109/iwqos.2012.6245975
4. Abramson, D., Jackson, J., Muthrasanallur, S., Neiger, G., Regnier, G., Sankaran, R., Schoinas, I., Uhlig, R., Vembu, B., Wiegert, J.: Intel virtualization technology for directed I/O. Intel Technol. J. **10**(03), 16 (2006). https://doi.org/10.1535/itj.1003. ISSN 1535-864X
5. Lowe, S., Natarajan, S., Unityvsa, E.M.C., Xenserver, C., Someswar, G.M., Kalaskar, H., Barham, P., Dragovic, B., Fraser, K., Hand, S., Harris, T., Ho, A., Neugebauer, R., Pratt, I., Warfield, A.: Comparative analysis for new virtualization customers. Security **2**(1), 369–381 (2016)
6. Victor, J., Savit, J., Combs, G., Hayler, S., Netherton, B.: Oracle Solaris 10 System Virtualization Essentials (2010)
7. Allot Communications. DDoS Attack Handbook (2017)
8. Intel. Intel Virtualization Technology Requirements (2017)
9. Kwiatkowski, J., Juszczyszyn, Krzysztof., Kolaczek, G.: An environment for service composition, execution and resource allocation. In: International Workshop on Applied Parallel Computing. Springer, Heidelberg (2012)
10. Fraś, M., et al.: Smart work workbench; integrated tool for IT services planning, management, execution and evaluation. In: International Conference on Computational Collective Intelligence. Springer, Heidelberg (2011)

Computer Network Architecture, Distributed Computer Systems, Quality of Service

Cost Effective Computing Unit Redundancy in Networked Control Systems Using Real-Time Ethernet Protocol

Jacek Stój[✉]

Silesian University of Technology, Gliwice, Poland
Jacek.Stoj@polsl.pl

Abstract. Designers of distributed computer systems have to take into account many requirements concerning the system operation and functionalities. In case of distributed real-time systems there are two characteristic features of great importance: reliability and temporal characteristics. High reliability level may be achieved by implementation of redundancy. However, every additional element in the system, also the redundant one, influences its temporal characteristic, one of which is the response time critical for the operation of Networked Control Systems. In the following chapter a system with cost effective computing unit redundancy operating in hot standby mode is described and analyzed from the point of view of its real-time operation. The possibility of implementation of synchronization routines between the computing units is also considered.

Keywords: Distributed computer system · Networked Control Systems
Real-time · Real-time system · Redundancy · Hot-standby redundancy
Synchronization · Real-Time Ethernet

1 Introduction

Industrial computer systems have some requirements that does not apply usually to regular distributed systems. Most important are real-time operation and reliability. For proper operation of industrial systems not only the result of their computation must be correct, but also it must be obtained in a specific time satisfying time constraints defined for a given system. Only then, the system may be considered as a real-time system, in this case referred as Network Control Systems NCS. Moreover, erroneous operation of NCS may have severe consequences like damage done to the environment or endangering life and health of people. Therefore, for safety critical systems, risk evaluation is done in order to decide on the required reliability of a given system.

To achieve appropriate reliability level according to the results of the risk evaluation, often redundancy is used to make the system invulnerable to faults. Redundancy should be considered as an extension of a given system by additional elements which do not provide any new functionalities, but allows the system to continue its operation in spite of a fault of its redundant components. An example is redundancy of computing units, e.g. Programmable Logic Controller PLC often implemented in industry. In that case, there are two PLCs, instead of one, both executing the same program. In a hot standby

L. Borzemski et al. (Eds.): ISAT 2018, AISC 852, pp. 43–53, 2019.
https://doi.org/10.1007/978-3-319-99981-4_5

operation mode, only one PLC is active, i.e. only one reacts to the events taking place in the system. When the active PLC becomes faulty, the other PLC becomes active and takes over the control of the system.

The most drawback considering redundancy is high cost of its application. Therefore, after referring to the related work in the following section, in the paper a cost effective computing unit redundancy is considered in the Sect. 3. It is also analyzed examined with some experimental research presented in Sect. 4. To make the cost effectiveness even higher, Sect. 5 describes some planned future work on the possibility of implementation synchronization routines into the redundant controllers.

2 Related Work

During previous research work, author analyzed and examined the temporal costs of redundancy. The research was focused on a couple of redundant architectures and on two states of systems with redundancy: steady state and transient state. The steady state was defined as the state of a given system with all the redundant in fault-free state [1]. Whereas the transient state was defined as starting on a fault of any component and lasting until a successful switchover to its corresponding redundant element [2].

This paper is a continuation of the above research with reference to Ethernet communication networks which are more and more widely used in industrial systems as relatively cheap in application and capable to support multiple services [3], In some of them the redundancy is possible and easy to implement, like in Profinet network [4]. In others redundancy may be applied, but not so freely, e.g. computing unit redundancy is hard to achieve in EtherCAT networks however ring architecture is quite feasible [5]. In any case the temporal characteristics of the networks should be always considered [6].

Worth noting it the IEC working group which created the IEC 62439 standard that defines redundancy methods applicable to most industrial networks, and which differ on the topology and the recovery time. The standard describes hot-standby switchover redundancy like the Media Redundancy Protocol MRP (IEC 62439-2, [7]), the High availability Seamless Redundancy HSR and Parallel Redundancy Protocol PRP (IEC 62439-3, [8]). The last two are active redundancy approaches that work without reconfiguration timeouts when a single failure in one of its two redundant network structures occurs [9].

3 Cost Effective Computing Unit Redundancy

Full redundancy of computing units may be considered as quite expensive as it would require duplication (or triplication in case of triple redundancy) of all the input/output modules connected to the redundant PLC. For that reason usually only the controller is made redundant, and all input/output modules are connected to the controllers as remote input/output station. The paper is focused on the later solution.

For realization of such architecture the communication interface has to be wisely chosen. It should give detailed information about its status and also provide means of controlling its operation, like exchanging the control on the physical outputs between the redundant controllers. Last but not least, the temporal characteristics of the data exchange should be considered by choosing one kind of networks or another.

While considering the cost effective architecture with computing unit redundancy, one should also adjust the solution to the need of exchanging vast amount of data between the redundant controllers without interfering the real-time communication with the remote input/output station. Data exchange between controllers is crucial to share the current status of the controllers and even sending requests, e.g. to perform control switchover between the redundant computer units.

4 Experimental Research

In the next subsection the testbed of the experimental research is presented. It is followed by description of obtained results and their analysis.

4.1 Testbed

The system, that was taken into consideration during the experimental research, consisted of one PLC (called later Process PLC) with one remote I/O station called RIO. Communication between the Process PLC and RIO station was based on Ethernet network and Real-Time Ethernet protocol by Beckhoff.

The Process PLC was a CX5140 embedded device by Beckhoff with no local I/O modules and no additional communication interfaces apart from Ethernet. The RIO station consisted of BK9100 Ethernet TCP/IP Bus Coupler by Beckhoff with the following I/O modules: 24VDC digital inputs KL1408 (2 pieces, 8 points each), 24VDC digital outputs KL2408 (2 pieces, 8 points each), 0–20 mA analogue inputs KL3448 (8 channels), 4–20 mA analogue outputs KL4424 (4 channels). That gives 36 bytes of user data, including status information about the state of the analogue inputs.

The above configuration will be referred as Basic System. The Basic System was then extended by redundant Process PLC, so it included two Process PLC and one RIO station. As the RIO station had two Ethernet ports, no additional switch was necessary. The later configuration will be referred as Redundant System in the paper and is presented in Fig. 1.

In the Redundant System both Process PLCs should receive data about the current state of RIO inputs, but only one Process PLC may control outputs of the RIO station. Usage of Real-Time Ethernet communication protocol gives the possibility to do so. It may be freely chosen during run-time whether a given PLC should only receive information about the state of inputs or may also control the outputs. The Process PLC monitors all the time the activity of the other PLC and the RIO station and makes decision about the required mode of operation.

For the needs of the experimental research of the systems in both configurations (Basic and Redundant), another PLC was provided (it is not depicted in Fig. 1.). It is called Diagnostic PLC. Some of its inputs were connected to the outputs of the RIO. Accordingly, some of its outputs were connected to the inputs of the RIO. It is described in more details in the following subsections.

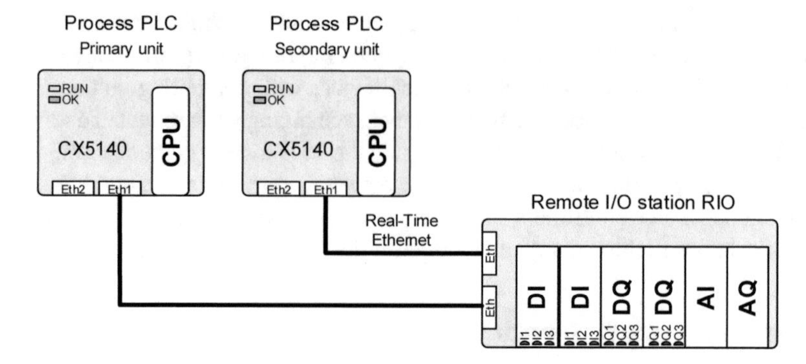

Fig. 1. The testbed – Redundant System

4.2 Square Wave in the Basic System

First goal of the experimental research was to determine how frequently outputs of the RIO may be updated in a timely manner. That way, it will be known how frequent outputs may be updated during the system operation. For this purpose a square wave was generated on one of the RIO outputs with different frequencies (the square wave output was updated every 1 ms). The output with the square wave signal was connected to the input of the Diagnostic PLC, which measured the time of the high and low level of the signal. The results for 3 square wave frequencies, are presented in the Fig. 2. On the *x*-axis the duration of the high or low level signal is given in milliseconds scaled logarithmically. The *y*-axis shows the number of measurements for a given signal duration. For example, there were over 40,000 measurements with the time equal 20 ms for 25 Hz square wave.

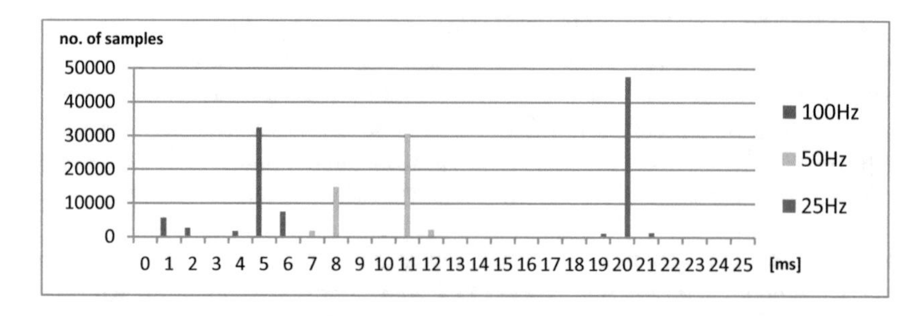

Fig. 2. Square wave measurement in a single node system – Basic System

In case of the 100 Hz square wave, the high and low level signal lasts 5 ms. In the graph in Fig. 2 it may be noticed that apart from the dominant value 5 ms there were also another values registered. The value of 6 ms suggests that some of the output updates were late, which was caused by the 1 ms square wave update period – the PLC program was being executed with 1 ms cycle time.

Interestingly, there are also values of 1 ms and 2 ms – occurrence of these values is caused by the 3 ms input filtering time of the Diagnostic PLC (it is described in more details in the next section). Moreover, after the 2 ms value there is a 2 ms gap, and then the 4 ms value. The 50 Hz square wave is a similar case – a gap of 3 ms is visible. It suggests that the outputs of the RIO may be updated not more frequent than every 2–3 ms.

In case of 25 Hz signal the measured values are mainly of the expected value of 20 ms. It means that every change of state of the square wave of that frequency was updated on the RIO output in time.

That above observations are important for the analysis of the request-response experiment results described in the following section.

4.3 Request-Response Experiment

After the basic square wave experiments, the response time was measured for both Basic and Redundant Systems. One of the outputs of the Diagnostic PLC was being energized cyclically. The output was connected to the input of the RIO and its state was logically copied in the user program of the Process PLC to the output of the RIO station which was connected to the input of the Diagnostic PLC. The Diagnostic PLC was measuring the time from output activation, to the change of the state of the input. The results are presented in Fig. 3 for Basic System and Fig. 4 for Redundant System.

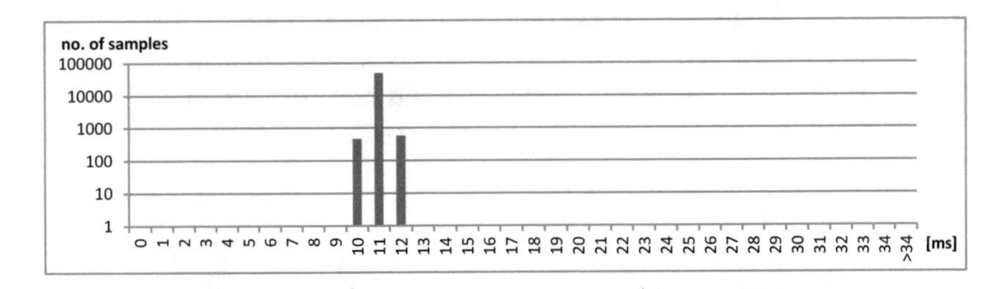

Fig. 3. Response time of the Process PLC in the Basic System

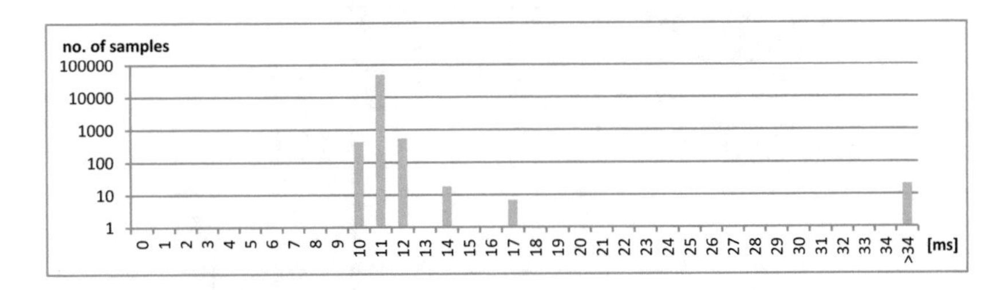

Fig. 4. Response time of the Process PLC in the Redundant System

As shown in Figs. 3 and 4 the response time in Basic and Redundant System is very similar. Only in Redundant System some measurements were greater than 12 ms, which it the longest response time registered for Basic Configuration. For Redundant System some responses were received after 14 ms, 17 ms or even more than 34 ms (the precise value of the latest measurement are not know because of the way the measurements were stored by the Diagnostic PLC). It seems that sometimes there were some problems with

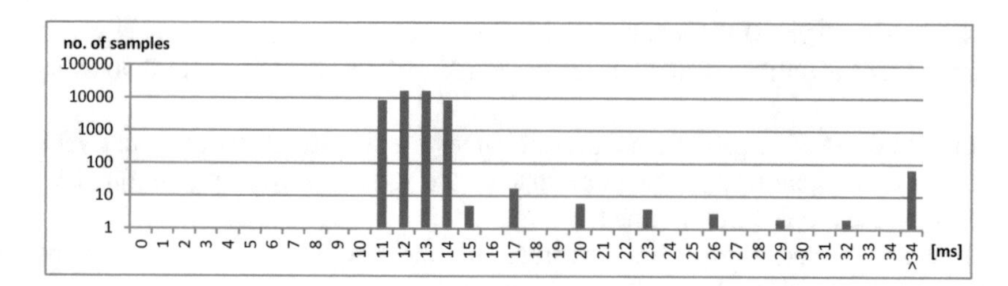

Fig. 5. Response time of the Process PLC in the Redundant System while switching of the controlling control source

communication between the Process PLC and RIO station. It was also noticed that during the experiment the "Error" diode on the BK9100 module was lit. That requires further analysis. It is worth noticing however, that the number of measurements with the value greater than 12 ms was 52, while total the number of measurements was 50,000. So the erroneous measurements occurred in about 0,1% of total number of measurements.

After measuring the response time in Redundant System another experiment was conducted concerning the redundant architecture. Again the response time was measured, but apart from that, the Diagnostic PLC was sending request to the Process PLC to perform a control switchover between the PLCs. It was sending the switchover request by its outputs connected to the RIO inputs. It received confirmation of successful switchover on its inputs. Every new switchover was requested after the previous one had been realized. The diagram of the testbed is presented in Fig. 6.

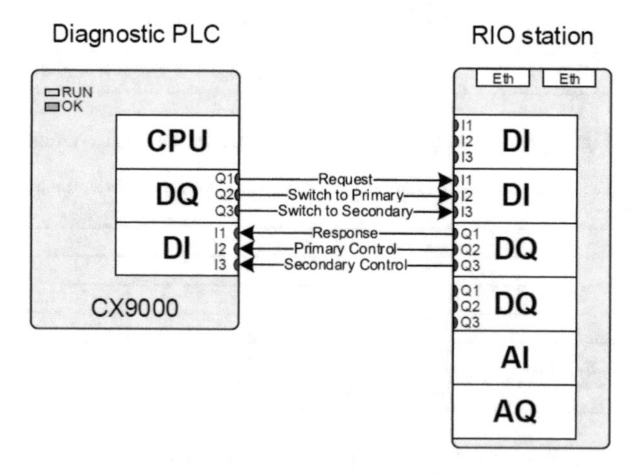

Fig. 6. Connections between Diagnostic PLC and RIO station for experiments with the control switchover

The results of the experiment are presented in Fig. 5. The response time is greater by about 1 ms while realizing the switchover routines than in the Redundant System without switchover (compare with Fig. 5). Moreover, the values are more equal, i.e. without one "outstanding" value with much greater number of occurrences like 11 ms in Fig. 4. There are also some values much greater than the rest, e.g. 17 ms, 20 ms, 23 ms etc. That again suggests some problem with communication. The number of those values was 103, while the total number of measurements was 50,000. So the erroneous measurements occurred in about 0, 2% cases. Nevertheless, occurrence of such delays requires further analysis and experiments.

Please note that the gaps between measured values in case of Redundant System (both with switching of control and without), that are longer than the most common values, like the ones mentioned above, have always the width of 3 ms. That corresponds to the update interval recognized during the square wave experiment (see: Sect. 4.2).

4.4 Temporal Considerations

The greatest response time measured in Basic System was 12 ms (see: Fig. 3) and in Redundant System, while realization of controller switchover, it was 15 ms (see: Fig. 5; the values which suggests communication faults as described in previous section, are not considered here). According to the worst case principle [10], the response time should be not more than the sum of the following elements:

- T_{IF} – request filtering time,
- T_{RIOi} – acquisition of the request (input state) by the RIO station,
- T_{NETi} – sending of the request to the controller,
- T_{CPU} – processing of the request by the controller,
- T_{NETo} – sending of the response to the request to the RIO station,
- T_{RIOo} – reception of the response by the RIO station and output update,
- T_{DIAG} – generation of the requests, reception of the responses and storing of the measurements,

The input filtering time T_{IF} is a delay in recognition of new input states. The filtering is included in the input module and is implemented for noise reduction. In case of the modules used in the experiment it has the value of 3 ms,

The acquisition time of the request T_{RIOi} and the output update time T_{RIOo} have the same values as they are included in the same RIO routines. It is described in the vendor's manual as "reaction time" which is declared to be not more than K-BUS backplane communication cycle plus 1 ms. The K-BUS cycle time depends on the hardware configuration and was equal 1.74 ms.

The value T_{NETi} and T_{NETo} may be considered similarly to T_{RIOi} and T_{RIOo} as equal. Moreover, the time needed for receiving and sending data between controllers and RIO station in case of Ethernet interface could me omitted, e.g. considering it from the point of view of input filtering time $T_{IF} = 3$ ms which is much greater than transmission of one Ethernet datagram.

To process the request and prepare the response by the controller at most 2 PLC cycles are needed. One cycle is the maximum time of the request acquisition, the other cycle includes the request processing. One cycle is realized every 1 ms.

The time needed for performing the experiment T_{DIAG} is equal to the response time in a single node system, which is a sum of filtering time and two diagnostic PLC cycle times. The filtering time and cycle time of the Diagnostic PLC and Process PLC are equal and have the value of 3 ms and 1 ms accordingly.

According to the above remarks, the response time T_{Res} is as follows:

$$\mathrm{T}_{\mathrm{Res}} = T_{IF} + T_{RIOi} + T_{NETi} + 2 \cdot T_{CPU} + T_{NETo} + T_{RIOi} + T_{DIAG} \quad (1)$$

and equals T_{Res}= 3 + (1 + 1.74) + 0 + 2 1 + 0 + (1 + 1.74) + (3 + 2·1) = 15.48 ms. The value corresponds to the measured response time which was not more than 15 ms. As already mentioned, at this point the values of 17 ms, 20 ms etc. visible in the Fig. 5 are not considered, as they are recognized as a consequence of some communication errors which requires further examination.

5 Future Works

During the switchover of control of redundant PLCs operating in hot standby mode, some unwanted changes of outputs states may occur. It is the result of independent execution of user programs implemented in both PLCs. Let's assume that two redundant PLCs called PLC1 and PLC2 are sending a square wave signal to the outputs. The output is set according to the state of the one of the PLCs, called Active PLC. Therefore, only the Active PLC is realizing the actual control of the outputs. The other PLC is a Backup PLC (see: Fig. 7).

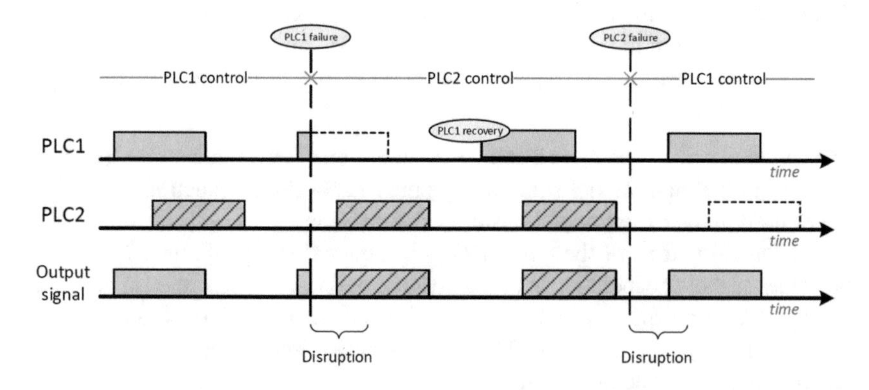

Fig. 7. Output state disruption during control switchover

At the beginning the PLC1 is the Active one. After a failure of the Active PLC (described as "PLC1 failure" in Fig. 7), the Backup PLC becomes active and starts controlling the outputs. The output signal changes to the state provided by that PLC – the PLC2. After the switchover, disruption on the output is present, as the square wave in PLC1 and PLC2 were not in the same phase (were shifted in time). Another disruption is present on the output after recovery of the PLC1 (which starts up as a Backup PLC) and failure of the PLC2.

To provide bumpless switchover, i.e. without disruptions like presented in Fig. 7, synchronization of the redundant PLCs is necessary, so that (among other things) the requested state of the outputs was corresponding to each other in both redundant PLCs. There would be not phase shift of the output signals in that case.

There are many hardware modules available on the market for synchronization of redundant PLCs. An example is Redundancy Memory Exchange RMX module for PacSystems Controllers by General Electric. The module provides fiber optic communication interface for data exchange between PLCs. The controllers state is synchronized in two points – input and output transfer point, as depicted in Fig. 8.

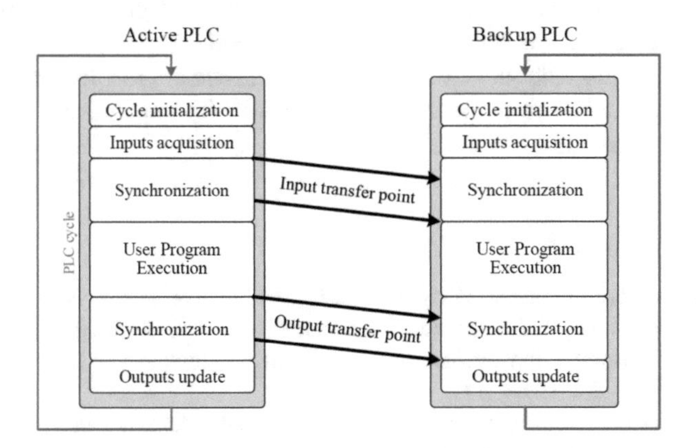

Fig. 8. Synchronization of redundant PLC in RMX modules by General Electric

The goal of future work is to realize the routines of synchronization of redundant PLCs using regular Ethernet connection, which would give much more cost effective solution in comparison to application of expensive dedicated modules. The idea is to implement the input and output synchronization points inside the user program executed by redundant PLCs. Typically, the user program includes some control algorithms that analyze input signals states and react to their state by updating output signals state.

While realizing the cost effective synchronization, in every PLC cycle only one out of three task would be executed: inputs synchronization, user program, outputs synchronization. It means that in consecutive cycles:

- Firstly, only inputs synchronization will be realized,
- Secondly, after successful synchronization, the user program execution will occur,
- Finally, after program execution, outputs synchronization will be performed and the PLC will start inputs synchronization again.

The question is, how much time would be needed for this kind of synchronization, of course taking into consideration the size of the system and the length of the synchronized memory area. Another problem is defining the best reaction of the Active PLC to an unsuccessful synchronization, which may be caused by for example communication link failure. The above and other problems will be analyzed in future works.

6 Final Remarks

As presented in the paper, it is possible to create a cost effective NCS architecture with computing unit redundancy using remote input/output station and Real-Time Ethernet protocol. It is a feasible task even if the vendor does not provide any "of the shelf" hardware or software modules for this kind of system. Of course, it should be expected that the additional, redundant elements in the system introduce additional delay which lengthens the response time. However, the decrease isn't great and in case of the analyzed system should not be considered as significant.

The presented research work will be continued. Synchronization of the redundant controllers, as described in the previous section, will be implemented and examined during more experimental research.

References

1. Kwiecień, A., Stój, J.: The cost of redundancy in distributed real-time systems in steady state. In: Kwiecień, A., Gaj, P., Stera, P. (eds.) Computer Networks. CN 2010. Communications in Computer and Information Science, vol. 79. Springer, Heidelberg (2010). https://doi.org/10.1007/978-3-642-13861-4_11
2. Stój, J., Kwiecień, A.: Temporal costs of computing unit redundancy in steady and transient state. In: Kosiuczenko, P., Madeyski, L. (eds.) Towards a Synergistic Combination of Research and Practice in Software Engineering. Studies in Computational Intelligence, vol. 733. Springer, Cham (2018). https://doi.org/10.1007/978-3-319-65208-5_1
3. Toral-Cruz, H., Mihovska, A.D., Gaj, P., et al.: Advances and challenges in convergent communication networks. Wirel. Pers. Commun. **96**, 4919 (2017)
4. Flatt, H., Schriegel, S., Neugarth, T., Jasperneite, J.: An FPGA based HSR architecture for seamless PROFINET redundancy. In: 2012 9th IEEE International Workshop on Factory Communication Systems, Lemgo 2012, pp. 137–140 (2012). https://doi.org/10.1109/wfcs.2012.6242555
5. Maruyama, T., Yamada, T.: Spatial-temporal communication redundancy for high performance EtherCAT master. In: 2017 22nd IEEE International Conference on Emerging Technologies and Factory Automation (ETFA), Limassol, pp. 1–6 (2017)
6. Gaj, P.: Pessimistic useful efficiency of EPL network cycle. In: Conference: 17th International Conference Computer Networks Location. CCIS, Ustron, Poland, 15–19 June 2010, vol. 79, pp. 297–305. Springer, Heidelberg (2010). https://doi.org/10.1007/978-3-642-13861-4_31
7. Zuloaga, A., Astarloa, A., Jiménez, J., Lázaro, J., Araujo, J.A.: Cost-effective redundancy for ethernet train communications using HSR. In: 2014 IEEE 23rd International Symposium on Industrial Electronics (ISIE), 1–4 June 2014, pp. 1117–1122 (2014)
8. Rentschler, M., Heine, H.: The parallel redundancy protocol for industrial IP networks. In: 2013 IEEE International Conference on Industrial Technology (ICIT), 25–28 February 2013, pp. 1404–1409 (2013)

9. Hoga, C.: Seamless communication redundancy of IEC 62439. In: 2011 International Conference on Advanced Power System Automation and Protection (APAP), 16–20 October 2011, vol. 1, pp. 489–494 (2011)
10. Wideł, S., Flak, J., Jestratjew, A., Gaj, P.: Analysis of real-time systems using convolution of probability mass function. In: Proceedings of 2012 IEEE 17th International Conference on Emerging Technologies & Factory Automation (ETFA 2012), pp. 1–4, Krakow (2012). https://doi.org/10.1109/etfa.2012.6489718

Bottleneck Utility-Based Rate Allocation for Autonomic Networks with Quality of Service

Dariusz Gasior(✉)

Faculty of Computer Science and Management,
Wroclaw University of Science and Technology, Wroclaw, Poland
dariusz.gasior@pwr.edu.pl

Abstract. In this paper, we consider the rate allocation problem in autonomic networks with Quality of Service guarantees. We assume that each link is managed by independent agent. Therefore, we consider our problem in the terms of the Game Theory. We introduce the algorithm which provable finds weak Pareto-optimal pure Nash Equilibrium. Such an approach may be implemented according to the autonomic networking paradigm. The initial simulation experiments, which have been performed for small backbone networks, indicate that the proposed method is a very promising one.

Keywords: Game theory · Nash Equilibrium · Quality of Service
Capacity allocation · Utility

1 Introduction

Modern applications, like e.g. UHD television, virtual and augmented reality, make it essential to manage networks in the way that enables appropriate values of transmission parameters like delay, jitter, loss probability, and transmission rate. That is why Quality of Service (QoS) networks attract so much attention.

While fulfilling the QoS requirements makes the application useful, the allocation of the additional resources allows improving the utility of the transmission [1,2]. Even though a lot of works has been devoted to the rate allocation problem over the last years, there is still a lack of effective resource allocation methods for such demanding applications like those mentioned earlier.

However, contemporary computer networks become very complex, difficult to manage and complicated to configure systems. This results in the rapid grow of operational costs. The idea of self-organizing, self-managing or autonomic networking seems to be the remedy for this problem ([3]). It is assumed that each network component may monitor, optimize and configure itself without human operator intervention. These local activities lead to effective management of the entire network.

In this paper, we propose to apply the autonomic networking concept which enables self-managing [4] to solve the rate allocation problem in the networks

L. Borzemski et al. (Eds.): ISAT 2018, AISC 852, pp. 54–64, 2019.
https://doi.org/10.1007/978-3-319-99981-4_6

with QoS guarantees. We assume that there is an agent associated with every link that wants to maximize its total utility and to ensure that QoS constraints are fulfilled for all transmission demands traversing this link. Moreover, we assume that the transmission routes are given in advance. We also limit the QoS parameters under consideration to the transmission rate, while it is believed to be the most crucial one [5]. The other QoS requirements may be satisfied with other network mechanisms. Since each link is managed independently, the problem under consideration is formulated as a game. The allocation algorithm is introduced and it is shown that it finds pure Nash Equilibrium.

2 Mathematical Model and Problem Formulation

We assume that the network consists of L links. Each link l is characterized by its capacity C_l and is managed independently by an agent. This parameter reflects the maximal amount of data which may be sent between nodes in the unit time. There are also R transmissions (which are called also flows). Each flow is described by the route. A route is a sequence of links used to transmit data with a particular flow. The routes are defined with the routing variable a_{rl} which indicates if l link is used for rth transmission. The capacity allocation x_{rl} is made on each link l traversed by the rth flow. The capacity allocations are calculated in each link by its agent. The rth transmission's rate allocation $\overline{x}_r$ results from the capacity allocations made on all links along its route. This rate must fulfill the QoS requirements expressed in terms of the minimum acceptable value $x_{r,\min}$. On the other hand, the links' capacities must not be exceeded. Like in [1], we also assume that there is a utility function $f(\overline{x}_r; w_r)$ associated with every transmission r which reflects , i.e. satisfaction level perceived when the transmission rate is $\overline{x}_r$ and the rth flow priority is w_r. The aim of each link's agent is to maximize the total utility from all transmissions traversing the corresponding link. The summary of notation is given in Table 1.

We introduce this problem as a noncooperative game [6]. The associations with the game theory are as follows. The links' agents are players (there are L players - one for each link). The feasible allocation for transmissions traversing particular domain $\mathbf{x}_l$ is a strategy. The allocation matrix $\mathbf{x}$ is a strategy profile. The objectives $Q_l(\mathbf{x}_d, \mathbf{x}_{-d})$ are the players' payoffs. We refer to this game as the Quality of Service Capacity Allocation Game (QCAG).

Formally, each player l (link's agent) solves the following optimization problem:

Given:

$$R, L, a_{rl}, C_l, w_r, \beta_r, \alpha, x_{\min,r}$$

Find:

$$\mathbf{x}_l^* = \arg\max_{\mathbf{x}_l} Q_l(\mathbf{x}_l, \mathbf{x}_{-l}) \quad (1)$$

such that:

$$\textstyle\sum_{r=1}^{R} a_{rl}x_{rl} \le C_l$$

$$\forall_{r=1,2,\ldots,R} \quad x_{rl} \ge a_{rl}x_{\min,r}$$

Table 1. Notation

L	Number of links
R	Number of flows (transmissions)
C_l	lth link's capacity ($l = 1, 2, \ldots, L$)
$\mathbf{a} = [a_{rl}]_{r=1,2,\ldots,R;l=1,2,\ldots,L}$	Routing matrix: $a_{rl} = 1, r$th flow traverse lth link; $a_{rl} = 0$, otherwise
w_r	rth flow priority parameter
$x_{r,\min}$	rth flow minimal transmission rate (QoS paramter)
$x_{rl} \geq 0$	lth link's capacity allocation for rth flow
$\hat{\mathbf{x}}_l$	lth link's capacity allocation vector: $\hat{\mathbf{x}}_l = [x_{rl}]_{r=1,2,\ldots,R}$
$\overline{x}_r \geq 0$	rth flow transmission rate: $\overline{x}_r = \min_{l:a_{rl}=1} x_{rl}$
$f(\overline{x}_r; w_r) = w_r\varphi(\overline{x}_r)$	rth flow utility function $\varphi(\overline{x}_r) = \begin{cases} \frac{\overline{x}_r^{(1-\alpha)}}{(1-\alpha)} & \alpha \geq 0 \ \wedge \ \alpha \neq 1 \\ \ln \overline{x}_r & \alpha = 1 \end{cases}$
$Q_l(\mathbf{x}_l, \mathbf{x}_{-l})$	The objective of the dth domain (payoff): $Q_l(\mathbf{x}_l, \mathbf{x}_{-l}) = \sum_{r=1}^{R} a_{rl}\beta_r f(\overline{x}_r; w_r) = \sum_{r=1}^{R} a_{rl}\beta_r w_r \varphi(\min_{j:a_{rj}=1} x_{rj})$
$\beta_r \geq 0$	given coefficient for rth flow e.g. $\beta_r = 1$ lub $\beta_r = (\sum_{l=1}^{L} a_{rl})^{-1}$
$\mathbf{x}_{-l}$	Allocation matrix $\mathbf{x}$ without lth link allocation, it is assumed: $\mathbf{x} = [\mathbf{x}_l, \mathbf{x}_{-l}]$

3 Game Properties

Since the considered problem is a noncooperative game, it is crucial to determine if the stable solutions (i.e. equilibria) exist. There are many concepts of equilibria. In this paper, we mainly focus on the pure Nash Equilibrium (PNE) [7]. If all agents choose PNE as their strategy, it ensures that no entity will change the strategy if the network state changes (e.g. when a new transmission appear or when a transmission disappear or when a new link is established or when a link is broken).

Remark 1. There exists at least one PNE for the considered game (QCAG). This remark may be justified with Rosen theory [11] since the formulated game is convex. We may also formulate the sufficient condition for the strategy to be PNE. Let us introduce the following auxiliary notation. A set of local transmissions $R^{(loc)} = \{r \in \{1, 2, \ldots, R\} : \sum_{l=1}^{L} a_{rl} = 1\}$ (i.e. set of transmissions traversing only one link). A set of transmission traversing more than one link $R^{(glob)} = \{r \in \{1, 2, \ldots, R\} : \sum_{l=1}^{L} a_{rl} > 1\}$. Finally, let $\hat{\mathbf{x}}_l(\overline{C}; \overline{R})$ be an optimal solution of the problem (1) for one link ($L = 1$) and the following problem parameters: $C_1 = \overline{C}, a_{r1} = \begin{cases} a_{rl} & \text{if } r \in \overline{R} \\ 0 & \text{otherwise} \end{cases}$ (for each $r = 1, 2, ..., R$), where $\hat{x}_{rl}(\overline{C}; \overline{R})$ is rth component of $\hat{\mathbf{x}}_l(\overline{C}; \overline{R})$.

Theorem 1. *If there exists an allocation* $\mathbf{x}$ *that fulfills the following conditions:*

- $\forall_{l_1,l_2} \forall_{r \in R^{(glob)}} \quad a_{rl_1} \max\{\hat{x}_{rl_1}(C_{l_1}; \{1, 2, \ldots, R\}), x_{\min,r}\} \geq a_{rl_1} x_{rl_1} = a_{rl_1} x_{rl_2} \geq a_{rl_1} x_{\min,r}$
- $\forall_{l=1,2,\ldots,L} \forall_{r \in R^{(loc)}} \quad x_{rl} = \hat{x}_{rl}(C_l - \sum_{r \in R^{(glob)}} a_{rl} x_{rl}; R^{(loc)})$

then $\mathbf{x}$ *is the PNE for QCAG.*

Before, we prove the above theorem, we must introduce four lemmas.

Lemma 1. *For each pair of feasible allocations* $\mathbf{x}$ *and* $\mathbf{x}'$ *such that* $\exists_{r \in R^{(glob)}} x_{rl} \leq x'_{rl}$ *the following equality holds* $Q_l([x'_{1l}, x'_{2l}, \ldots, x'_{(r-1)l}, x_{rl}, x'_{(r+1)l}, \ldots, x'_{Rl}], \mathbf{x}_{-l}) = Q_l(\mathbf{x}'_l, \mathbf{x}_{-l})$.

Proof (Proof of the Lemma 1*).* Since the following equality is met:

$$\min_{j:a_{rj}=1} x_{rj} = \min\{x'_{rl}, \min_{j \neq l:a_{rj}} x_r j\},$$

thus we obtain:

$$Q_l(\mathbf{x}_l, \mathbf{x}_{-l}) = \sum_{r=1}^{R} a_{rl} \beta_r w_r \varphi(\min_{l:a_{rl}=1} x_{rl})$$

$$= \sum_{r=1}^{R} a_{rl} \beta_r w_r \varphi(\min\{x'_{rl}, \min_{j \neq l:a_{rj}=1} x_{rj}\}) = Q_l(\mathbf{x}'_l, \mathbf{x}_{-l}).$$

□

Lemma 2. *Let us denote:* $\hat{x}_{r_1 l} \triangleq \hat{x}_{r_1 l}(C_l; \{1, 2, \ldots, R\})$ *i* $\hat{x}_{r_2 l} \triangleq \hat{x}_{r_2 l}(C_l; \{1, 2, \ldots, R\})$ *For any* l *and for any feasible allocation* $\mathbf{x}_l$*, if* $\exists_{r_1,r_2} \; x_{r_1 l} \leq \hat{x}_{r_1 l} \;\wedge\; x_{r_2 l} \geq \hat{x}_{r_2 l}$ *then* $\forall_{\delta>0:\; x_{r_1 l}-\delta \geq x_{\min,r_1}} \; f(x_{r_1 l}; w_{r_1}) + f(x_{r_2 l}; w_{r_2}) \geq f(x_{r_1 l}-\delta; w_{r_1}) + f(x_{r_2 l}+\delta; w_{r_2})$ *equivalently:* $\forall_{\delta>0:\; x_{r_1 l}-\delta \geq x_{\min,r_1}} \; f(x_{r_1 l}; w_{r_1}) + f(x_{r_2 l}; w_{r_2}) - (f(x_{r_1 l} - \delta; w_{r_1}) + f(x_{r_2 l} + \delta; w_{r_2})) \geq 0$.

Proof (Proof of the Lemma 2*).* The following two statements results from the lemma's assumption and the concavity of the objective function: Since $\hat{x}_l$ is optimal solution of (1) for $L = 1$, and taking into account lemma's assumption, we obtain:

$$f(\hat{x}_{r_1 l}; w_{r_1}) + f(\hat{x}_{r_2 l}; w_{r_2}) - (f(\hat{x}_{r_1 l} - \delta; w_{r_1}) + f(\hat{x}_{r_2 l} + \delta; w_{r_2})) \geq 0. \quad (2)$$

From the concavity of the utility functions, we obtain:

$$\begin{aligned} & f(x_{r_1 l}; w_{r_1}) - f(x_{r_1 l} - \delta; w_{r_1}) \quad (3) \\ & \geq f(\hat{x}_{r_1 l}; w_{r_1}) - f(\hat{x}_{r_1 l} - \delta; w_{r_1 l}) f(\hat{x}_{r_2 l} + \delta; w_{r_2}) - f(\hat{x}_{r_2 l}; w_{r_2}) \\ & \geq f(x_{r_2 l} + \delta; w_{r_2}) - f(x_{r_2 l}; w_{r_2}). \end{aligned}$$

Summing ineqaulties resulting from (2) and applying (3) $f(x_{r_1l}; w_{r_1}) + f(x_{r_2l}; w_{r_2}) - (f(x_{r_1l} - \delta; w_{r_1}) + f(x_{r_2l} + \delta; w_{r_2})) \geq f(\hat{x}_{r_1l}; w_{r_1}) + f(\hat{x}_{r_2l}; w_{r_2}) - (f(\hat{x}_{r_1l} - \delta; w_{r_1}) + f(\hat{x}_{r_2l} + \delta; w_{r_2})) \geq 0$ □

Lemma 3. *For any l, $\forall_{r\in\{1,2,\ldots,R\}}$ $\hat{x}_{rl}(C_l;\ \{1,2,\ldots,R\})$ is nondecreasing againts C_l.*

Proof (Proof of the Lemma 3 (indirect)).
Indirect proof assumption:

$$\exists_{i\in\{1,2,\ldots,R\}}\ \exists_{C_l\geq 0\ \wedge\ \Delta>0}\ \ \hat{x}_{il}(C_l;\{1,2,\ldots,R\}) > \hat{x}_{il}(C_l+\Delta;\{1,2,\ldots,R\})$$

It leads to the following results:

$$\begin{aligned}\exists_{\overline{R}_l\subseteq\{1,2,\ldots,R\}}\ (\forall_{r\in\overline{R}_l}\ \exists_{\delta_r>0}\ \hat{x}_{rl}(C_l+\Delta;\{1,2,\ldots,R\}) &= \hat{x}_{rl}(C_l;\{1,2,\ldots,R\}) + \delta_r)\\ \wedge\ \exists_{n\notin\overline{R}_l}\ \exists_{\delta_n>0,\epsilon_n\geq 0}\ \hat{x}_{nl}(C_l+\Delta;\{1,2,\ldots,R\}) &= \hat{x}_{nl}(C_l;\{1,2,\ldots,R\}) + \delta_n + \epsilon_n\\ \wedge\ \hat{x}_{il}(C_l+\Delta;\{1,2,\ldots,R\}) &= \hat{x}_{il}(C_l;\{1,2,\ldots,R\}) - (\sum_{r\in\overline{R}_l}\delta_r + \delta n)\end{aligned}$$

From Lemma 2 we have

$$\begin{aligned}&f(\hat{x}_{il}(C_l;\{1,2,\ldots,R\}); w_i) + \sum_{r\in\overline{R}_l} f(\hat{x}_{rl}(C_l;\{1,2,\ldots,R\}); w_r)\\ &+ f(\hat{x}_{nl}(C_l;\{1,2,\ldots,R\}) + \epsilon_n; w_n) \geq f(\hat{x}_{il}(C_l;\{1,2,\ldots,R\})\\ &-(\sum_{r\in\overline{R}_l}\delta_r + \delta n); w_i) + \sum_{r\in\overline{R}_l} f(\hat{x}_{rl}(C_l;\{1,2,\ldots,R\}) + \delta_r; w_r)\\ &+ f(\hat{x}_{nl}(C_l;\{1,2,\ldots,R\}) + \delta_n + \epsilon_n; w_n)\end{aligned}$$

It contradicts with indirect proof assumption, because f is nondecreasing. □

Lemma 4. *For any l, let $\mathbf{x}_l$ be the feasible allocation and let $\overline{R}_l = \{r : x_{rl} \leq \hat{x}_{rl}(C_l;\{1,2,\ldots,R\})\}$ then*

$$\hat{x}_{rl}(C_l;\{1,2,\ldots,R\}) \leq \hat{x}_{rl}(C_l - \sum_{r\in\overline{R}_l} a_{rl}x_{rl};\{1,2,\ldots,R\}\setminus\overline{R}_l)$$

Proof (Proof of the Lemma 4). Since $\forall_{r\in\{1,2,\ldots,R\}\setminus\overline{R}_l}$ $\hat{x}_{rl}(C_l,\{1,2,\ldots,R\}) = \hat{x}_{rl}(C_l - \sum_{r\in\overline{R}_l} a_{rl}\hat{x}_{rl}(C_l,\{1,2,\ldots,R\}),\{1,2,\ldots,R\}\setminus\overline{R}_l)$.
From the Lemma 3, we obtain: $\hat{x}_{rl}(C_l - \sum_{r\in\overline{R}_l} a_{rl}\hat{x}_{rl}(C_l,\{1,2,\ldots,R\}), \{1,2,\ldots,R\}\setminus\overline{R}_l) \leq \hat{x}_{rl}(C_l - \sum_{r\in\overline{R}_l} a_{rl}x_{rl},\{1,2,\ldots,R\}\setminus\overline{R}_l)$. □

Now, we can give a proof of the Theorem 1.

Proof (Proof of the Theorem 1). Let $\mathbf{x}_l$ fulfills the assumptions of the Theorem 1, then:

1. From Lemma 1: for any $r \in R^{(glob)}$ increasing allocation cannot improve payoff.
2. From Lemma 2 taking into account Lemma 4 decreasing of the allocation for any transmission $r \in R^{(glob)}$ cannot increase payoff.
3. For any $r \in R^{(loc)}$ allocation $\mathbf{x}_l$ is optimal allocation of the remaining capacity. Therefore, we cannot increase payoff by alternating this allocation.

Every allocation $\mathbf{x'}_l$ may be obtained from $\mathbf{x}_l$ with a sequence of changes considered in above three points.
Thus, we cannot change the allocation $\mathbf{x}_l$ in such a way that the payoff is increased. It means that $\mathbf{x}_l$ is PNE of the considered game. □

To evaluate the equilibria points the concepts of Social Welfare, Price of Anarchy (PoA) and Price of Stability (PoS) should be introduced ([8]). First, we define the following Social Welfare function:

$$SW \triangleq Q(\mathbf{x}).$$

Then the Price of Anarchy (PoA) is defined as:

$$PoA = \frac{SW^*}{\min_{\mathbf{x} \in SNE} SW(\mathbf{x})}$$

where: SNE - set of all PNE, $SW^* = \max_{\mathbf{x} \in \hat{D}_{\mathbf{x}}} SW(\mathbf{x})$, $\hat{D}_{\mathbf{x}}$ - set of all feasible strategy profiles.
Finally, the Price of Stability (PoS) is defined as:

$$PoS = \frac{SW^*}{\max_{\mathbf{x} \in SNE} SW(\mathbf{x})}$$

The PoA measures the deterioration of the SW when each player makes decision independently (non-cooperative game) in comparison to the situation when all players act for common good (maximal social welfare) for the worst case. While, PoS measures the deterioration of the SW for the best case.

Remark 2. If the Social Welfare is defined as: $SW \triangleq Q(\mathbf{x})$ then POA may be unbounded while PoS may be equal to 1.

4 Rate Allocation Algorithm for Autonomic QoS Networks

To solve the considered problem, we propose the BURA-Q algorithm (Algorithm 2) which should be implemented by each agent (on each link).

The main rate allocation algorithm (BURA-Q) uses the procedure called SQRA (Algorithm 1), which is introduced firstly. The presented method finds potentially the best allocation from the local link's point of view, while the QoS requirements are preserved.

Algorithm 1. Single link QoS Rate Allocation Algorithm (SQRA)

Obtain: $\tilde{R}, \tilde{C}_l, \mathbf{x}_l$
Ensure: $\tilde{\mathbf{x}}_l$
1: $\hat{R} \leftarrow \tilde{R}$, $\forall_{l \in \tilde{L}} \hat{C}_l \leftarrow \tilde{C}_l$, $qosviolated \leftarrow$ True
2: **for** $r \in \{1, 2, \ldots, R\}$ **do**
3: **if** $r \notin \hat{R}$ **then**
4: $\tilde{x}_{rl} \leftarrow x_{rl}$
5: **end if**
6: **end for**
7: **while** $\hat{R} \neq \emptyset \ \wedge \ qosviolated =$ True **do**
8: $\forall_{r \in \hat{R}}$ Update: $\tilde{x}_{rl} = \frac{a_{rl}(w_r \beta_r)^{1/\alpha}}{\sum_{q \in \hat{R}} a_{ql}(w_q \beta_q)^{1/\alpha}} \hat{C}_l$
9: **for** $r \in \hat{R}$ **do**
10: $qosviolated \leftarrow$ False
11: **if** $\tilde{x}_{rl} < x_{\min,r}$ **then**
12: Update: $\tilde{x}_{rl} \leftarrow x_{\min,r}$, $\hat{R} \leftarrow \hat{R} \setminus r$, $\hat{C}_l - x_{\min,r}$
13: Update: $qosviolated \leftarrow$ True
14: **end if**
15: **end for**
16: **end while**
17: Return $\tilde{\mathbf{x}}_l = [\tilde{x}_{1l}, \tilde{x}_{2l}, \ldots, \tilde{x}_{Rl}]$

Now, we can utilize the introduced procedure. Initial allocations are computed with the use of the Algorithm 1 (SQRA). Then, all sources start transmitting data and rates of all flows increase. The rate of a flow stops accelerating when some link on the path becomes a bottleneck. It means that those link's capacity is exhausted. Each flow that traverses such a bottleneck is called a saturated flow. In consequence, its rate cannot be increased any more. Nevertheless, other links on the path of such flow may have spare capacity. So, it is possible to assign this capacity among the non-saturated flows, increase their rates (using SQRA algorithm) and repeat that until all the flows become saturated.

Algorithm 2. Bottleneck Utility-based Rate Allocation with QoS Algorithm (BURA-Q)

1: $n \leftarrow 1$, $\tilde{R} \leftarrow \{1, 2, \ldots, R\}$, $\tilde{L} \leftarrow \{1, 2, \ldots, L\}, \forall_{l \in \tilde{L}} \tilde{C}_l \leftarrow C_l$
2: **while** $n \leq L$ **do**
3: $qos \leftarrow False$
4: **while** $qos = False$ **do**
5: $qos \leftarrow True$
6: $\forall_{l \in \tilde{L}}$ Calculate: $\tilde{\mathbf{x}}_l$ with SQRA algorithm (Algorithm 1) for $\tilde{R}, \tilde{C}_l, \mathbf{x}_l$.
7: $\forall_{r \in \tilde{R}} \forall_{l \in \tilde{L}: a_{rl}=1}$ Update: $x_{rl} = \min_{l: a_{rl}=1} \tilde{x}_{rl}$
8: $R_{qos} \leftarrow \{r \in \tilde{R} : x_{rl} = x_{\min,r}\}$
9: **if** $R_{qos} \neq \emptyset$ **then**
10: Update: $\tilde{R} \leftarrow \tilde{R} \setminus R_{qos}$ and $qos \leftarrow False$ and $\forall_{l \in \tilde{L}} \tilde{C}_l \leftarrow \tilde{C}_l - \sum_{r \in R_{qos}} a_{rl} x_{rl}$.
11: **end if**
12: **end while**
13: Find: $l' = \min\{l \in \tilde{L} \sum_{r \in \tilde{R}} a_{rl} x_{rl} = \tilde{C}_l\}$
14: Update: $\tilde{L} \leftarrow \tilde{L} \setminus \{l'\}$ and for all $l \in \tilde{L}$ update: $\tilde{C}_l \leftarrow \tilde{C}_l - \sum_{r \in \tilde{R}} a_{rl'} a_{rl} x_{rl'}$
15: Update: $\tilde{R} \leftarrow \tilde{R} / \{r \in \tilde{R} : a_{rl'} = 1\}$
16: Update $n \leftarrow n + 1$.
17: **end while**
18: Return $\mathbf{x}$.

5 Algorithm Properties

The following properties of the Algorithm BURA-Q occur.

Theorem 2. *Algorithm BURA-Q (Algorithm 2) finds PNE.*

Proof (Proof of the Theorem 2). Allocation obtained in 2 (BURA-Q) fulfills the necessity condition for PNE - see: Theorem 1. □

Theorem 3. *Algoritm BURA-Q (Algorithm 2) finds weak Pareto-optimal PNE.*

Proof (Proof of the Theorem 3). The allocation obtained for the link saturated in the first iteration ($n = 1$) cannot be improved since it is optimal solution of the allocation problem for those link (obtained with SQRA algorithm). □

6 Simulation

In this paper, we present results of some preliminary simulation experiments. The main objective was to initially evaluate the elaborated allocation algorithm. The algorithm was implemented using Python 2.7 programming language and the experiments were conducted using the computer with Intel Core M 1.1 processor and 8 GB RAM.

The simulations was run for $\beta_r = 1$ and the following problem parameters:

- network's parameters:
  - $N \in \{15, 16, 17, 18, 19, 20\}$ - number of nodes,
  - $\rho \in \{0.2, 0.3, 0.4, 0.5, 0.6\}$ - network density, reflecting number of links: $L = \frac{1}{2}\rho N(N-1)$,
  - $C_l \sim U(1000, 10000)$ - lth link capacity (Mbps).
- transmissions' parameters:
  - $R \sim U(\frac{1}{2}\underline{\theta}n(n-1); \frac{1}{2}\overline{\theta}n(n-1))$ - number of flows, $\underline{\theta} = 0.3, \overline{\theta} = 0.7$,
  - for each flow r the pair of nodes (i, j) (source, destination) was randomly selected,
  - a_{rl} - calculated as the shortest path (from source to destination),
  - $w_r \sim U(1, 10)$,
  - $\alpha = 0.5$,
  - $x_{\min,r} \sim U(0.3\hat{C}_r; 0,5\hat{C}_r)$, where $\hat{C}_r = \min_{l:a_{rl}=1} \frac{C_l}{\sum_q a_{ql}}$,

The uniform distribution on the interval $[c, d]$ is denoted by $U(c, d)$. The network topologies were generated with the method described in [10]. The simulation was run three times for each set of parameter values (N, ρ). The considered network instances correspond to the small real-life backbone networks [9].

The quality criterion is defined as follows:

- $\gamma = \frac{SW(\mathbf{x}^{BURAQ})}{\max_{\mathbf{x}\in\hat{D}_{\mathbf{x}}} SW(\mathbf{x})}$, where $SW(\mathbf{x}) = \sum_{r=1}^{R} f(\overline{x}_r; w_r), \hat{D}_{\mathbf{x}} = \{\mathbf{x} : \forall_l \sum_{r=1}^{R} a_{rl}x_{rl} \leq C_l \;\wedge\; \forall_r \; x_{rl} \geq a_{rl}x_{\max,r}\}$,

where $\mathbf{x}^{BURAQ}$ is the solution found by the algorithm. One may notice that $PoS \leq \frac{1}{\gamma} \leq PoA$. The optimal value of the social welfare function $SW(\mathbf{x})$ was found with primal-dual projected gradient method.

The statistical results of the simulation experiments are given in Table 2. The representative chart of the average value of quality criterion for a different number of network nodes and different number of network domains is given in Fig. 1.

Table 2. Simulation results

Statistics for γ	BURA-Q
Average	0.9
Median	0.92
Variance	0.006
Min	0.67
Max	1.0

For the considered networks instances, BURA-Q algorithm finds solution with Social Welfare near the optimal one. It is easily seen that the higher network density, the better algorithm performance. It seems that there is no strict dependence between algorithm quality and the size of network (number of nodes - N). In about 10% cases BURA-Q finds the solution maximizing Social Welfare.

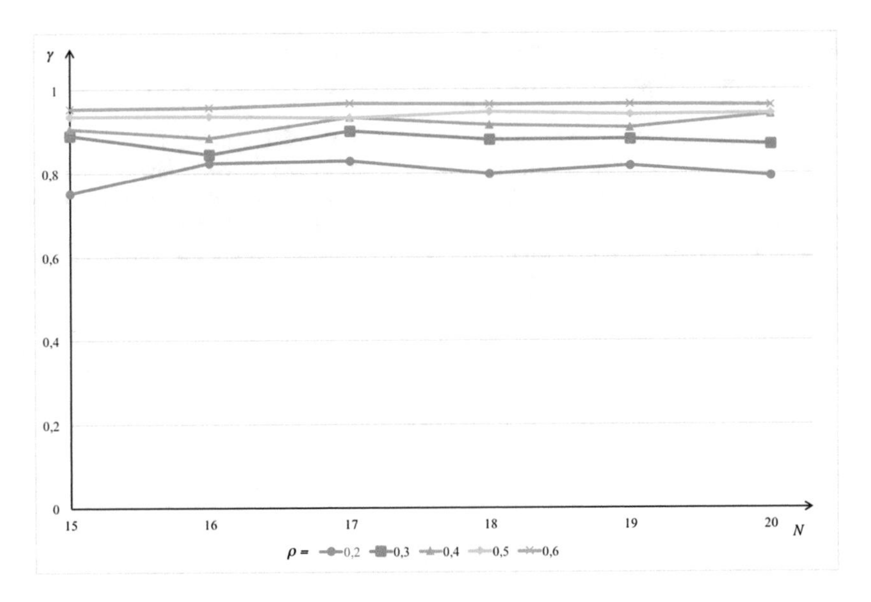

Fig. 1. The average value of quality criterion depending on number of nodes for different network densities.

7 Final Remarks

In this paper, the Quality of Service rate allocation problem in the autonomic network was formulated in terms of Game Theory. The introduced algorithm BURA-Q finds the weak Pareto-optimal PNE for the considered game. The performed simulation for the small backbone networks indicates a very good quality of the proposed method.

References

1. Kelly, F.P., Maulloo, A.K., Tan, D.K.: Rate control for communication networks: shadow prices, proportional fairness and stability. J. Oper. Res. Soc. **49**(3), 237–252 (1998)
2. Gasior, D.: QoS rate allocation in computer networks under uncertainty. Kybernetes **37**(5), 693–712 (2008)
3. Kephart, J.O., Chess, D.M.: The vision of autonomic computing. Computer **36**(1), 41–50 (2003)
4. Gasior, D., Drwal, M.: Pareto-optimal Nash equilibrium in capacity allocation game for self-managed networks. Comput. Netw. **57**(14), 2817–2832 (2013)
5. Wydrowski, B., Zukerman, M.: QoS in best-effort networks. IEEE Commun. Mag. **40**(12), 44–49 (2012)
6. Nisan, N., Roughgarden, T., Tardos, E., Vazirani, V.V.: Algorithmic Game Theory. Cambridge University Press, Cambridge (2007)
7. Nash, J.: Non-cooperative games. Ann. Math. **54**, 286–295 (1951)

8. Koutsoupias, E., Papadimitriou, C.: Worst-case equilibria. In: Annual Symposium on Theoretical Aspects of Computer Science, pp. 404–413. Springer, Heidelberg (1999)
9. Reference Networks. http://www.av.it.pt/anp/on/refnet2.html
10. Bu, T., Towsley, D.: On distinguishing between Internet power law topology generators. In: Proceedings of the Twenty-First Annual Joint Conference of the IEEE Computer and Communications Societies. vol. 2, pp. 638–647 (2002)
11. Rosen, J.B.: Existence and uniqueness of equilibrium points for concave n-person games. Econ. J. Econ. Soc. **33**, 520–534 (1965)

Use of Ising Model for Analysis of Changes in the Structure of the IT Network

Andrzej Paszkiewicz[1(✉)] and Kamil Iwaniec[2]

[1] The Faculty of Electrical and Computer Engineering, Department of Complex Systems, Rzeszow University of Technology, Rzeszow, Poland
andrzejp@prz.edu.pl
[2] Ropczyce, Poland

Abstract. The paper presents the potential of using the Ising model to analyze changes in IT networks. These systems belong to the group of complex systems, like other natural systems, therefore, the processes occurring in them can be modeled using similar algorithms. The authors presented a modification of Ising algorithm based on the sensitivity factor. It allowed to take into account the individual susceptibility of separate nodes in the network structure to occurring changes. Due to the wide range of possibilities of using Ising model and, above all, its modifications, only a certain area of application has been presented in individual algorithms. Additionally, the use of the order parameter enabled presentation of phase transitions in the network structure.

Keywords: Ising model · Complex networks · Social networks
Phase transitions

1 Introduction

Network structures allow to describe a lot of dependencies occurring both in the natural and artificial environment. Examples of such relationships may be social networks, road and airport networks, computer networks, professional networks, river networks, links between web pages and websites, material structures, citation networks, etc. In all these cases, the network structure can be used to model behaviors, transfer information or other relationships. Therefore, their analysis should constitute an important area of scientific research. Of course, the research in this area has been conducted for many years, discrete mathematics and the entire theory of graphs was applied [1, 2]. This refers to the arrangement of the elements themselves, as well as the creation of a connection structure between them, both in a physical and logical context. These issues, however, are well known and described in numerous publications [3, 4]. Heuristic, genetic, accurate, etc. solutions serve this purpose. However, these solutions do not always allow to map real processes occurring in network structures, and in particular the dynamics of their changes. Therefore, in order to improve existing models, as well as to create new ones that better reflect real relations in network structures, the theory of complex systems is used [5]. It has its source in biology. Mechanisms and processes occurring in natural systems make it impossible to treat them as simple systems. The development of system

L. Borzemski et al. (Eds.): ISAT 2018, AISC 852, pp. 65–77, 2019.
https://doi.org/10.1007/978-3-319-99981-4_7

theory [6], and then system engineering, allowed for gradual inclusion of elements of complex systems in technical and later economic sciences. One of the ways to describe complex systems is fractal geometry [7], which perfectly describes hierarchical network structures, including their scalability.

In this section, we will mainly refer to one of the properties of complex systems, namely phase transitions [8]. This property is associated with a rapid change in the value of the relevant parameters of a given object or a physical body. As a result of phase transitions, a given system (object, body, structure) also changes its properties. Of course, these changes can be considered on a micro and macroscopic scale. Therefore, changing values of local parameters may lead to a change in the phase of the whole system, and thus changes in the value of global parameters. Thus, phase transitions, perfectly reflect significant structural changes in network systems, including relationships between individual elements (nodes) of these structures. This applies to both physical and logical structures. Therefore, they relate to phenomena of a critical nature accompanying the processes taking place in these structures. The theory of phase transitions is classified as the field of physics on the borderline of thermodynamics, material physics, field theory, physical chemistry, etc. However, for some time now, phase transitions have also been studied in other areas such as social, cultural or economic relations [9–11]. Moreover, such phenomena may occur in IT networks, e.g. social network structures, wireless networks, Ad-Hoc networks, and even the network of relations between software modules, etc.

One of the well-known solutions for describing phase transitions is Ising model [12]. It is mainly used in statistical physics (originally as a mathematical model of ferromagnetism), but has also found application in other areas, e.g. social networks. The basis for the choice of this model by the authors was its universal character. It allows for taking into account the dynamic of changes occurring in the different network structures and identification of phase transitions. Thus, it is perfect for modeling changes of opinion in social networks [13], in which the environment of a given actor (node) can exert pressure on it and thus change its attitude or opinion on a given topic. Both classic Ising model and its modifications can be used for this type of analysis.

Another area in the field of IT network structures is the progress (expansion) of the epidemic of viruses or other types of malicious software. Ising model allows to analyze the course of infection in computer networks, the susceptibility of individual nodes, as well as the possible regression process. In this respect, modifications of the basic Ising model built on susceptibility parameters and threshold values may have particular application.

Of course, the above examples of the use of Ising model for analyzing changes that occur in the network of IT structures presented in the paper do not exhaust the full spectrum of possibilities. In the future, models may be created and developed i.e. in the area of communication in wireless networks, IoT, etc.

2 Ising Model

One of the exemplary solutions used to model phase transitions in network structures is Ising magnetism model [12]. The classic model is based on the assumption that each node (spin) can take one of two values +1 or −1, i.e. $\vec{S}_i \in \{-1, 1\}$. The energy of any state is given by the Hamiltonian function [13]:

$$H = -\sum_{\langle i,j \rangle} J_{ij} \vec{S}_i \vec{S}_j - B \sum_i \vec{S}_i, \tag{1}$$

where: J_{ij} is the coupling constant between *i*-th, and *j*-th node, while *B* is the outer magnetic field.

However, to describe the network structures, we can use any number of states representing e.g. different values, opinions (groups of opinions) referring to a specific topic. Then, $S_i \in \{s_1, \ldots, s_k\}$. Thanks to this, it is possible to model the behavior of the network structure depicting the state adopted by the individual nodes that currently dominates in their neighborhood. In the case of systems in which more than 2 states/opinions are considered, the Potts model can be used [14]. However, it is a generalization of Ising model. Therefore, the paper only refers to Ising model as the original one.

The idea of this algorithm is precisely described in Fig. 1. According to the assumptions of the model, even a slight advantage of one of the states (opinions) in the environment of a given node may cause a change in its status (opinion). In practice, the advantage of one "voice" calls for a change in the state of a given node and may cause further changes in its neighborhood, and then start a chain reaction that changes the state of the whole system, that is, the network structure, until the moment of balance. This moment may be associated with the adoption of 100% of one state or a balance between *n* states from the set of all available states.

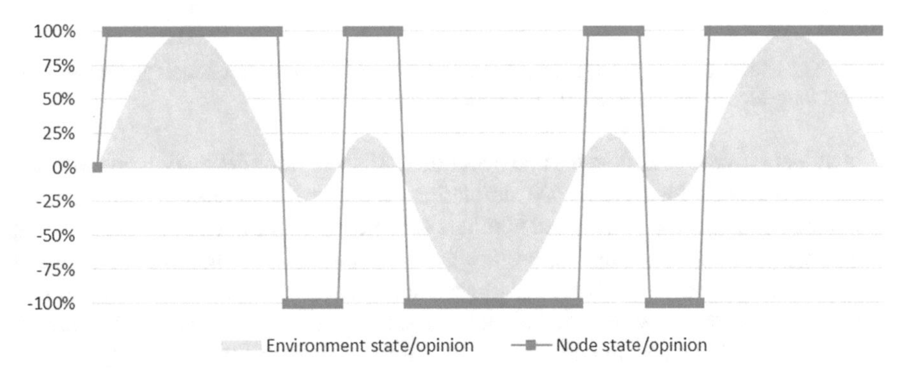

Fig. 1. The idea of changing the states (opinions) of a node depending on the distribution of states (opinions) in the environment for standard Ising model

The classic algorithm describing the operation of phase transitions for Ising model, used for modeling opinions or states, has the following structure:

1. `Defining the studied population/system.`
2. `Generating random opinions/states among the population/systems under study.`

```
3. Calculating opinions/states of neighboring units for
   each unit.
4. Calculating new opinions/states based on dominant opin-
   ions/states among neighbors for each unit.
5. Repeating steps 3 and 4 until no change occurs in the
   structure.
```

When examining random simulations using the presented algorithm, the most important changes occur at the beginning of the algorithm's operation, and the distribution is smoothed, which leads to the stabilization of the system (Fig. 2). This stabilization is achieved when the system is in balance.

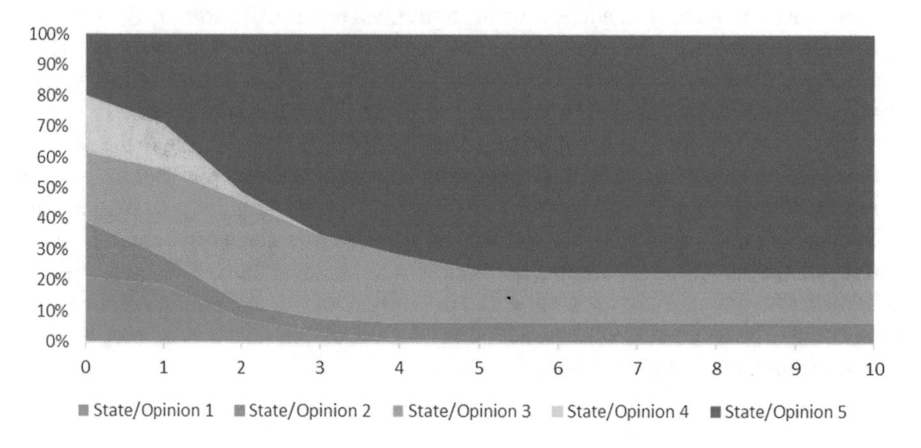

Fig. 2. The course of changes in the distribution of states in subsequent simulation steps using standard Ising algorithm (for exemplary 5 initial states)

In order to investigate changes that occur in the distributions depending on the initial number of states in the system, 100 simulations were performed based on the above-mentioned algorithm for 10 different numbers of initial opinions (Fig. 3). Thus, referring to the original Ising algorithm, there are as many possible opinions/states as many possible values of the spin. It was also assumed that the weight of each connection between nodes is the same, and the relations between neighbors affect the final opinion/state adopted by the node. In addition, it was assumed that the limit of the algorithm's operation is the stabilization of the system/network.

Analyzing the obtained results, it can be noticed that the statistics concerning the number of state changes by particular nodes increase with the increase of the number of initial states. In order to obtain more information about the state distributions in the studied network, the average number of final states in relation to the initial states was also examined and in what proportions they occur most frequently (Table 1).

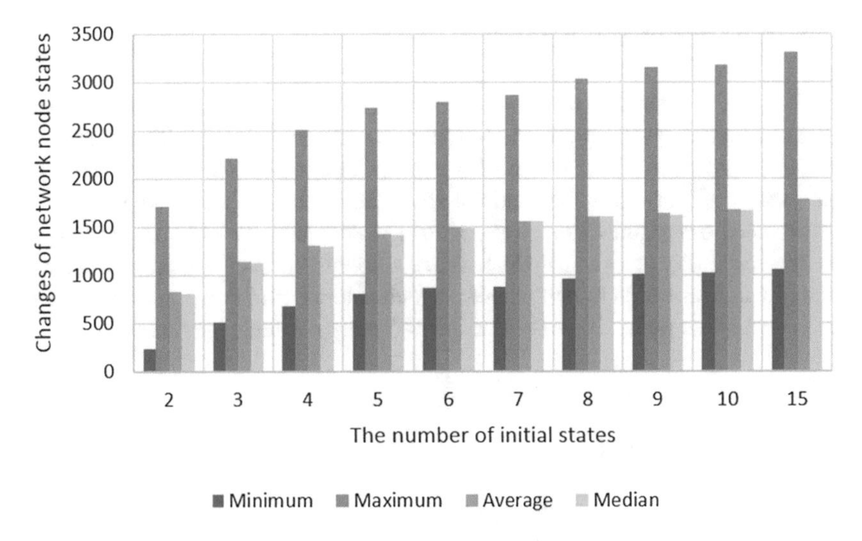

Fig. 3. Changes in the states of individual nodes depending on the number of states in the network after 1000 simulations using the standard algorithm

Table 1. The distribution and the average number of final states depending on the number of initial states after 1000 simulations using a standard algorithm.

		The number of initial states									
		2	3	4	5	6	7	8	9	10	15
The number of final states	**1**	105	33	13	5	6	1	2	0	1	1
	2	895	372	222	137	106	67	54	41	37	15
	3		595	538	445	374	335	272	238	210	151
	4			227	353	413	409	434	465	442	374
	5				60	96	173	202	213	260	352
	6					5	15	36	40	48	98
	7						0	0	3	2	9
	8							0	0	0	0
	9								0	0	0
	10									0	0
	Avg	**1,895**	**2,562**	**2,979**	**3,326**	**3,502**	**3,731**	**3,888**	**3,982**	**4,075**	**4,391**

Based on the obtained results, it can be noticed that the average number of final states increases with the number of initial states and, as in the change statistics, this increase is logarithmic. Additional fact is that, there can be a distribution in which one state dominates completely in a given system for almost any number of initial states tested.

3 Sensitivity in Ising Model

The stability of the system, and in this case the network structure, is as important as the ability to adapt the system (network) to changes occurring in it resulting from internal and external processes. In the case of the basic Ising model, there is a risk of quick changes resulting only from the minimal advantage of one of the states in the vicinity of a given node or possible anomalies occurring in the network structure caused e.g. by unjustified instability of a given node. In order to counteract such situations, a **sensitivity** parameter was introduced to Ising model, which determines the threshold for accepting changes by a given node. Then, the change of the node state is the following:

$$S_i^k \rightarrow S_i^l, \tag{2}$$

with the condition of transition:

$$\sum_m S_m^k \ll \sum_n S_n^l \tag{3}$$

for:

$$\sum_n S_n^l = \sum_m S_m^k (1 + d), \tag{4}$$

where: $m, n \in V_i$ and $m \neq n$, and V_i is the set of nodes located in the environment of v_i node, while d is the threshold value set for the given network. The idea of this approach is presented in Fig. 4. As can be seen, the node changes its state when the minimum threshold is exceeded.

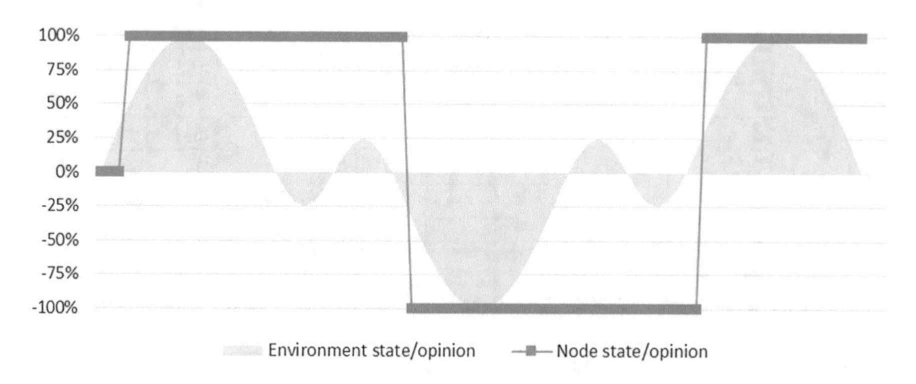

Fig. 4. The idea of changes in states (opinions) of the node depending on the distribution of states (opinions) in the environment for Ising model threshold

In order to investigate the influence of the sensitivity parameter on Ising algorithm, a series of simulations were carried out for different values of this parameter. Figure 5 presents selected results for the value of 0% of the sensitivity parameter (Ising's classic algorithm), 30% and 50%. Analyzing the results obtained, it can be concluded that the introduction of this parameter allows to reduce the number of changes in the states of particular nodes, and thus the system (network) works more stably.

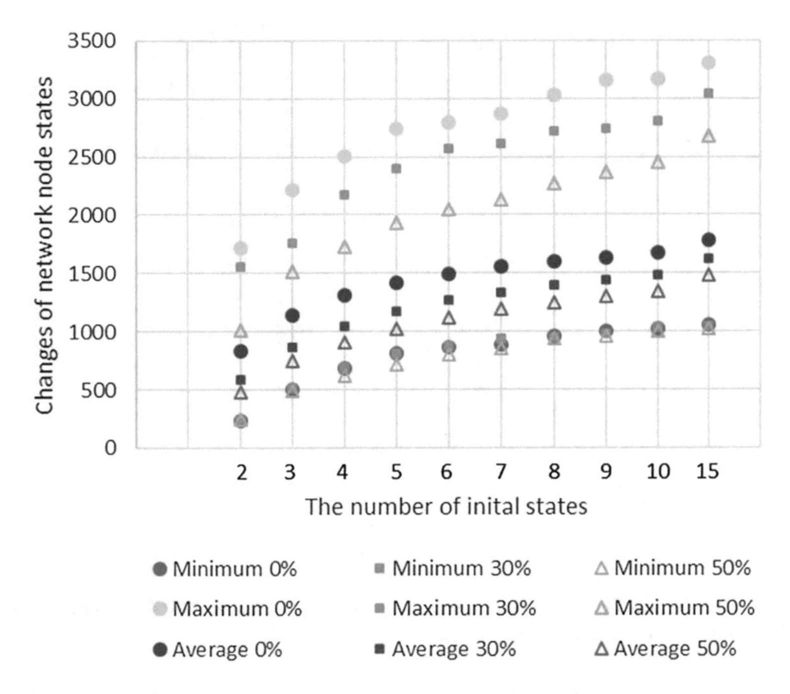

Fig. 5. Changes in the states of individual nodes depending on the number of states in the network after 1000 simulations using the algorithm including the node sensitivity parameter (for sensitivity values 0%, 30% and 50%)

In the above case, the constant sensitivity of the entire system (network) was considered. However, it is possible to take into account the individual sensitivity of separate nodes. Then the condition changes in the scope of:

$$\sum_{n} S_n^l = \sum_{m} S_m^k (1 + d_n), \tag{5}$$

where: $d_n \in D$, and $D = \{d_i : i \leq |V|\}$, where V is a set of all nodes.

4 An Order Measure and Phase Transitions

In order to determine the current phase in which a given system (network) can be found, the parameter of order [15] can be used:

$$M = \sum_{S=1}^{q} \left| N_S - \frac{N}{q} \right|, \tag{6}$$

where: N is the number of all nodes in the network structure, N_S is the number of nodes being in the S state, and q is the number of all states in the network. Of course, for a special case where $q = 2$ Eq. (3) can be reduced to a form $|N_1 - N_2|$.

If all the initial node states are considered during the whole simulation, the value of the order parameter increases to the maximum value, and then the system goes into a stable state (ordered). This is shown in Fig. 6. The changes only occur at the initial stage of the algorithm's operation. This is due to the nature of Ising algorithm.

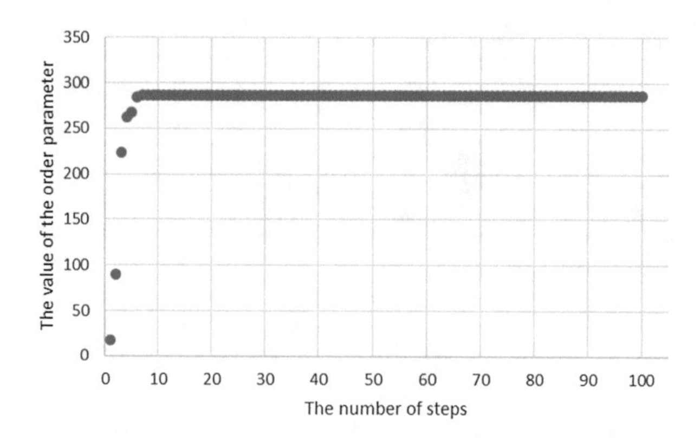

Fig. 6. Values of the order parameter for the classic Ising algorithm (including all initially available states during the entire simulation)

In case only currently available node states are included in the calculations, the graph presenting the values of the order parameter looks different. As shown in Fig. 7, before the stabilized phase, the values of the order parameter increase, followed by their correction and the system goes into a stable state. It results from the fact that before the transition of the system into a stable state, the final number of states (opinions) currently

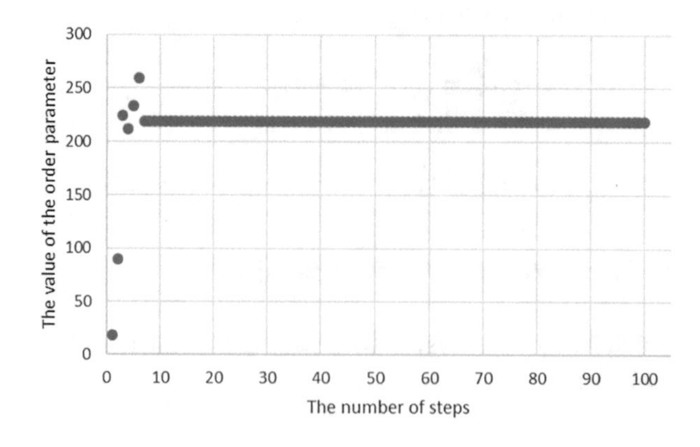

Fig. 7. The values of the order parameter for the classic Ising algorithm (during the individual simulation steps - only those states that are currently remaining in the system are taken into account)

assigned to individual nodes is limited, and this number directly affects the value of the order parameter.

In the case of the classic Ising algorithm, the network converts quite quickly into an orderly state, even with a larger number of the initial states available. However, taking into account the system sensitivity parameter, the stabilization phase takes place later (Fig. 8). This depends on the value of the sensitivity parameter, the number of nodes and the number of initial states.

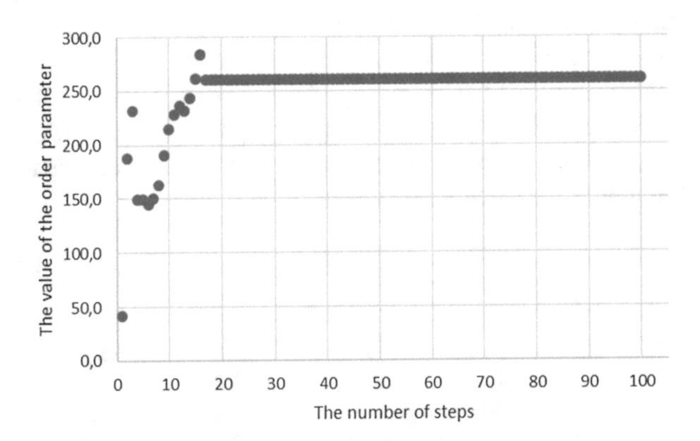

Fig. 8. The values of the order parameter for Ising algorithm considering the sensitivity parameter (during the individual simulation steps only those states that are currently remaining in the system are taken into account)

The graphs above show the two phases of the network (system) work, the disordered phase and the ordered phase. However, in reality, the system can be in three phases in this case, two already mentioned and furthermore, the initially ordered one, which is not visible on the charts. The phase initially ordered when the algorithm starts to work immediately goes to the disordered phase, therefore the values for it are not saved.

5 The Examples of Use

A big threat to the proper functioning of IT networks is all kinds of malicious software. Therefore, modeling and studying the processes of their dissemination help to get to know them better, and thus predict the consequences and scope of their impact.

In the model proposed that is based on the assumptions described in the previous sections, it was assumed that each node v_i can take one of the states:

$$S = \begin{cases} 0 & \text{uninfected node} \\ 1 & \text{infected node} \end{cases}. \tag{7}$$

In order to take into account different levels of security for individual nodes, the following sensitivity parameters were adopted:

$$d_j = \begin{cases} 0\% & \text{unsecured system} \\ 25\% & \text{low security system} \\ 50\% & \text{medium security system} \\ 75\% & \text{high security system} \\ 100\% & \text{fully safe system} \end{cases}. \quad (8)$$

Thus, there are different groups of nodes with different susceptibilities. Taking into account the randomness of the sensitivity in the system, incidental network structures consisting of 250 nodes with a randomly assigned level of sensitivity were examined. Figure 9 shows an example of the frequency of state changes in the whole network. Of course, these values depend on the values of input parameters such as: the initial number of infected nodes, the distribution of node sensitivity levels, and the capability to "cure" a given node, i.e. return to the *uninfected* state.

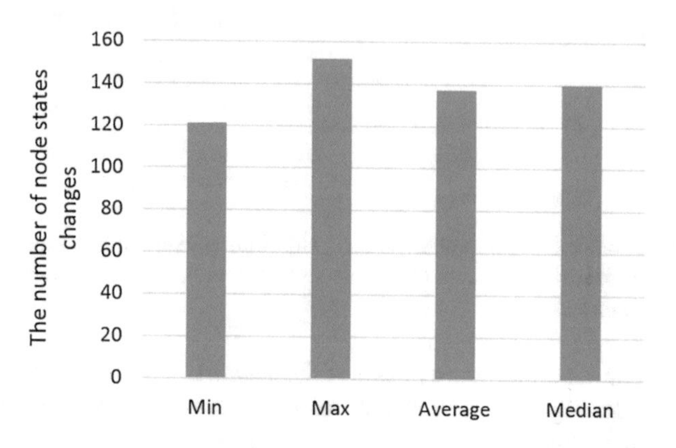

Fig. 9. Number of node state changes in the network for 250 nodes with a random sensitivity value

An exemplary distribution of the sensitivity parameter value in the structure of 250 nodes is presented in Fig. 10. These distributions correspond to the input data for the 10 simulation series carried out. Of course, it is possible to manually define this parameter and thus carry out the assumed scenario of the spread of malicious software for the defined network infrastructure.

Analyzing the results obtained, it can be observed that the course of changes in the value of the order parameter for the conducted simulations corresponds to the assumptions of the previously adopted Ising model. The phase of dynamic changes taking place in the network is clearly visible followed by the phase in which the network remains in a stable state (Fig. 11).

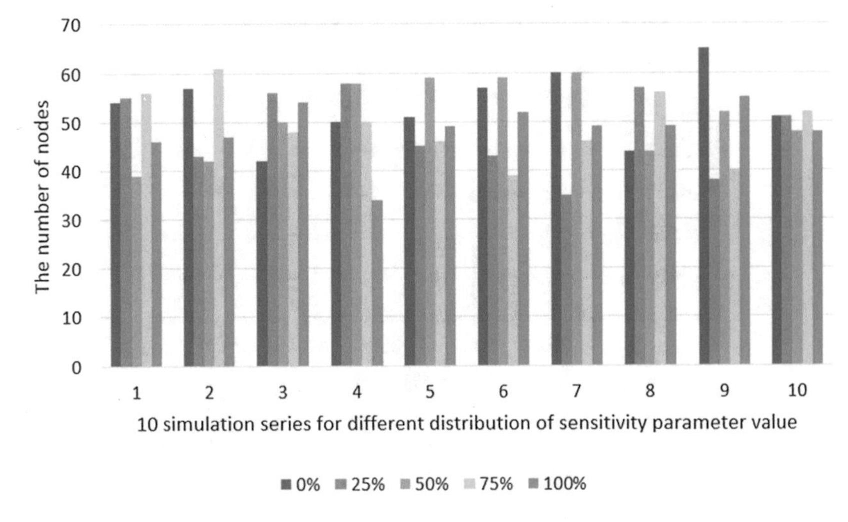

Fig. 10. Distribution of the sensitivity parameter values assigned for 250 network nodes

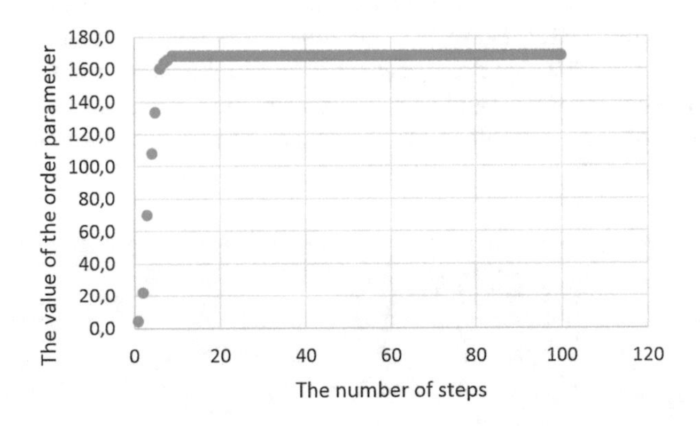

Fig. 11. An example value of the order parameter for the system with 2 states, taking into account the randomness of threshold values of each individual

In the case of analysis of changes taking place in the structures of social networks, the following situation can be presented. Node states correspond to the opinion of a given person to a given topic. Then the number of opinions can be much larger than it was in relation to the analysis of the infection in the network structure. For example, the assumption was made that there are 20 opinions in the system, and each individual (node) has different threshold for susceptibility to opinion change in accordance with the values defined in expression (8). On this basis, the values of the system order parameter were obtained (Fig. 12).

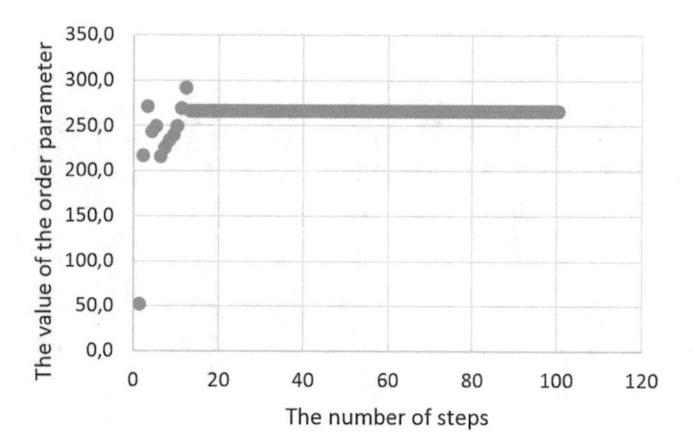

Fig. 12. Values of the order parameter for a system with 20 initial opinions, taking into account the randomness of threshold values of each individual

As can be observed in Fig. 12, the value of the order parameter is low only at the beginning of the process of changes taking place in the social network structure, and then increases significantly. These results correlate with the results shown in Fig. 13. It can be observed there that significant changes in the number of opinions in the simulated system change between steps 2 and 12. Later, the structure of the social network remains in a stabilized state. As a result of the algorithm, the number of opinions from 20 is ultimately reduced to 5.

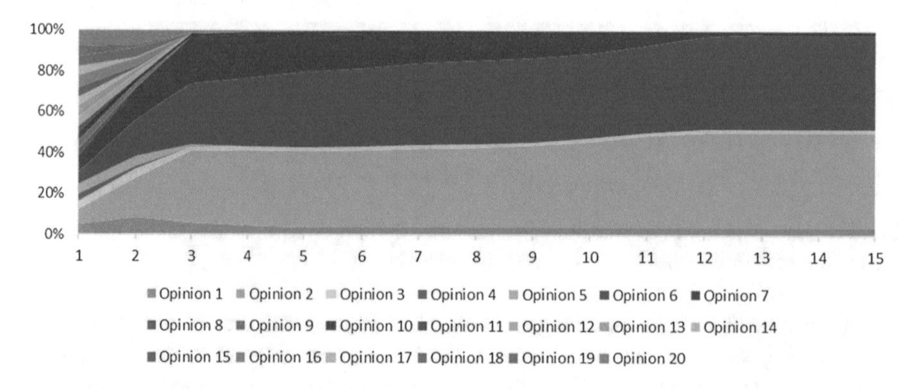

Fig. 13. The course of changes in the distribution of opinions in the social network in the next steps of the simulation (initial number of opinions 20, final 5)

6 Conclusions

A lot of network structures can be observed in the world. Understanding the processes occurring in them has a significant impact on learning about the accompanying phenomena. For this purpose, both real structures should be investigated and they should be subjected to modeling and simulation taking different conditions and operating parameters into account. Therefore, it is very important to look for the right models and algorithms. This model may be Ising model, which may be used in many areas of IT networks. Moreover, the examples of using Ising model to simulate e.g. the spread of malicious software in a network infrastructure, or modeling the processes of changing opinions in a social network environment, can be further developed. One should consider the implementation of mechanisms introducing anomalies occurring in this type of network structures.

The use of Ising model also allows to analyze processes that have the character of phase transitions, in which rapid changes in the value of significant parameters of a given network structure take place. Thanks to this, changes can be clearly seen in the trends of opinions in social networks, the level of infection of the network structure, or the area of overloading the computer network. Thus, the definition of phases in the system can allow i.a. to define its parameters so that it works optimally, using the least possible resources.

References

1. Diestel, R.: Graph Theory. Springer, New York (2000)
2. Bondy, J.A., Murty, U.S.: Graph Theory with Applications. Springer, New York (2008)
3. Johari, R., Tsitsiklis, J.N.: A scalable network resource allocation mechanism with bounded efficiency loss. IEEE J. Sel. Areas Commun. **24**(5), 992–999 (2006)
4. Chang, S., Gavish, B.: Telecommunications network topological design and capacity expansion: formulations and algorithms. Telecommun. Syst. **1**(1), 99–131 (2005)
5. Auyang, S.Y.: Foundation of Complex-System Theories. Cambridge University Press, Cambridge (1999)
6. Von Bertalanffy, L.: General System Theory. George Braziller, New York (1993)
7. Song, C., Havlin, S., Makse, H.A.: Origins of fractality in the growth of complex networks. Nat. Phys. **2**(4), 275–281 (2006)
8. Stanley, H.E.: Introduction to Phase Transitions and Critical Phenomena. Oxford University Press, New York (1989). ISBN 9780195053166
9. Fronczak, P., Fronczak, A., Hołyst, J.A.: Phase transitions in social networks. Eur. Phys. J. B **59**(1), 133–139 (2007)
10. Wang, B., Han, Y., Chen, L., Aihara, K.: Multiple phase transitions in the culture dissemination. In: Complex Sciences, pp. 286–290. Springer, Heidelberg (2009)
11. Novotny, J., Lavička, H.: Non-equilibrium phase transitions of economic systems under exogenous structural shocks (2014). https://doi.org/10.13140/rg.2.1.3472.4962
12. Selinger, J.V.: Ising model for ferromagnetism, introduction to the theory of soft matter. In: Soft and Biological Matter, pp. 7–24. Springer International Publishing (2015). https://doi.org/10.1007/978-3-319-21054-4_2
13. Ishii, M.: Analysis of the growth of social networking services based on the Ising type agent model. In: Pacific Asia Conference on Information Systems, Proceedings, p. 331 (2016)
14. Wu, F.Y.: The Potts model. Rev. Mod. Phys. **54**(I), 235–268 (1982)
15. Toruniewska, J., Suchecki, K., Hołyst, J.A.: Unstable network fragmentation in co-evolution of Potts spins and system topology. Phys. A **460**, 1–15 (2016)

A Generic System for Automotive Software Over the Air (SOTA) Updates Allowing Efficient Variant and Release Management

Houssem Guissouma(✉), Axel Diewald, and Eric Sax

Karlsruhe Institute of Technology, Engesserstr. 5, 76131 Karlsruhe, Germany
{houssem.guissouma,axel.diewald,eric.sax}@kit.edu

Abstract. The introduction of Software Over The Air (SOTA) Updates in the automotive industry offers both the Original Equipment Manufacturer and the driver many advantages such as cost savings through inexpensive over the air bug fixes. Furthermore, it enables enhancing the capabilities of future vehicles throughout their life-cycle. However, before making SOTA a reality for safety-critical automotive functions, major challenges must be deeply studied and resolved: namely the related security risks and the required high system safety. The security concerns are primarily related to the attack and manipulation threats of wireless connected and update-capable cars. The functional safety requirements must be fulfilled despite the agility needed by some software updates and the typically high variants numbers.

We studied the state of the art and developed a generic SOTA updates system based on a Server-Client architecture and covering main security and safety aspects including a rollback capability. The proposed system offers release and variant management, which is the main novelty of this work. The proof of concept implementation with a server running on a host PC and an exemplary Electric/Electronic network showed the feasibility and the benefits of SOTA updates.

Keywords: Connected vehicles · SOTA updates · Variant management · Security Safety · Release management · Electronic control unit

1 Introduction

Electric/Electronic (E/E) architectures include nowadays up to 150 ECUs with various safety and real-time demands and over 100 million lines of code [1]. The increasing integration of electronics and software in modern vehicles in form of embedded systems raises the error probability of ECU's code. These errors cause the program to perform in a way that produces an unintended outcome [2], which can lead to system failures needing to be fixed by an adequate software update. A lot of efforts are spent to detect these errors before the final production [3]. But more and more often errors occur during use. In this case, updates are urgent and the Original Equipment Manufacturer (OEM) must develop a bug fix, which is usually updated during global recall campaigns.

In 2020, 75% of cars shipped globally are expected to have wireless connectivity [4]. One key benefit of the rising vehicle communication is the deployment of software or

L. Borzemski et al. (Eds.): ISAT 2018, AISC 852, pp. 78–89, 2019.
https://doi.org/10.1007/978-3-319-99981-4_8

firmware updates over the air. In this work, we describe both kinds (software and firmware) as SOTA considering firmware as a special case of software. The Over The Air (OTA) approach would minimize the customer inconvenience and allow faster updates, since the software could be downloaded, as soon as the release is ready [5]. Otherwise, it will save important costs compared to the traditional update process in the workshop. According to IHS Technology, the total worldwide OEM cost savings from SOTA updates are forecasted to grow to over $35 billion in 2022 [6]. For these reasons, many institutions in the automotive industry are currently working on the introduction of SOTA updates as core feature of their vehicles and agree that this kind of updates will gain more importance in the upcoming years [7].

However, before licensing SOTA updates for ECUs and including them in the fleet's life-cycle management, multiple challenges must be further studied. In addition to security risks, such as manipulating the update's content by an unauthorized third party, the existing enormous number of system variants needs new processes, methods and tools to achieve efficient and safe SOTA updates [7]. This variants abundance, rising from the multitude of customer's wishes and the numerous configuration's possibilities, makes the validation of software releases for whole product lines a difficult task.

This paper is structured as follows: in Sect. 2, we studied the state of the art of software updates in research and industry. Then, in Sect. 3, we gave a brief overview of some related works. Thereupon, the architecture and main parts of the implemented generic system for SOTA Updates is explained in Sect. 4. Thereafter, a system for variant and release management based on a dedicated database is introduced in Sect. 5. Section 6 includes a conclusion and a description of future work.

2 State of the Art

2.1 Traditional Software Updates for Vehicles

A system software release is usually defined for each vehicle and flashed at the end of the production line. New system releases during after sales phase are implemented depending on the product nature with different frequencies, for example once per six months. One important use case for such releases is recall campaigns conducted by OEMs after discovering a severe bug, which could influence the safety of the passengers or lead to a problem with respect to a national or an international licensing norm. In order to install these releases on the concerned target ECUs, updates are deployed traditionally in dealer workshops over cables. A sequence diagram describing the typical steps of such an update process is illustrated in Fig. 1. In this lengthy and costly process, the OEM must not only manage the complexity of multiple software versions, vehicle variants and configurations, but also manage the distribution to dealerships. The update is usually done by connecting through the On-Board Diagnostics (OBD) port [9], and can take about 15 to 90 min [10]. This can block other garage activities during the update process, and requires special equipment, which is not always available in the workshop. For the consumers, who must visit the dealer and wait for the update to be completed, this presents an inconvenience factor, which reduces their satisfaction [10].

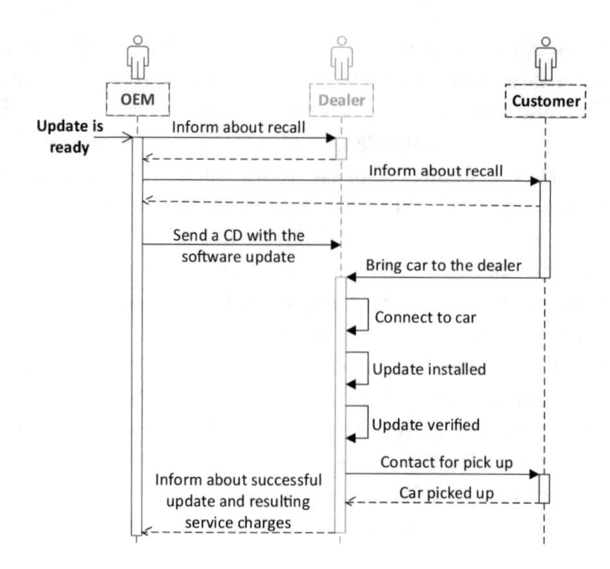

Fig. 1. Sequence diagram of a typical update's process in a dealer's workshop (according to [8])

Another possible update alternative is the customer implemented update, which is already offered by some OEMs for updating the radio/infotainment system. In this case, the customer receives a CD with the software installation files or has to manually download the software on a USB drive [8]. This approach is also lengthy and inconvenient for the customer, and induces a high risk by putting much responsibility on the customer's side. From a security point of view, it increases the possibility of reverse engineering and uploading a hacked version of the installation file by having the software in the hands of many customers [8].

2.2 Updates for Mobile Devices and Computers

Software updates are today an important part of the lifecycle management of computer systems and mobile devices such as smartphones or tablets. Diagnostic, version and configuration management, updates and other life-cycle activities need to be integrated in an efficient framework to keep an overview of all devices and their evolution. Seen the rapid spread of wireless connected devices and the success they are having in our modern society, this framework usually also supports an OTA implementation.

One of the most important mobile devices management solutions is the Open Mobile Alliance (OMA) Device Management (DM), which is practically supported by all smartphone manufacturers [11]. It defines the management information (management objects) for mobile devices in a so called DM tree. The management objects are the entities for software setting that may exist as environment variables or firmware images [12]. The management is done remotely, for example over a WLAN connection, through the interaction between the OMA DM server and the management agents using the OMA DM protocol aimed at providing remote synchronization of mobile devices.

Although many requirements for automotive SOTA updates, such as higher safety levels, distribution of functions through dozens of connected ECUs and the wider multi-disciplinarity including mechanical aspects, are not included in the OMA DM standard, many aspects can be transferred and used in the automotive field.

2.3 Current Situation of Automotive SOTA Updates

SOTA updates are nowadays not only a topic of research, but also already implemented by different OEMs, like BMW, VW and Tesla for navigation maps. Total vehicles that have map OTA updates are projected to grow from approximately 1.2 million units in 2015 to nearly 32 million units by 2022 [13]. SOTA updates for Infotainment services like email or social media have also been realized.

At ECU's firmware level such as for chassis or power-train modules, SOTA is still at the very beginning of the road. These updates are very critical to the safety of the passengers and need to be secure and safe enough to avoid each unexpected ECU behavior. Only the automaker Tesla is known for offering SOTA capability for almost all domains, including updates for the Autopilot, which is the autonomous driver assistance system of the car [14]. However, the number of vehicle variants in this case is still relatively low, which makes the management of the fleet easier.

3 Related Works

Considering the significance of SOTA updates for the future of the automotive industry, different research works have been conducted in the recent years.

In [15], a computationally efficient protocol with low memory overhead for secure firmware updates over the air is presented.

Mansour et al. introduced a system in [16] called AiroDiag for diagnosing embedded systems and updating ECU's software of vehicles through the Internet using a client-server structure.

In [17], a generic SOTA system covering safety and security aspects for an Electronic Brake Control (EBC) system has been presented. The safety and security of the vehicle as well as the availability of the car for the passengers are introduced as the main criteria for the acceptance of SOTA by legal authorities and customers.

Most of the existing works focused clearly on the security aspect of SOTA updates. The variant management challenge and its inclusion in a generic SOTA system allowing the validation of software updates for whole fleets throughout their life-cycles hasn't been studied deep enough yet. And one of the main goals of this work is to cover this gap and propose convenient methods to manage this complexity.

4 A Safe and Secure SOTA System

The developed generic system is realized using a classic Server-Client approach composed of the OEM's SOTA Server and the vehicle as the SOTA Client. The SOTA Server acts as the administration unit of the system and therefore manages the update process. The responsibilities of the SOTA Client include the reception, validation and distribution of update files to the respective ECUs within the vehicle. The features introduced in this section are based on other works and merged into a generic system.

4.1 System Overview

The SOTA Server's three main objectives are:

- update process control
- data management
- administrator/client access.

In order to perform organized SOTA updates, the system needs to keep track of software images, i.e. the binary files to be programmed on the Flash memory of ECUs, the status of updates and the amount and type of ECUs in the different vehicles. Furthermore, vehicles are grouped into fleets, so the software updates can be applied to only a specific set of cars. A fleet in this sense can contain all the vehicle variants of an OEM or only a well-defined group of variants managed by a separate department depending on the release development strategy.

Software management is key to introduce proper software version control and to prevent compatibility mismatches. Therefore, software images are characterized by their name, version and compatible ECU. An update management component is responsible for the administration of software releases. Software updates are issued for a specific ECU type and therefore create a unique relation between a software image and its matching ECU type. These updates are then combined into so-called Deployment Packages which contain all information and software images needed to update a certain type of vehicles of the selected fleet. All necessary information and data, as current software versions of cars in the field, pending software updates and validated software images, are stored in a local database within the server.

The client's and administrator's access to the SOTA server are granted through RESTful Application Programming Interfaces (APIs). These allow system administrators to manage the distributed ECUs and their software details in cars and also their affiliation to fleets, moreover to issue software updates and initiate them. All interactions with the SOTA client such as software image downloads and status reports by the vehicle are tunneled through the client API.

The SOTA client is divided into a Telematics Unit, a Central Gateway including a local database and the various ECUs. This architecture is shown in Fig. 2. The Telematics Unit represents the vehicle's access point and provides a mobile communication link via e.g. WiFi or cellular networks. It is also in charge of the encryption and decryption of any outgoing or incoming transmissions. In order to orchestrate SOTA updates within the vehicle, the Central Gateway's tasks include periodically checking for new updates within the server, downloading and unzipping pending Deployment Packages and the distribution to the ECUs. After a successful installation, the Central Gateway refreshes its database regarding the affected ECUs and reports it back to the SOTA Server.

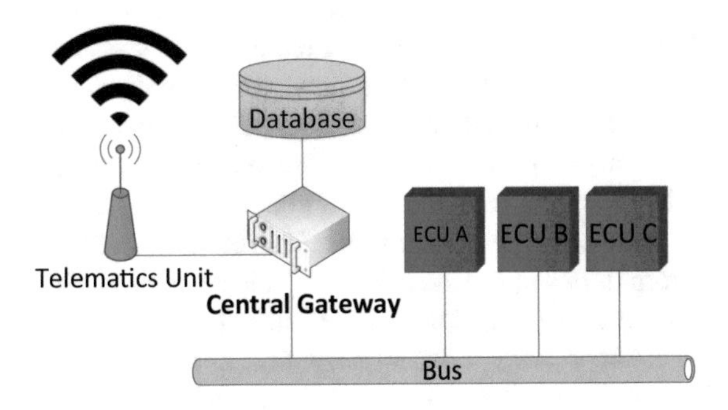

Fig. 2. SOTA Client architecture

For the implementation, a Node.js [18] server is used as the SOTA server, a Raspberry Pi 3 represents the Central Gateway as well as the Telematics Unit, and the various ECUs are realized by Texas Instruments' Tiva C LaunchPads all connected to the Central Gateway via a common Controller Area Network (CAN) bus.

4.2 Security

Due to the flexible data connectivity of a vehicle to another vehicle or an infrastructure, new risks may occur during the data transmission. As a risk can be defined as a likelihood of an attack caused by a threat, risks of unauthorized access of vehicle's software via the interconnectivity has to be predicted, and tackled by providing some additional features in the data transmissions.

The possible attacks to a SOTA system might include identity theft and manipulation or repetition of transmitted messages which may result in gaining the vehicle's control by an unauthorized third party. This encompasses intellectual property theft of software images or sensitive vehicle data. An attacker in possession of OEM server's identity could infiltrate the electronic system and install fraudulent software on ECUs of thousands of vehicles. The interception of messages that contain clear text using a man-in-the-middle attack could unveil confidential information to unauthorized users through unauthorized access.

In order to reduce the risk of a possible identity theft, a strong authentication process has to be implemented. The client's identity is verified using an open standard protocol called OAuth2 [19]. For this matter, the client retrieves an Access Token from an authorization server providing a unique and confidential secret to proof its identity. It then uses this Access Token to make an API call to the SOTA server which verifies the received token with the authorization server and sends back the requested information. The process of the server's authentication is realized using a self-signed certificate which gets transmitted to the client during the establishment of a secure communication channel using the Transport Layer Security (TLS) handshake. The client then verifies the integrity of that certificate with a Certificate Authority.

The confidentiality of the data exchanged between the SOTA server and client is secured using reliable symmetric encryption methods. The greatest risk that endangers the integrity of any data that is being sent draws from the disclosure of the keys used for the de- and encryption procedures. Therefore, these keys need to be stored securely all the time and may only be transmitted via a secured connection established by a TLS handshake. For this, the client asks the server about which TLS version is intended to be used and about its known cipher algorithms. The server answers providing the chosen cipher algorithm from the selection, its public key and certificate. The client then verifies the certificate with the Certificate Authority. Furthermore, it generates the session key which will later be used for the symmetric encryption, encrypts it with the server's public key and sends it back. Being the owner of the respective private key, the server then decrypts the session key and acknowledges its reception. From then on, the session key is used for both encryption and decryption of any messages transmitted. An overview of methods and issues related to encryption key management can be found for example in [20].

Besides, it is important to insure the freshness of data in order to prevent unauthorized third parties from interfering with the communication by intercepting messages and sending them again at a later point in time. This risk has been prevented by appending a consecutive package number to every message.

Since security was not the main focus of this work, further research dealing with other security topics such as public keys management and certificate revoking should be carried out. As a general framework for prototyping and developing Cybersecurity in connected vehicles, the standard SAE J3061, published in January 2016 by SAE International and described for example in [21], could be used as a reference.

4.3 In-Vehicle Update Strategy

The software architecture of ECUs is designed to separate the bootloader from the actual application according to the AUTOSAR standard. This allows for the maintainability of the application while the bootloader is left intact and operational. As it comes to update strategies, there are three decisions to be made which each has an impact on the cost, design, time consumption and feature capabilities of SOTA updates. Figure 3 shows the resulting main realization alternatives.

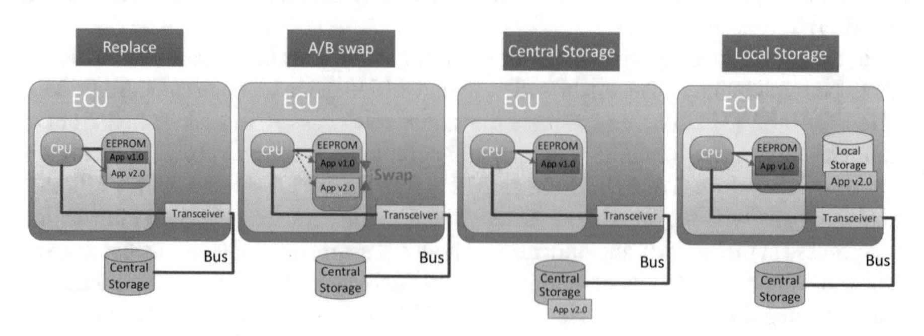

Fig. 3. Update Strategies (according to [17])

SOTA updates for ECUs can either be performed using Full Binary or Delta Updates. For Full Binary Updates, the application is being replaced entirely, whereas for Delta Updates only the application parts that differ from the new version are being edited. The implemented system uses for simplification the Full Binary approach.

Another conceptual choice that needs to be made is whether an incoming software update should replace its predecessor in its memory location (Replace Approach) or should be saved in a separate memory block to which the program control is shifted, thus being called A/B Swap. Despite doubling the memory space and therefore increasing the hardware cost of the ECU, the A/B Swap enables the functionality of rollbacks, i.e. shifting back to the old software version in case of an update's failure.

In addition, the placement of the storage device, from which the new software image is retrieved, must be negotiated. Using only a central storage comes with the advantage that no changes have to be applied to the ECU's design, but will also result in long vehicle downtimes in the case of multiple software updates coming in at once. Software images can be retrieved much quicker from a local storage, which also enables the system to perform multiple updates in parallel as the required files can be transmitted while the car is in operation.

The resulting solutions are compared in Table 1 considering: vehicle downtime, extra cost, rollback time and necessary ECU design changes.

Table 1. Comparison of the three update approaches

	Down time	Extra cost	Rollback time	ECU design changes
Replace & Central storage	Long	Minor	Long	None
A/B & Central storage	Long	Medium	Very short	Minor
Replace & Local storage	Short	Medium	Medium	Medium

The A/B Swap with a central storage causes additional hardware costs because the size of internal flash memory of the microcontroller has to be doubled and also results in a long downtime of the vehicle during the update process due to the fact that software images can only be transmitted to the ECUs one at a time. Nevertheless, the downtime can be reduced by using a faster bus system, e.g. Ethernet, or running the ECU applications from an external flash memory. This creates the possibility to perform simultaneous updates of multiple ECUs. The big advantages this approach has over the Replace Approach with a local storage, are the significantly shorter rollback time and the fact that only minor design changes have to be applied to the ECUs. Because of these reasons, it has been adopted in the system of this paper.

4.4 Rollback Function

Establishing automotive SOTA updates will increase the frequency of software upgrades being rolled out to vehicles and therefore elevate the error probability during the update process. The option to recover from an ECU deadlock due to a failed update is in this case a very important safety feature. It can be realized through a Rollback function.

One of two possibilities to perform a Rollback is to flash a previous version of the application running on the ECU back to its program memory from either a central or local storage device. Depending on the location of the storage device, this procedure may take several seconds or even minutes to finish for one ECU.

A second option is to reverse an already executed A/B Swap and shift the program control back to block A.

5 Variant and Release Management

5.1 Variant Management in the Automotive Field

Through the offered freedom to configure vehicles according to the customer needs and wishes, a huge variant space arises comprising millions of unique products with differences at various granularity levels. This leads to a high degree of complexity inside automotive product lines. Within each vehicle model, the customer has normally the choice between different equipment variants. For each equipment variant, there is the possibility to choose different kinds of motors, gearboxes and bodies. We named these first variability parameters Variants Features, as shown in Fig. 4. Then the numerous ECU functions such as airbag, radio or driver assistance systems can be defined as further configuration features, which increase the variants number considerably.

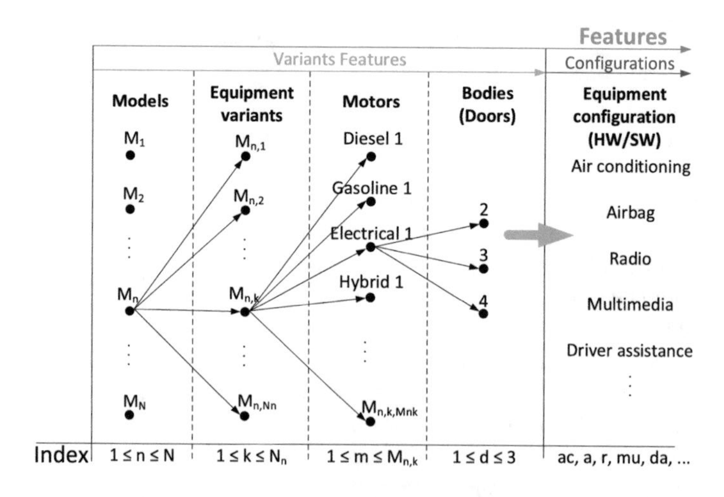

Fig. 4. Features determining the vehicle variant

Conducting an evaluation of the number of variants of e.g. Volkswagen cars using the online model configurator [22] and considering only the variants features of Fig. 4, we found a sum of 1164 variants. By including the ECU configuration features, this number will increase exponentially. By introducing optional SOTA updates, another variability dimension in time is added to the already existing variant space. This means that the same software modules can exist for the same product with different versions depending on whether or not optional updates have been installed.

One of the most used notations to model variability is the feature-based representation, similar to the model in Fig. 4 [23]. It documents the features of a product line and their relationships and specifies the set of valid products.

5.2 Realization of Variant and Release Management

At the back-end (server) side of the system, each vehicle is defined by a unique ID differentiating it from all other vehicles. All its relevant management information is saved according to the feature-based variant modeling in a structured database, which has been implemented using the SQL language. This data structure serves as the reference for deploying the updates into the fleet and keeping diagnostic information up-to-date. It should also allow for the management of the development and validation of SOTA updates under consideration of the existing variant space. This can be achieved by traceability implementation to the static and dynamic system models at different abstraction levels.

At the beginning of a new SOTA release, an administrator selects a fleet and the software updates to be distributed. In the next step, the management system creates a Deployment Package for every unique vehicle configuration of the fleet in question. The Deployment Package contains all necessary software images for a particular vehicle configuration. The selection of the relevant software images is made based on a comparison of two subsets of ECU types and their resulting cut set. One subset contains the ECU Types of the vehicle configuration present in the fleet in question. The other subset

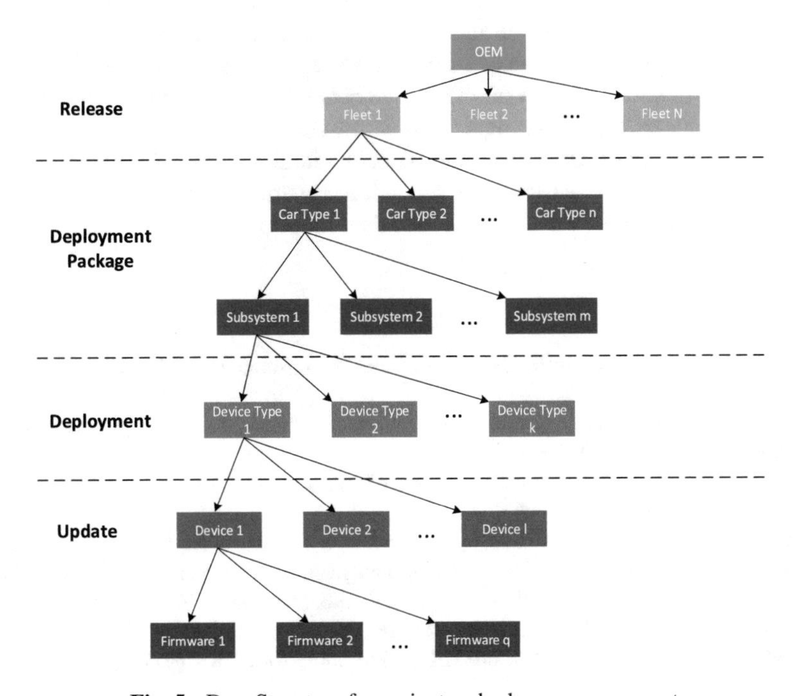

Fig. 5. Data Structure for variant and release management

contains the ECU Types to which the selected software updates are compatible to. To track the progress of the installation of the Deployment Packages on every vehicle of the fleet, the management system creates a Deployment entity for every car that receives a Deployment Package and an Update entity for every ECU that is being supplied with a new software image. This management system based on the implemented database is represented is Fig. 5. With this procedure, the OEM can manage the evolution of its fleet and the distribution of the Deployment Packages derived from the main software ECU updates.

6 Conclusion

After studying the state of current automotive updates, such as during global recall actions, the main benefits as well as the challenges of SOTA could be identified. The state of the art of common SOTA updates for mobile devices and computers as well as the already existing first SOTA systems in the automotive industry have also been described. Besides, some of the related works have been mentioned and the boundary to this work has been explained.

A generic SOTA system has been designed and implemented in a proof of concept framework. Security mechanisms, different update strategies combined with various rollback function realizations have been introduced as main characteristics of the system. The needed resources and structures for efficient variant and release management have been included as principal part of the generic system.

Due to the generic aspect of this work and the diversity of the involved parts in the system, different subjects will be further investigated in future works. Experimental work to validate the release management procedure for realistic update use cases will be carried out. In addition, the variant management of software updates for distributed functions should be deeper investigated and validated by considering the dependencies between the involved ECUs. Consistency checks, simulation and virtualization techniques need to get involved in a concrete validation scenario, where dependencies require different update versions between various variant models.

References

1. Staron, M.: Automotive Software Architectures: An Introduction, 1st edn. Springer, Cham (2017)
2. Hobbs, C.: Embedded Software Development for Safety-Critical Systems. Auerbach Publications, Boston (2015)
3. Sax, E.: Automatisiertes Testen Eingebetteter Systeme in der Automobilindustrie. Hanser-Verlag, München (2008). ISBN 978-3-446-41635-2
4. Khurram, M., Kumar, H., Chandak, A., Sarwade, V., Arora, N., Quach, T.: Enhancing connected car adoption: Security and over the air update framework. In: 2016 IEEE 3rd World Forum on Internet of Things (WF-IoT), pp. 194–198, December 2016
5. Nilsson, D.K., Sun, L., Nakajima, T.: A framework for selfverification of firmware updates over the air in vehicle ECUs. In: 2008 IEEE Globecom Workshops, pp. 1–5, November 2008

6. Bird, E., Colin, J.: Improving software, reliability & innovation - executive summary. IHS Technology, Technical report (2015)
7. Sax, E., Reussner, R., Guissouma, H., Klare, H.: A survey on the state and future of automotive software release and configuration management. KIT, Technical report, November 2017
8. Dakroub, H., Cadena, R.: Analysis of software update in connected vehicles. SAE Int. J. Passeng. Cars Electron. Electr. Syst. **7**(2), 411–417 (2014). https://doi.org/10.4271/2014-01-0256
9. Els, E.: The hackers holy grail - the OBD has manufacturers worried. In: Automotive Diagnostic Systems. CarTech Inc., Denver, June 2017
10. Odat, H.A., Ganesan, S.: Firmware over the air for automotive, fotamotive. In: IEEE International Conference on Electro/Information Technology, pp. 130–139, June 2014
11. Liu, L., Moulic, R., Shea, D.: Cloud service portal for mobile device management. In: 2010 IEEE 7th International Conference on E-Business Engineering, pp. 474–478, November 2010
12. Shin, J., Chung, Y., Ko, K.S., Eom, Y.I.: Design and implementation of the management agent for mobile devices based on OMA DM. In: Proceedings of the 2nd International Conference on Ubiquitous Information Management and Communication, ICUIMC 2008, pp. 575–579. ACM (2008)
13. Culver, M.: Over-the-air software updates to create boon for automotive market, IHS says. IHS Automotive, September 2015
14. Tesla: Software updates (2017). https://www.tesla.com/software
15. Nilsson, D.K., Larson, U.E.: Secure firmware updates over the air in intelligent vehicles. In: ICC Workshops - 2008 IEEE International Conference on Communications Workshops, pp. 380–384, May 2008
16. Mansour, K., Farag, W., ElHelw, M.: AiroDiag: a sophisticated tool that diagnoses and updates vehicles software over air. In: 2012 IEEE International Electric Vehicle Conference, pp. 1–7, March 2012
17. Freiwald, A., Hwang, G.: Safe and secure software updates over the air for electronic brake control systems. SAE Int. J. Passeng. Cars Electron. Electr. Syst. **10**(1), 71–82 (2016)
18. Tilkov, S., Vinoski, S.: Node.js: using JavaScript to build high-performance network programs. IEEE Internet Comput. **14**(6), 80–83 (2010). https://doi.org/10.1109/mic.2010.145
19. Internet Engineering Task Force (IETF): OAuth 2.0 Authorization Framework, May 2018. https://tools.ietf.org/html/rfc6749
20. Tse, D.W.K., Chen, D., Liu, Q., Wang, F., Wei, Z.: Emerging issues in cloud storage security: encryption, key management, data redundancy, trust mechanism. In: Wang, L.S.-L., June, J.J., Lee, C.-H., Okuhara, K., Yang, H.-C. (eds.) Multidisciplinary Social Networks Research. Springer, Heidelberg (2014)
21. Schmittner, C., Ma, Z., Reyes, C., Dillinger, O., Puschner, P.: Using SAE J3061 for Automotive Security Requirement Engineering
22. Volkswagen: Online configuratro, February 2016. https://www.volkswagen.de/app/konfigurator/vw-de/de
23. Berger, T., Rublack, R., Nair, D., Atlee, J.M., Becker, M., Czarnecki, K., Wasowski, A.: A survey of variability modeling in industrial practice. In: Proceedings of the Seventh International Workshop on Variability Modelling of Software-intensive Systems (2013)

Internet Traffic Management Under EU Regulation 2015/2120 – The Need for the Co-operation of Legal and Computer Scientists

Andrzej Nałęcz(✉)

Faculty of Management, Warsaw University, Szturmowa 1/3, 02-678 Warsaw, Poland
ANalecz@wz.uw.edu.pl

Abstract. EU Regulation 2015/2120 introduces a network neutrality regime into European electronic communications law. The application and enforcement of the abstract rules established by the regulation influences all the stakeholders in the internet value chain. The regulation's provisions are ambiguous and inconclusive. Legal and computer scientists must co–operate to provide an interpretation of those provisions that would, first, respect the rights of the end–users of internet access services and provide a user experience in line with the regulation, and second, respect the technical realities of managing traffic on the internet. The paper presents the legal framework of traffic management under the regulation and exposes problems which may only be resolved by a synthesis of the legal and computer science approaches. These include: the classification of traffic management measures into basic, reasonable and exceptional categories, recognized by the regulation; the admissibility of traffic classification; the definition of congestion befitting the regulation; the role of traffic management measures in congestion control. If these issues are not resolved, the exercise of the rights of end–users of internet access services may be limited, and the providers of these services may be subjected by national regulatory authorities to unreasonable obligations.

Keywords: Basic traffic management measures
Reasonable traffic management measures
Exceptional traffic management measures · Internet congestion
Classes of traffic

1 Introduction

Internet traffic management was not traditionally regulated by EU law. However, that changed with the entering into force of the Regulation 2015/2120 of the European Parliament and of the Council of 25 November 2015 laying down measures concerning open internet access and amending Directive 2002/22/EC on universal service and users' rights relating to electronic communications networks and services and Regulation (EU) No 531/2012 on roaming on public mobile communications networks within the Union (further referred to as Regulation 2015/2120). The regulation introduced a network

L. Borzemski et al. (Eds.): ISAT 2018, AISC 852, pp. 90–99, 2019.
https://doi.org/10.1007/978-3-319-99981-4_9

neutrality regime into EU electronic communications law, establishing the rights of end-users of internet access services and requiring the providers of those services to manage internet traffic respecting the rules set by the law.

The purpose of this article, written by a legal scientist, is to determine which problems related to internet traffic management have been solved by the law and which should be solved by computer scientists. While due to the principle of technology neutrality [1] the Regulation 2015/2120 does not specify any concrete technical standards and relies solely on abstract rules, in the end it is applied in the real world, where suppliers of network devices and providers of internet access services face the problem of providing and utilizing solutions that are both effective and not illegal under the Regulation 2015/2120. Legal scientist and computer scientists should come together to make clear what these solutions should be.

2 The Terminology of the Regulation 2015/2120

2.1 Internet Access Services and Specialized Services

Article 2 of the Regulation 2015/2120 defines internet access services as publicly available electronic communications services that provide access to the internet, and thereby connectivity to virtually all end points of the internet, irrespective of the network technology and terminal equipment used. From the point of view of analyzing traffic management measures, the most important conclusion arising from the interpretation of the cited provision is the fact that the Regulation 2015/2120 pertains only to publicly available electronic communications services – that is those that are offered to the public, either on the consumer or business markets, or on both, rather than only to a select group of end-users. Thus, traffic management in private networks is not subject to the requirements of the Regulation 2015/2120 and is not mentioned or discussed further in the paper.

Article 3 (5) of the Regulation 2015/2120 allows providers of internet access services and providers of content to offer services other than internet access services which are optimized for specific content, applications or services, or a combination thereof, where the optimization is necessary in order to meet the requirements for a specific level of quality. These are the so–called specialized services. Unlike internet access services, they come with Quality of Service guarantees. The distinction is important inasmuch that it shows that while traffic management of internet traffic may involve traffic classification and different treatment of the various classes of traffic (which is discussed below), it may only distinguish between classes of best–effort traffic types, since QoS optimization is limited to specialized services.

2.2 The Lack of a Legal Definition of "Traffic Management" in the Regulation 2015/2120

The Regulation 2015/2120 does not define the term "traffic management". However, traffic management was discussed in the debate leading up to the regulation being legislated. The Body of European Regulators for Electronic Communications (BEREC), which was tasked by the European Commission with investigating issues related to network neutrality and coordinated the discussion and presently coordinates the regulation's enforcement, leans towards a very broad understanding of traffic management, describing it as "the way the traffic is forwarded in the networks (…), which may include both regular first-come-first-serve management of traffic and more advanced ways of shuffling traffic through the networks" [2].

Such a broad understanding of traffic management differs from the one presented in computer science literature. In [3], Constantine and Krishnan describe the traffic management capabilities of network devices as those pertaining to: traffic classification, traffic policing, traffic scheduling, traffic shaping and active queue management. This approach, then, excludes the simplest technical solutions, such as passive queue management, from the scope of the term "traffic management".

The broad approach to traffic management in interpreting the Regulation 2015/2120 is necessitated by the requirements of network neutrality and by the need to safeguard the rights of end-users – as shaped by the regulation. Only a broad approach may prevent the providers of internet access services from circumventing the network neutrality rules by claiming a practice they are engaging in does not in fact constitute traffic management subject to the Regulation 2015/2120. Thus, traffic management under the Regulation 2015/2120 should be understood to include all the technical solutions, be it simple or complex, used to measure, analyze, classify and forward traffic in the network.

3 Traffic Management Measures Under the Regulation 2015/2120

3.1 Primal and Dual Congestion Control Under the Regulation 2015/2120

One of the objectives of utilizing traffic management measures under the Regulation 2015/2120 is congestion control. In the networking literature [4], "primal congestion control refers to the algorithms executed by the traffic sources for controlling their sending rates or window sizes. (…) TCP algorithms fall in this category. Dual congestion control is implemented by the routers through gathering information about the traffic traversing them. A dual congestion algorithm updates, implicitly or explicitly, a congestion measure or congestion rate and sends it back, implicitly or explicitly, to the traffic sources that use that link." While the Regulation 2015/2120 itself does not mention these two types of congestion control, they are referenced by BEREC and in the legal literature as "end–point based" and "network–internal" congestion control [5, 6]. The former does not contravene the provisions of the Regulation 2015/2120 on traffic management [5] and is in fact outside their scope [6]. Thus, the regulation applies only to the latter, network–internal measures.

3.2 Classification of Traffic Management Measures Under the Regulation 2015/2120

The Regulation 2015/2120 implies the existence of basic traffic management measures, and explicitly mentions two other types of measures – reasonable traffic management measures, regulated by Article 3 (3) second subparagraph of the regulation, and those going beyond the reasonable, regulated by Article 3 (3) third subparagraph. The latter are called exceptional traffic management measures by BEREC and in the legal literature [5, 6]. Basic measures are the simplest ones and do not involve any traffic classification. Reasonable measures may rely on traffic classification and may involve different treatment of specific categories of best–effort traffic. Exceptional measures may be employed only under exact circumstances, enumerated in the Regulation 2015/2120, and they may go as far as traffic blocking. All types of measures are described in further detail below.

3.3 Basic Traffic Management Measures

Article 3 (3) first subparagraph of the Regulation states that Providers of internet access services shall treat all traffic equally, when providing internet access services, without discrimination, restriction or interference, and irrespective of the sender and receiver, the content accessed or distributed, the applications or services used or provided, or the terminal equipment used. The cited provision implies the recognition of a type of traffic management measures, unnamed in the regulation, and known in the legal literature as basic traffic management measures [6]. The key phrase in the provision is that all traffic should be treated equally. Since only reasonable and exceptional traffic management measures allow traffic classification and the differentiated treatment of specific classes of traffic, the equal treatment of traffic in this context must be understood in its most narrow, linguistic meaning. Thus, basic traffic management measures must entail the identical treatment of every data packet traversing the network. As has been explained in the legal literature, they involve a transmission of data packets without any interference and completely independent of the class of traffic they belong to, and are therefore completely neutral from the point of view of the end-users [6].

Handling traffic using Passive Queue Management (PQM), involving first–in–first–out and tail drop algorithms, and giving all data packets the same priority undoubtedly constitutes basic traffic management measures. However, the status of classless Active Queue Management (AQM) as either a basic or reasonable traffic management measure is unclear under the Regulation 2015/2120. BEREC seems to indicate that all AQM (be it class-based or classless) should be assessed under the requirements for reasonable rather than basic measures [5]. This is unconvincing. As will be explained below, reasonable traffic management measures rely on traffic classification. Since classless AQM, such as Random Early Detection (RED) do not distinguish between traffic classes, they should be considered basic traffic management measures. RED with In and Out (RIO) and Weighted RED (WRED) algorithms allow a differentiated treatment of packets and therefore should be considered reasonable traffic management measures under the Regulation 2015/2120 (described in the following section), rather than basic ones. This is one of the areas where the application and enforcement of the Regulation 2015/2120 would greatly benefit from the input of computer scientists.

3.4 Reasonable Traffic Management Measures

Requirements for Reasonable Traffic Management – Introduction. Article 3 (3) second subparagraph of the Regulation 2015/2120 introduces the concept of reasonable traffic management measures and sets out the requirements which must be met by a measure for it to be considered reasonable. Under the cited provision, such measures must be transparent, non-discriminatory and proportionate, and must not be based on commercial considerations but on objectively different technical quality of service parameters of the specific categories of traffic. They must not monitor the specific content and must not be maintained for longer than necessary. Article 3 (3) third subparagraph, which generally pertains to exceptional traffic management measures, also exemplifies what kinds of the treatment of traffic may not be considered reasonable. All these requirements are discussed in detail below.

Transparency. The transparency of a traffic management measure pertains to its disclosure to both the end-users and the national regulatory authorities. While the end-users must only very broadly be informed of traffic management measures by means of the contract including internet access services, the information provided to the national regulatory authority must be detailed enough for the regulator to be able to determine whether a given measure satisfies the requirement of reasonableness.

Non-discrimination. Traffic Classes. Reasonable traffic management measures must be non-discriminatory. Non-discrimination is a well–known concept in EU law. It "is not equivalent to the distinction as such but refers to the disadvantageous treatment of an individual based on a ground prohibited by law [and] it occurs when comparable situations are treated differently and different situations are treated in the same way", when it is not objectively justified [7]. In the case of internet traffic management, the requirement of non-discrimination pertains not to individuals but rather to data packets transferred over the internet. All packets must be transmitted according to the same rules, e.g. using the first-in-first-out algorithm, unless there is an objective justification for differentiation. The Regulation 2015/2120 unequivocally states in Article 3 (3) second subparagraph that specific categories of traffic may be treated differently based on their objectively different quality of service requirements (the regulation uses the term "traffic categories" rather than "traffic classes", used in computer science). This harmonizes with the objective of reasonable traffic management, described in recital 9 of the Regulation 2015/2120 as contributing to an efficient use of network resources and to an optimization of overall transmission quality. Treating all data packets in the same way regardless of the circumstances, especially when network congestion occurs, would lead to an inefficient allocation of network resources, such as capacity, and it would also adversely affect the user experience. Thus, the key issues of interest to both legal and computer scientists are: establishing classes of traffic with objectively different quality of service requirements, identifying data flows as belonging to a specific class of traffic (traffic classification), and specifying the circumstances under which the treatment of the various classes of traffic should be differentiated. The Regulation 2015/2120 does not specify which network devices may be involved in classifying traffic flows as belonging to specific categories of traffic – it may therefore occur in any device, e.g. in routers or in specialized middleboxes.

Proportionality. Reasonable traffic management measures must also be proportionate. Proportionality is a well established general principle of EU law. It "requires that there be a reasonable relationship between a particular objective to be achieved and the means used to achieve that objective" [8]. When an action is undertaken that is subject to a proportionality review, a test may be carried out to evaluate whether the action satisfies several specific requirements: the pursuit by that action of legitimate aims; the suitability of the means, referring to its ability to further the achievement of the objective; the necessity of the means, that is whether it affects the rights of individuals as little as possible; and finally, proportionality in the narrow sense, that is the weighing of the reduction of the enjoyment of rights against the level of realization of the objective [8]. A traffic management measure may be considered reasonable only when it passes the test – that is when it cumulatively satisfies all its four criteria, described below in further detail.

Legitimate Aim. A traffic management measure pursues a legitimate aim when it is not based on commercial considerations (which is discussed below) but seeks to achieve an efficient use of network resources and contributes to the user experience, while not infringing on the rights of the end-users enshrined in the Regulation 2015/2120. E.g. a measure giving precedence to traffic generated by a specific application (rather than to a class of traffic generated by all the applications sharing similar quality of service requirements) would not be considered as pursuing a legitimate aim, even if it contributed to the efficient use of network resources, since it would discriminate between specific applications, which the regulation forbids to protect the end–users' rights.

Suitability. A measure is suitable when it indeed effectively serves the purpose for which it was introduced. E.g. if the purpose of employing a specific measure is the control of congestion, it will be considered suitable only if it can be proved that it in fact contributes to limiting congestion.

Necessity. A measure is necessary when the aim it pursues is not achievable by using another measure, which would influence the right of the end-user to access any internet content to a lesser extent. E.g. if the efficient use of network resources and a satisfying user experience may not be achieved by using classless traffic management measures, solutions distinguishing between several classes of best–effort traffic types may be utilized.

Proportionality in the Narrow Sense. Finally, a traffic management measure is proportionate in the narrow sense when its introduction does not adversely affect the end-user's rights to an excessive extent, e.g. a measure introduced to provide efficient access to one category of internet content (translating into a specific class of traffic) may not lead to another category of content becoming effectively unusable by the end–user.

Commercial Considerations. Reasonable traffic management measures may not be based on commercial considerations. This means that a provider of internet access services may not make using or refraining from the use of specific traffic management measures conditional upon receiving monetary or non-monetary remuneration from any party, be it an internet content provider, an end–user, or anybody else.

Monitoring Specific Content. Under Article 3 (3) second subparagraph of the Regulation 2015/2120, reasonable traffic management measures may "not monitor the specific content". This is understood by BEREC as a prohibition on monitoring the specific content of a data packet, i.e. the transport layer protocol payload [5]. On the other hand, the information contained in the IP packet header or transport layer protocol header is considered by BEREC as generic rather than specific content and therefore reasonable traffic management measures may rely on it [5]. This includes such common identification methods as those based on the connection's port numbers [9]. In practice, the prohibition on monitoring specific content means that reasonable traffic management may not under any circumstances involve deep packet inspection techniques (DPI). This may prove problematic in practice, as such solutions are often used in middleboxes [10]. However, the interpretation of the Regulation 2015/2120 does not preclude the use of machine learning solutions, if they do not rely on DPI. Thus, the utilization of pattern recognition algorithms described in [9] is allowed.

How Long Reasonable Traffic Management Measures May be Maintained? The Congestion Control Controversy. Under the Regulation 2015/2120, reasonable traffic management measures may not be maintained for longer than necessary, which relates to the proportionality principle. Both BEREC [5] and legal scientists [6] distinguish between trigger functions and employing traffic management measures as such. A trigger function may be "implemented and in place (…) on an ongoing basis inasmuch as the traffic management measure only becomes effective in times of necessity" [5]. Thus, employing a measure which continuously or regularly affects traffic is questionable from the point of view of the requirement of necessity [5, 6].

Technical solutions involving class-based AQM algorithms that trigger specific treatment of data packets to control congestion, but which allow for first–in–first–out, class–less functionality when there is no congestion seem to be in line with the Regulation 2015/2120 and fall within the category of reasonable traffic management measures. However, this sensible conclusion is weakened to the extent of being muddled by BEREC's stance and the imprecise wording of the regulation. BEREC indicates that class–based measures preventing "impending congestion" constitute exceptional traffic management measures and are subject to the limitations of Article 3 (3) subparagraph three rather than subparagraph two [5]. If that interpretation were to be accepted, then there would appear to be no circumstances under which any reasonable traffic management measures could ever be used. Why else reasonably manage traffic than to prevent and mitigate congestion, if class–based differentiation based on commercial considerations is banned altogether? Computer scientists' input on the issue would be invaluable.

What Treatment of Traffic May not be Considered Reasonable. Article 3 (3) third subparagraph of the Regulation 2015/2120 exemplifies what may not be involved in reasonable traffic management. The actions mentioned include, but are not limited to: blocking, slowing down, altering, restricting, interfering with, degrading or discriminating between specific content, applications or services, or specific categories thereof. Such actions are limited to exceptional traffic management measures, as described by BEREC [5] and in the legal literature [6]. The interpretation of the cited provision is problematic, especially with respect to the slowing–down of traffic. Best–effort, class–based AQM involves priority schemes for data packets, effectively slowing some packets down in relation to others. This should be considered as belonging to reasonable traffic management measures (as long as traffic classes are based on objective and genuine QoS requirements), rather than exceptional ones, as the Regulation implies. Computer scientists' elaboration on the issue would be greatly appreciated.

3.5 Exceptional Traffic Management Measures

Exceptional Traffic Management Measures – Introduction. Under Article 3 (3) third subparagraph of the Regulation 2015/2120, in certain, very specific circumstances, enumerated in the regulation and discussed below, traffic management measures going beyond those deemed reasonable may be used. In the literature and BEREC documents, these measures are called exceptional traffic management measures [5, 6]. Even though it is not clearly stated in the Regulation 2015/2120, a legal interpretation of the provisions leads to the conclusion that these measures – just like the reasonable ones – must be transparent and proportional, may not be based on commercial considerations and may not be maintained for longer than necessary. Unlike reasonable measures, however, the exceptional ones may under some circumstances monitor the specific content. They may also involve the actions mentioned in the previous paragraph above, including the blocking of traffic.

Circumstances under Which Exceptional Traffic Management Measures May be Used. The circumstances under which exceptional traffic management measures may be used have been enumerated in Article 3 (3) (a)–(c) of the Regulation 2015/2120 and are discussed below.

Compliance with Legislation. Under Article 3 (3) (a) of the Regulation 2015/2120, exceptional traffic management measures may be employed to ensure compliance with EU or national legislation, and with acts that comply with that legislation, e.g. decisions by courts or other public authorities. This involves inter alia the blocking of illegal content, such as child pornography under national laws resulting from the transposition of Article 25 of the Directive 2011/92/EU of the European Parliament and of the Council of 13 December 2011 on combating the sexual abuse and sexual exploitation of children and child pornography, and replacing Council Framework Decision 2004/68/JHA. The relevant legislation usually determines what specific traffic management measures should be employed to enact it, e.g. Domain Name Service blocking. EU and national legislators would be advised to cooperate with computer scientists in choosing the most appropriate, specific measures to be used.

Preservation of Network Integrity and Security. Under Article 3 (3) (b) of the Regulation 2015/2120, exceptional traffic management measures may be employed to preserve the integrity and security of the network, of the services provided via that network, and of the terminal equipment of end-users. This is rather uncontroversial and has been convincingly explained by BEREC in [5].

Preventing and Mitigating Congestion. Under Article 3 (3) (c) of the Regulation 2015/2120, exceptional traffic management measures may be employed to "prevent impending network congestion and to mitigate the effects of exceptional or temporary network congestion, provided that equivalent categories of traffic are treated equally". The cited provision is problematic for two reasons: first, the Regulation 2015/2120 does not define network congestion, and second, it does not contain provisions that would sufficiently differentiate between cases when reasonable or exceptional traffic management measures should be employed to control congestion.

Understanding what the term "congestion" used in the Regulation 2015/2120 is supposed to mean is not as straightforward as it may seem. In networking literature congestion is "defined as a state or condition that occurs when network resources are overloaded, resulting in impairments for network users as objectively measured by the probability of loss and/or delay" [4]. However, other definitions of congestion are also presented in the literature [11]. Legal and computer scientists should work together to provide a definition of congestion best suited to the application and enforcement of the Regulation 2015/2120.

Taken literally, Article 3 (3) (c) of the Regulation 2015/2120 implies that any "impending network congestion" justifies the use of exceptional traffic management measures. Such an interpretation however is unacceptable. As has been mentioned above, under the Regulation 2015/2120 solutions implemented in the network to aid in congestion control, such as Active Queue Management, generally fall into the category of reasonable traffic management. The use of exceptional rather than reasonable traffic management measures in order to control impending congestion is certainly justified in cases when a congestion collapse of the network is predicted based on the analysis of the network traffic. Other cases should be elaborated upon by computer scientists, considering both the need to safeguard the end–users' rights enshrined in the Regulation 2015/2120, and the technical realities of network traffic management.

4 Conclusions

Since the Regulation 2015/2120 came into force, all traffic management measures used by providers of internet access services in the EU must meet the requirements introduced by its provisions. Thus, the law has become one of the factors that have to be considered when designing and applying internet traffic management measures employed in networks used to provide publicly available internet access services in Europe. The proper interpretation of the Regulation 2015/2120, serving its sensible application and enforcement, necessitates the co-operation of legal and computer scientists. The former may not disregard the technical realities of managing internet traffic, which should be elaborated on by the latter, in turn taking legal considerations into account.

Computer scientists' input would be invaluable in clearing up the many ambiguities created by the Regulation 2015/2120 with respect to the various types of traffic management measures categorized by it, namely the basic, reasonable and exceptional ones. The distinction between these three categories of measures is of the utmost importance to the application of the Regulation 2015/2120, and yet it was not made clear enough by the legislators or BEREC. Computer scientists undoubtedly may help in determining which concrete, technical solutions or types thereof fall within each of the three categories. Specifically, even though the need to control congestion in the network is apparently one of the main reasons of introducing any traffic management measures, the Regulation 2015/2120 is unclear on when reasonable or exceptional measures may be used to control congestion. Computer scientists' input in resolving this conundrum, as well as interpreting how the term "congestion" should be understood for the purpose of applying and enforcing the regulation would be appreciated.

Computer scientists may also greatly assist in resolving the issues related to the treatment of various classes of traffic. The Regulation 2015/2120 allows the objectively justified differentiation in the treatment of traffic classes, based on their different technical quality of service requirements. Computer scientists may delineate classes of traffic with objectively different quality of service requirements, determine the means of identifying data flows as belonging to a specific class of traffic (traffic classification), and specify the circumstances under which the treatment of the various classes of traffic should be differentiated.

References

1. Maxwell, W.J., Bourreau, M.: Technology neutrality in internet, telecoms and data protection regulation. Comput. Telecommun. Law Rev. **21**(1), 1–4 (2015). https://www.hoganlovells.com/~/media/hogan-lovells/pdf/publication/201521ctlrissue1maxwell_pdf.pdf
2. BEREC. https://berec.europa.eu/eng/netneutrality/traffic_management/. Accessed 13 May 2018
3. Constantine, B., Krishnan, R.: Traffic Management Benchmarking. RFC 7640 (2015). https://tools.ietf.org/html/rfc7640
4. Papadimitrou, D., Welzl, M., Scharf, M., Briscoe, B.: Open Research Issues in Internet Congestion Control. RFC 6077 (2011). https://tools.ietf.org/html/rfc6077
5. BEREC: BEREC Guidelines on the Implementation by National Regulators of European Net Neutrality Rules. BoR (16) 127. https://berec.europa.eu/eng/document_register/subject_matter/berec/download/0/6160-berec-guidelines-on-the-implementation-b_0.pdf
6. Piątek, S.: Rozporządzenie UE Nr 2015/2120 w zakresie dostępu do otwartego internetu, 1st edn. Wydawnictwo CH Beck, Warszawa (2017). (in Polish)
7. Maliszewska-Nienartowicz, J.: Direct and indirect discrimination in European union law – how to draw a dividing line? Int. J. Soc. Sci. **III**(1), 41–55 (2014). http://www.iises.net/download/Soubory/soubory-puvodni/pp041-055_ijoss_2014v3n1.pdf
8. Klatt, M., Meister, M.: The Constitutional Structure of Proportionality, 1st edn. Oxford University Press, Oxford (2012)
9. Dainotti, A., Pescapé, A., Claffy, K.C.: Issues and future directions in traffic classification. IEEE Netw. **26**(1), 35–40 (2012). https://doi.org/10.1109/MNET.2012.6135854
10. Bremler-Barr, A., Harchol, Y., Hay, D., Koral, Y.: Deep packet inspection as a service. In: CoNEXT 2014 Proceedings of the 10th ACM International Conference on Emerging Networking Experiments and Technologies, pp. 271–282. ACM, New York (2014). https://doi.org/10.1145/2674005.2674984
11. Bauer, S., Clark, D., Lehr, W.: The Evolution of Internet Congestion (2009). https://groups.csail.mit.edu/ana/Publications/The_Evolution_of_Internet_Congestion_Bauer_Clark_Lehr_TPRC_2009.pdf

Ranking Composite Services

Vangalur Alagar[1(✉)], Ammar Alsaig[1], and Mubarak Mohammad[2]

[1] Concordia University, Montreal, Canada
alagar@cs.concordia.ca, a_alsaig@encs.concordia.ca
[2] Software Consultant, Toronto, Canada
mubarak.sami@gmail.com

Abstract. Service compositions are a means to create value-added services for service provider, as well as for service requester. From the consumer-centric point of view services that meet the predefined preferences of service requester add both economic value and satisfaction. In order to achieve consumer satisfaction in a market place it is necessary for service providers to design fair-algorithms for service composition and ranking of services, so that consumers will have less difficulty in selecting their most preferred services without much difficulty. This consumer-oriented approach has the potential to add value-added trust for business, leading to profitable business and healthier consumer relation. Motivated by this theme this paper proposes a fair ranking algorithm for composite services.

Keywords: Fair ranking · Simple and composite services
Algorithm performance

1 Introduction

Services can be broadly classified into *atomic* (simple) services and *complex* (composite) services. An atomic service cannot be split in terms of either its functionality or its mode of existence. A complex service is obtained by "putting together" one or more services that are either atomic or complex. Service composition methods refer to different ways in which services are put together as a "package" [6]. The survey article [8] describes the different frameworks for service compositions for cloud, mobile, and web services. The main drivers for service compositions are (1) to create value-added services, (2) enhance usage simplicity, (3) improve efficiency of reusability, and (4) to create composition mechanisms as building blocks for service-oriented system management. Most of the existing composition designs can be broadly classified as *hierarchical* and *conversational*. In the former approach, the choice of services in a composition, and how they are composed are controlled solely by *Service Provider* (SP), and as such it is completely *opaque* (black box) to *Service Requester* (SR). This design approach is used by many enterprises, such as *Tour Operators*, *Power Suppliers*, and *Insurance Companies*. In a conversational design SR is given access to

L. Borzemski et al. (Eds.): ISAT 2018, AISC 852, pp. 100–110, 2019.
https://doi.org/10.1007/978-3-319-99981-4_10

certain stages in the composition process. This grey box composition approach allows SR to "negotiate" with SP on certain service quality choices. We comment that hierarchical compositions are *static* (cannot be changed by any one other than SP) because they are done at service publication time, whereas conversational compositions are *dynamic* because they are done at service execution time with input/consultation of SR. Clearly, static compositions are driven by the "value-added" goals of SP, while dynamic compositions are partly driven by user's demand and are done by execution engine in the "service provision framework". In this paper we look at user-centric *ranking* of composite services, and hence the compositions are driven solely by the set of *preferences* specified by SR only.

User-centric service compositions are desired in many areas of social importance, such as travel, health, consumer-oriented retail industries. In a social setting composite services arise and are in great demand. As an example, a travel package offered by a SP is a (static) complex service, which typically includes air travel, hotel accommodations, car rental and other simple services such as "tour guides". A consumer can buy only the whole package offered by the SP, without control over changing the features of any service in the package and in general cannot "meaningfully compare" two different tour packages offered by different SPs to know which best meets their preferences. However, a SR can select simple services individually and make a tour package. Given the variety of services that have the same functionality with differing qualities are available on-line it becomes an arduous task for a SR to put together different packages and compare them. So, there is a need to provide a medium which (1) receives a list of functional specifications from SR, each function related to one simple service, (2) for each function fetches all services from SP that match the functional requirement of SR, (3) receives preferences from SR for selecting services that are acceptable to SR, and (4) ranks all possible compositions and displayers a ranked list of composite services. We assume that SP selects and preprocesses simple services of a specific functionality into vector models (of quality attributes) and publishes the services [6]. The ranking algorithm for composite services presented in this appear is a generalized version of the ranking algorithm [1] for atomic services. The complete theory behind the presented algorithm appears in [2].

2 Background Material

Because the composition algorithm presented in this paper is an extension of the composition algorithm for atomic services [1,2], it is essential to understand the basic conceptual models, notations, and algorithmic structure. So, in this section we explain the basic material in order that the algorithm in Sect. 3 can be easily followed.

2.1 Service Model and Query Structure

A SP publishes atomic services, where each atomic service is a first class object having a "service part" and a "contract part" [6]. The service part describes the functional behavior of the service, described through its functional specification, attributes, and other quality of service properties. The contract part describes legal and context information for service execution and delivery. The essence is that in general a service functionality may exist with different quality of service attributes and contracts. So, a SR will find many possible service options while matching exactly with one service functionality. Ranking arises only here, namely to rank these services, all of which have the same functionality, in decreasing order of relevance to the set of preferences of SR. Hence, for algorithmic level abstraction we let the number of "quality/trustworthy/contract" (QTC) attributes to be fixed for all services in the system and model them as vectors of finite length. That is, every service functionality s_f is associated with a finite number of vectors $\{\langle a_1, a_2, \ldots, a_n \rangle\}$, where a_i, $i = 1 \ldots n$, is a QTC attribute. A user, having specified functionality, will specify a query $\langle V_1, V_2, \ldots, V_n \rangle$ where V_i is a value preferred by her for attribute a_i. The value V_i is either a numerical value or a range of numerical values or a non-numerical for attribute a_i depending upon the type of attribute a_i. Our algorithm [1] requires "semantics" for user specified values in order that the user's preferences are properly enforced in the algorithm. The semantic part, explained below, along with the query vector $\langle V_1, V_2, \ldots, V_n \rangle$ is the user input to the ranking algorithm).

1. *Consumer Query Vector:* $\{V_1, \cdot, V_n\}$, where V_i is a value of a_i.
2. *Level of Importance, modeled by a Weight Vector:* $\{W_1, \cdot, W_n\}$, where W_i indicates the level of importance of a_i.
3. Feature Preferences: A feature can be modeled by a range of values. A feature value may be compared with a query value for that feature in mode "B" (best match) or "E" (exact match). We call "B" and "E" as *modes.* A user is empowered to state whether or not a specific service feature is "essential", by setting the *Essential Option* of that feature to either "T" (True, meaning essential) or "F" (False meaning non-essential).
4. *Semantics for Best Match:* We want our algorithm to suggest to user services which are "better suited" than what was really stated in the preferences. As an example, assume that a SR wants apartments whose rent do not exceed a certain amount and whose locations are within a specific area of city. Clearly, the mode of search must be "B" and it is advantageous to the user to use "Less is Better" (LB) semantics while matching rents. If the user includes "living area" (or lease term) as a requirement then "More is Better" (MB) is the semantics while comparing living areas. So, we have considered the two possible semantics 'LB and MB'. As an example, if the consumer query includes the value "3 year warranty" for "warranty" attribute, the ranking algorithm will find services that offer warranty for *at least* 3 years and place them on top of the ranked list.
5. *Normalization and Accuracy Levels:* In many similarity calculations, the computation may throw us into unbounded domains or in a situation where the differences are so small that within a tolerance limit the difference may not be distinguishable. So, we do have a "normalization" scheme, and a user-centric determining factor for "accuracy level". We have introduced in our earlier algorithm [1] "Algorithmic Accuracy" and "Essential Accuracy" that are to

be set by users in order to fine tune the algorithm, provide stability, and getting sharper ranking results.

6. *AllBest Option:* It is introduced to force higher ranking for certain services. It can be set to either "T" or "F". If "T" is specified, the X-Algorithm is forced to prefer services that meet *all requirements for all features* over other services in which some features may offer "better values" although not all requirements are met. The algorithm relaxes this strict constraint when the value "F" is set for this option.

An example of user query that is structured to include the above information is shown below.

Query (Feature Values)	[100, 20, 10, 120]
Weights	[Critical, Normal, Normal, Critical]
Mode	[B, E, E, B]
Essential	[T, T, F, T]
Semantics	[MB, LB, LB, MB]
AllBest	TRUE

2.2 A Brief Overview of Ranking Algorithm for Simple Services

The algorithm proposed for ranking composite services in Sect. 3 uses the X-Algorithm [1] which ranks each simple service in the composition. We explain here the three steps involved in the X-Algorithm, the function used to calculate the similarity measure in Step 1 of the X-Algorithm, and give an example. The first step is "Pre-processing phase (PP)" in which the input query vector $\langle V_1, V_2, \ldots, V_n \rangle$ and the stated preferences/semantics are compared against every vector of the quality attributes $\{\langle a_1, a_2, \ldots, a_n \rangle\}$ of the service function, and a matrix PM of feature ranks is prepared. The rank PM_{ij} is the *normalized rank* of ith attribute comparison of jth quality vector against the query vector $\langle V_1, V_2, \ldots, V_n \rangle$. The calculation of the rank is based on the similarity measure [1] shown in Eqs. 1, 2, 3, and 4. An extensive discussion justifying the merits of this similarity measure appears in [1,2]. The second step is "Multiplication Phase (MP)" that calculates the "Unsorted Rank (UR)" $UR = Weight \star PM$, where $Weight = User\ Weights + Essential\ Weight$. At this stage the "AllBest" requirement of the user is applied on the unsorted ranks to filter the unsorted ranks of services that meet all user requirements. In the final step, a fast sorting algorithm is applied on the vector UR to produce the ranked list of services wherein services appear in decreasing order of their ranks.

The function that calculates the similarity measure for a feature is given below. A detailed justification for this choice and its technical merits appear in [2].

$$SM_X = \begin{cases} SM_{EB} & X = 0 \\ RC_X & (BestMode \text{ or } ExactMode) \text{ and } (X = -1 \text{ or } X = 1) \end{cases} \quad (1)$$

where RC_X is defined in Eq. 4 and SM_{EB} is defined in Eq. 2.

$$SM_{EB} = \begin{cases} +1 & s = q \ , \ s \in q \\ 0.5 & s \subset q, q \subset s \\ 0 & s \nsubseteq q \end{cases} \quad (2)$$

$$X = \begin{cases} -1 & BestMode, semantic = MB \\ +1 & BestMode, semantic = LB \end{cases} \quad (3)$$

$$RC_X = \begin{cases} |X| & s = q \\ |X + \frac{|q-s|}{q}| & s < q \\ |X - \frac{|q-s|}{s}| & s > q \end{cases} \quad (4)$$

*Example 1 (*Ranking Simple Services using the X-Algorithm*).* Assume that all services have four features. The user query profile is $[f_1(Price) = 1000; LB, BestMode]$, $[f_2(Rating) = 4*, MB, BestMode]$, $[f_3(Colour) = [Read, Black], EB, BestMode]$, and $[f_4(Weight) = 4, LB, ExactMode]$. The four available services are $S_1 = \{900, 9*, RED, 6\}$, $S_2 = \{1500, 7*, WHITE, 2\}$, $S_3 = \{400, 3*, BLACK, 3\}$, and $S_4 = \{1000, 4*, BLACK\&WHITE, 2.7\}$. When the above functions are applied to calculate the scores (in three steps as explained earlier) we get the scores 4.33 for S_1, 2.6 for S_2, 4.1 for S_3, and 3.175 for S_4. Hence, the ranked list of services produced by the X-Algorithm is S_1, S_3, S_4, S_2.

3 Ranking Composite Services

A composition is performed on many services, some of which may be complex. It is necessary to enable the user construct a composite query interface, called *Extended Simple Request Structure* ($ESRS$) as shown in Fig. 1. The SR field in $ESRS$ is the user request to rank simple services by the X-Algorithm that fit the attribute values and semantics defined in it. The value *weight* defined by the user for the attribute SRW in $ESRS$ is the weight assigned to *all* services ranked in response to the *Simple Request* part. If the attribute SRE in $ESRS$ is set to "ON" ("OFF") then all services ranked by the X-Algorithm is set to "ON" ("OFF"). In the two fields SRE and SRW the user is required to assign preference and level of importance (regular or essential) to the service within the composition, which will lead to determining whether a simple service in a composition can be more important than another simple service. Eventually, this will affect the ranks produced for the different compositions. Thus, a user wishing to compose m different services for a composition will construct a $ESRS$ for each service type. Once the $ESRS$ for all m requests are submitted to the composition phase, our composition algorithm will automatically construct a *Composite Request Structure* (CRS) [2]. Our X-Algorithm is extended to take CRS as user "preferences", create "composite service alternatives", and then rank them according to CRS.

$Simple\ Request(SR)$				SRW	SRE
-	$Feature_1$	...	$Feature_n$	$weight$	ON/OFF
$Query$	$value_1$	...	$value_n$		
$Semantics$	$semantic_1$	...	$semantic_n$		
$Weight$	$weight_1$	...	$weight_n$		
$Mode$	$mode_1$	...	$mode_n$		
$Range$	$range_1$	...	$range_n$		
$Essential$	ess_1	...	ess_n		
$AllBest$	ON/OFF				

Fig. 1. The structure of $ESRS$ request

3.1 Constructing Composite Request Structure (CRS)

Let $RqC = \{Rq_1, Rq_2, \ldots, Rq_n\}$, where Rq_i denotes the $ESRS$ for i_{th} service. In the X-Algorithm every attribute will get a bounded score. We need to extend this concept to service level. So, we provide a method for calculating the "maximum rank" $maxRank$ for each Rq_i. The calculation is done based on the regular ranks, and other options of the attributes within Rq_i definition in $ESRS$. So, this maximum rank can be calculated prior to the ranking process. We define $maxRank = MRR + MEss + MAllBest$, where (1) MRR is the maximum rank that the X-Algorithm achieves when both $Essential$ and $AllBest$ options are not set, (2) $MEss$ is the maximum rank when $Essential$ option is set to "ON", and (3) $MAllBest$ is the maximum when $AllBest$ option is set to "ON". Since MRR, $MEss$, and $MAllBest$ are all bounded, the value $maxRank$ will be bounded. MRR is calculated in the multiplication of the weights by the scores of service features, when all scores are set to the maximum reward ($maxReward$) and all the weights are set to the maximum weight(w_h). Thus, $MRR = n * w_h * maxReward$, and $MEss = maxReward * ((w_h * 1000 * w_e) + w_h) * nef$, where n is the number of features defined in Query, w_h and w_e are the highest regular weight and the essential weight respectively, and nef is the number of essential features defined in the user request. The maximum impact caused by this option has been observed [1] to be on the final rank of Rq_i due to the addition of the number of features defined in Rq_i to the final rank of the results of Rq_i. Thus, $MAllBest = n * AllBestFlag$, (where n is the number of features in query and $AllBestFlag$ is 0 if $AllBest$ option is "OFF" and $AllBestFlag$ is 1 if $AllBest$ option is "ON"). The $maxRank_i$ for $ESRS$ Rq_i is included in CRS (Table 1) in the row CRQ (Complex Request Query). The other two rows in Table 1 are Complex Request Weight (CRW), Complex Request Essential (CRE) that are directly available in $ESRS$ (Fig. 1).

3.2 Responding to a Composite Request

X-Algorithm is extended to do the following steps.

Step 1: Services for each Simple Request(Rq_i) are ranked using X-Algorithm. Corresponding to Rq_i let TR_i denote the total ranks of services calculated in this step.

Table 1. The structure of CRS request for m $ESRS$

-	Rq_1	...	Rq_m
CRQ	$maxRank_{Rq_1}$	...	$maxRank_{Rq_m}$
CRW	SRW_{Rq_1}	...	SRW_{Rq_m}
CRE	SRE_{Rq_1}	...	SRE_{Rq_m}

Table 2. $ESRS$ Hotel and Air Travel Services: Example 2

Simple Hotel Request(Rq_h)				SRW_h	SRE_h
-	$HotelPrice(h_1)$	$Stars(h_2)$	$Location(h_3)$	*Significant*	0(OFF)
Query	100$	4/5	*Downtown*		
Semantics	*LB*	*MB*	*EB*		
Weight	*Significant*	*Normal*	*Low*		
Mode	*Best*	*Exact*	–		
Range	–	–	–		
Essential	1	0	0		
AllBest	0(OFF)				

Simple AirTravel Request(Rq_f)			SRW_f	SRE_f
-	$TicketPrice(f_1)$	$Route(f_2)$	*Normal*	0(OFF)
Query	1000$	*Direct*		
Semantics	*LB*	*EB*		
Weight	*Significant*	*Normal*		
Mode	*Best*	–		
Range	–	–		
Essential	0	0		
AllBest	1(ON)			

Step 2: The Cartesian Products $TR = \{TR_1 \times TR_2 \times \cdots \times TR_m\}$ are computed and represented in $CPlan$ matrix $\{r_{ij}\}$, wherein each row shows the rankings (in decreasing order) of the simple services in the composition. That is, the i_{th} row is the ranking vector of i_{th} composition, such that $r_{ij} \in TR_j$ is the ranking of the simple services corresponding to Rq_j.

Step 3: The rows of $CPlan$ matrix are considered as alternates and are ranked with respect to "Exact Match" criteria of the ranks specified in CRQ, subject to the weights CRW and the *Essential* option CRE specified Table 1. We justify this step because CRS is in effect "the user request for composition" and $CPlan$ is the "available alternatives" of composition scores. Below we illustrate the steps of ranking composite services through an example.

*Example 2 (*Ranking the Composition of Hotel and Air Travel Simple Services*).* The user constructed $ESRS$ for Hotel and Air Travel are shown in Table 2. Hotel and Air Travel Services that are available are shown in Table 3. The extended X-Algorithm computes the rank for the composition of these two services as follows.

Table 3. Available Services: Example 2

Hotel	$Price(hs_1)$	$Stars(hs_2)$	$Location(hs_3)$	Air Travel	$Ticket\ Price(fs_1)$	$Route(fs_2)$
HS_1	60$	3.5/5	*Suburb*	FS_1	900$	*Indirect*
HS_2	250$	5/5	*Downtown*	FS_2	1200$	*Direct*
HS_3	100$	5/5	*Suburb*	FS_3	1000$	*Direct*

1. Calculates $maxRank_h$ for the Simple Hotel Request Rq_h, and $maxRank_f$ for the Simple AirTravel Request Rq_f:
   Calculation of $maxRank_h$ for Travel Request Rq_f:
   $maxRank_h = RR_h + Ess_h + AllBest_h$
   where,
   $MRR_h = nfq * w_h * maxReward = 3 * 0.005 * 3 = 0.045$
   $MEss_h = maxReward * ((w_e * 1000w_h) + w_h) * neq$
   $= 3 * ((0.005 * \frac{0.1+(3-1)*(5)}{0.1} + 1) * 5 + 0.005) * 1 = 22.59$
   $MAllBest = 3 * 0 = 0$
   Thus, $maxRank_h = 22.62$
   Similarly, the $maxRank_f$ for the Simple Request Rq_f is calculated: $MRR_f + MEss_f + MAllBest_f = 0.03 + 0 + 2 = 2.03$.

2. The CRS request is automatically structured by the extended X-Algorithm, as shown in Table 4.
3. The Total Ranks computed for Hotel Services TR_h and for Air Travel Services TR_f are shown in Table 5.
4. The $CPlan$ matrix computed from the product $TR = TR_h \times TR_f$ is shown in Table 6.
5. The X-Algorithm is executed for the request CRS on the available "composition" alternatives in $CPlan$. The Ranks of rows in $CPlan$ are listed in Table 7 (ordered in a non-increasing order). That gives ranked list of compositions.

Table 4. The structure of CRS request for the two simple requests Rq_h and Rq_f: Example 2

-	Rq_h	Rq_f
CRQ	$maxRank_{Rq_h}(22.62)$	$maxRank_{Rq_f}(2.03)$
CRW	$SRW_{Rq_h}(0.005)$	$SRW_{Rq_f}(0.003)$
CRE	$SRE_{Rq_h}(0)$	$SRE_{Rq_f}(0)$

3.3 Complexity of Ranking Composed Services

Let Rq_i denote the ith $ESRS$ in a Composite Request CRS. Let m denote the number of simple services in CRS. Let the size of composition plan be C. Based

Table 5. Ranking of Hotel and Air Services: Example 2

$HotelService$	$Rank$	$AirService$	$Rank$
HS_1	7.42	FS_1	2.009
HS_2	4.19	FS_2	1.804
HS_3	6.28	FS_3	1.724

Table 6. Different composition plans ($CPlan$) for Hotel and Air Travel Services: Example 2

$TR = TR_h \times TR_f$
$[HS_1\ ,\ FS_1]$
$[HS_1\ ,\ FS_2]$
$[HS_1\ ,\ FS_3]$
$[HS_2\ ,\ FS_1]$
$[HS_2\ ,\ FS_2]$
$[HS_2\ ,\ FS_3]$
$[HS_3\ ,\ FS_1]$
$[HS_3\ ,\ FS_2]$
$[HS_3\ ,\ FS_3]$

$$\Longrightarrow CPlan = \begin{bmatrix} HS_1 & FS_1 \\ HS_1 & FS_2 \\ HS_1 & FS_3 \\ HS_2 & FS_1 \\ HS_2 & FS_2 \\ HS_2 & FS_3 \\ HS_3 & FS_1 \\ HS_3 & FS_2 \\ HS_3 & FS_3 \end{bmatrix}$$

Table 7. Ranks of different compositions of services: Example 2

Ranked TR	*CPlan Ranks*
$[HS_1\ ,\ FS_3]$	0.0048
$[HS_1\ ,\ FS_2]$	0.0042
$[HS_3\ ,\ FS_3]$	0.0041
$[HS_1\ ,\ FS_1]$	0.0040
$[HS_3\ ,\ FS_2]$	0.0035
$[HS_3\ ,\ FS_1]$	0.0032
$[HS_2\ ,\ FS_3]$	0.0029
$[HS_2\ ,\ FS_2]$	0.0023
$[HS_2\ ,\ FS_1]$	0.0021

on the complexity of X-Algorithm [1], the time necessary to calculate the ranks of $CPlan$ is $O(C) + O(C * m) + O(C\ log(C))$, where, $O(C)$ is the time necessary to obtain composition plans, $O(C*m)$ is the time necessary to calculate the total ranks for m composition plans in $CPlan$, and $O(C\ log(C))$ is the time necessary to sort the total ranks of $CPlan$. We remark that C is essentially dependant on the number of simple services ranked and composed by the X-Algorithm and m is the number of $ESRS$ requests. There is no obvious correlation between C and m. So, for a fixed value of m the complexity is bounded by $O(C\ log(C))$.

4 Conclusion

The X-Algorithm [1] used for ranking "simple services" is extended in this paper for ranking "composite services". The key points in the extension are (1) the automatic construction of CRS request from the $ESRS$ input of clients, (2) creating "CPlan" from the ranked lists of simple services, and (3) executing the X-Algorithm again with "CRS as request and $CPlan$ as the list of available services. Our algorithm emphasizes user-centric ranking, an aspect of great importance in document retrieval, social ranking, and healthcare research. To the best of our knowledge there are only a few papers [3,4,12] that formally discuss ranking of composite services. Compared to these existing works, we claim the following advantages to our approach:

- The ranking method is fair, in the sense that only on-line ranking is done strictly respecting user preferences. Many algorithms [10] use off-line processing that might violate user preferences.
- The complexity of our algorithm is independent of the number of features included in the user query, a significant contrast to other algorithms [9] which restrict the number of features in user queries.
- Our method is independent of specific composition mechanisms. Sequential, parallel, and conditional composition methods have been discussed in [6]. These techniques can be applied once the $Cplan$ ranking is done. Hence, our method is more general than the ranking methods proposed in [3,4,12].
- After ranking each simple service type, the user can query that ranked list and choose only those services that are in the top k (for some k determined by the user) for composition with other services. The choice of k may be different for each service type. As an example, a user may choose the top 10 of hotel services to be composed with the top 15 of Airline services. If the composition is done before ranking the simple services to be composed, there is a potential explosion of "service space". In our approach both space and time complexity are optimized, the first due to "user specification of preferences" and the second due to the computational simplicity of the similarity function that we designed to handle user preferences with semantic clarity, and algorithmic efficiency justified by $O(C\ log(C))$ complexity.
- Instead of a SR constructing the $ESRS$ an SP can create one embedding her own criteria and semantics, and use X-Algorithm to rank services at service provision stage. It is very likely that SP can successfully publish service packages that meet her business goals.

We are currently exploring methods for using our algorithm for (1) ranking automatic service compositions in service networks, (2) ranking patient records in the context of drug or treatment similarity, and (3) criteria-based ranking of files that are dynamically created by data fusion in the study of Big Data Science and Engineering.

References

1. Alsaig, A., Alagar, V., Mohammad, M., Alhalabi, W.: A user-centric semantic-based algorithm for ranking services: design and analysis. Serv. Oriented Comput. Appl. **11**(1), 101–120 (2017). Accessed Sept 2016
2. Alsaig, A.: Semantic-based, multi-featured ranking algorithm for services in service-oriented computing. Master of Computer Science thesis, Concordia University (2013). http://spectrum.library.concordia.ca/978006/
3. Bawazir, A., Alhalabi, W., Mohammad, M.: A formal approach to matching and ranking context-dependent services. In: Proceedings of The IIER International Conference, pp. 6–18 (2016)
4. Brucker, A.D., Zhou, B., Malmignati, F., Shi, Q., Merabi, M.: Modelling, validating, and ranking of secure service compositions. Soft Pract Exper. **47**, 1923–1943 (2017)
5. De, S., Barnaghi, P., Bauer, M., Meissner, M.: Service modelling for the Internet of Things. In: Proceedings IEEE of the Federated Conference on Computer Science and Information Systems, Szczecin, Poland, pp. 949–955 (2011)
6. Naseem, I.: Specification, composition and provision of trustworthy context-dependent services. Ph.D. thesis, Department of Computer Science and Software Engineering, Concordia University (2012)
7. Ibrahim, N., Mohammad, M., Alagar, V.: Adaptable discovery and ranking of context-dependent services. In: Proceedings of 2011 IEEE Asia-Pacific Services Computing Conference, pp. 1–6. IEEE (2011)
8. Lemos, A.L., Daniel, F., Benatallah, B.: Web service composition: a survey of techniques and tools. ACM Comput. Surv. **48**(3), 33:1–33:41 (2015)
9. Milovanović, A., Mitričević, M., Mijalković, A., et al.: The Analytic Hirarchy Process (AHP) Application in Equipment Selection. The Growth of Software Industry in the World With Special Focus on Bosnia and Herzegovina (2012)
10. Schafer, J.B., Konstan, J., Riedi, J.: Recommender systems in e-commerce. In: Proceedings of the 1st ACM Conference on Electronic Commerce, pp. 158–166. ACM, New York (1999)
11. Zhang, R., Birukou, A., Kidawara, Y., Kiyoki, Y.: Web service ranking based on context. In: Second International Conference on Cloud and Green Computing, pp. 375–382 (2012)
12. Zhang, S., Dou, W., Chen, J.: Selecting Top-K composite web services using preference-aware domianance relationship. In: The IEEE 20th International Conference on Web Services, Santa Clara, California, USA, pp. 75–82. IEEE Computer Society Press (2013)

Effectiveness Metrics for the SelfAid Network, a P2P Platform for Game Matchmaking Systems

Michał Boroń(✉), Anna Kobusińska, and Jerzy Brzeziński

Poznań University of Technology, Poznań, Poland
mboron@cs.put.poznan.pl

Abstract. This paper describes effectiveness metrics for the SelfAid Network — quantified values providing information about the quality of functioning of the system. Proposed metrics capture, among others, the cost of running the application, user satisfaction, and resistance to failures. The metrics were specifically chosen for the SelfAid network, a platform for game matchmaking systems. Matchmaking, in online multiplayer games, refers to the process of finding players for online play sessions. It can function based on ranking (such as ELO in chess), matching players choosing the same difficulty level (for cooperation-based games), etc. The paper presents the results of applying the proposed metrics to a simulation of the SelfAid network.

Keywords: P2P · Peer-to-peer · Large scale systems
Edge computing · Fog computing · Measure · Metric

1 Introduction

Online multiplayer games market is popular and still growing. It inspires research of solutions to various unique problems, such as matchmaking players. SelfAid [1] is a P2P platform for building matchmaking systems with custom rules. This custom matchmaking system is built by defining multiple services. A service is an implementation of a network protocol. For each defined service, SelfAid creates a group (ring) of nodes which members run a copy of the service. Then the network traffic is automatically distributed (load balanced) among the members of the group. SelfAid increases the number of members of a group when the demand for their service rises. It also provides a mechanism for a client to find the nodes hosting a particular service. For example, a matchmaking system could consist of one service for beginner, one for intermediate and one for advanced players.

In order to properly evaluate the quality of the SelfAid system, it is necessary to take multiple factors into account. First, from the perspective of a user who has to contribute some of his resources in order to benefit from the network, it is important that the amount of those resources is reasonable. Also, the user wants to quickly receive responses for his requests, so that he is able to quickly

L. Borzemski et al. (Eds.): ISAT 2018, AISC 852, pp. 111–122, 2019.
https://doi.org/10.1007/978-3-319-99981-4_11

start playing the game. From the point of view of the system, it is important that it continues to properly function even when failures occur. Accordingly, this paper proposes to measure the values of the following features: resource usage, responsiveness, reliability.

The paper is structured as follows. First, in Sect. 2, the proposed metrics are described and rationale for their formulas is provided. Section 3 discusses the importance of considering various load scenarios in the domain of game matchmaking systems. Then, the simulation environment which was used for experiments is discussed in Sect. 4. Following, in Sect. 5, the results obtained by applying proposed metrics to the simulation of the SelfAid network are presented. Then, related work is listed in Sect. 6. Finally, conclusions and future work directions are presented in Sect. 7.

2 Concept

This section discusses the metrics proposed for the SelfAid network.

The system does not pose much requirements towards saving data on the hard drive, nor using RAM memory, nor performing intensive computation. Thus, network resources (mostly network bandwidth) are the bottleneck for system performance. An assumption was made that all nodes have the same preferences as to how much bandwidth (at most) they are willing to contribute. When the quantity of traffic exceeds preferences of a node, it ignores application requests (does not respond to them).

The following subsections present metrics for each of the following features: resource usage, responsiveness, reliability.

2.1 Measuring Resource Usage

Measuring resource usage aims to capture how demanding is the application on a participant (a node in the P2P network). The measurement is focused on the network bandwidth, since it is the main resource needed for SelfAid network operation.

One of the resource usage metrics is the metric of percent of time with load above average. That is, how much time (with respect to total simulation time) did a node spend, contributing more network bandwidth than planned. Let us define TLAA as the amount of Time a peer spent with Load Above Average. The equation representing this metric, named OTR (Overload Time Ratio), is shown below.

$$OTR = \frac{TLAA}{total\ time} \tag{1}$$

The metric mentioned in the previous paragraph gives a good idea about how often the nodes are overloaded, but does not provide insight how much are they overloaded. Some information about the amount of the overload can be inferred from simple metrics of total network traffic of a peer (for sent and received traffic respectively). This metric also provides a comparison of how fairly (uniform) the load is divided between peers.

2.2 Measuring Responsiveness

Responsiveness captures user satisfaction, i.e. how much time is needed to complete a user request. In SelfAid, the user request is the request for finding an opponent.

The corresponding metric can be simply expressed as:

$$responsiveness = \frac{1}{n}\sum_{i=1}^{n}(t_{response}(req_i) - t_{sent}(req_i)) \tag{2}$$

where req_i refers to a client request to find an opponent, $t_{sent}(req_i)$ is the moment of time (expressed in real time) in which the request was sent. $t_{response}(req_i)$ is the moment of time in which the user receives a response for his request req_i. In order to prevent skewing results by unanswered requests, they are assumed to count as a timeout value (a node stops waiting for the response after some time).

This metric reflects user experience or satisfaction while using the service. It is possible to adjust this metric in order to reflect the state of the system as a whole. The adjustment is required, because responsiveness may vary significantly based on latency between the user machine and the SelfAid service node. Let us define ELSN as Estimated Latency to the Service Node. In order to better express system effectiveness, the SR (system responsiveness) metric is proposed:

$$SR = \frac{1}{n}\sum_{i=1}^{n}(t_{response}(req_i) - t_{sent}(req_i) - ELSN) \tag{3}$$

From the point of view of the system, one can ask how many requests a service node got and how many of them were left unanswered. The equation representing this metric, named SNRF (Service Node Response Faults), is shown below.

$$SNRF = requests\ received - times\ responded \tag{4}$$

2.3 Measuring Reliability

Reliability says how well the system handles node failures. To measure how failures impact the functioning of the system, one can use responsiveness and resource usage metrics and compare them under different failure scenarios.

$$metric^{D} = metric^{FS} - metric^{OS} \tag{5}$$

where $metric^{D}$ is the deviation of the metric, defined as the difference between the metric observed in the failure scenario ($metric^{FS}$) and the metric observed in the optimal scenario ($metric^{OS}$). The optimal scenario is a scenario in which there are no node failures and the latency between the nodes is the same as in the failure scenario. The deviation is a signed value. As all of the defined metrics follow the rule of "the smaller the better", the same applies to the deviation metric. If the deviation metric results in a negative value, that means that the system achieved better performance in the failure scenario than in the optimal operation.

There are several SelfAid metric, which deviation gives meaningful results. The *system responsiveness*^{D} shows how much longer a client has to wait for his request to be completed. The deviation in the Service Node Response Faults ($SNRF^D$) shows the amount of additional overload, while its duration is reflected in the deviation of Overload Time Ratio (OTR^D).

3 Considering Various Load Scenarios

In online games the demand for the services changes with time. During the day, usually more people play in the afternoon, or in the evening. During the week, more players appear on fridays, weekends and holidays. Other events may also influence this, such as a game release (or a new content release), a successful marketing campaign, etc. Real world data for gaming traffic can be found, for example, on the webpage [21]. These scenarios result in many possible load distributions, which should be taken into account when measuring resource usage, reliability and responsiveness.

The SelfAid network should adapt to the increase in load by converting more nodes to being ring members. This process of conversion additional nodes will take some time, during which the responsiveness will decrease. Such a situation results from the issue that, temporarily there will not be enough service nodes to take care of all the requests. Directly connected to this issue is the resource usage. If there are not enough service nodes, the existing service nodes have to perform at maximum allowed load, which is more than the preferred amount.

The load scenarios that seem reasonable to test (i.e. seem to be properly reflecting nature of the traffic) are the following:

- uniform distribution, tests the behavior of the system under constant load, which is equivalent to a stable number of players. E.g. in the middle of an afternoon.
- step distribution (a uniform distribution up to a point and then another uniform distribution), tests the reaction to a quick and permanent increase of load, which can happen after a new game version is released.
- cone shaped distribution (linear increase up to a point and then linear decrease), tests the reaction to a steady growth and reduction of load. This is representative of a natural transition between different sizes of active player population, such as an increase starting in the afternoon until the evening and then decrease towards the night.

4 Environment

The simulation and analysis of SelfAid network were performed with the use of Peerfact.KOM simulator [2]. It was developed in Darmstadt university, Germany in 2011 and is actively updated. It has a layered and modular architecture. The layers include: network, transport, overlay, application. The network layer controls latency between hosts, bandwidth limitations, channel loss etc. Transport layer deals with the choice of TCP/UDP, provides API functions to send/receive messages. Overlay layer contains overlay-specific abstractions, such as: join, leave, store, lookup. Application layer is everything higher than overlay.

Multiple overlay implementations are available (Chord [3], Kademlia [4], Gnutella [5], CAN [6], Pastry [7]).

Simulation parameters may be controlled by setting appropriate values in the xml configuration file. It is possible to control, among others, packet sizes, packet fragmenting rules, jitter, packet loss, overlay choice, simulation time, random seed, number of hosts, churn model. It is also possible to schedule an action to be performed at a given time (e.g. at 30 min mark) by using action files. Actions are defined by specifying a group of nodes which are subject to the action, time at which the action is to be performed, and the action itself (e.g. joining the overlay).

There is also a built-in visualisation component. After a simulation is generated, the visualisation scenario may be saved as a separate file. When played, it is possible to see (in "real" time), the messages exchanged by the nodes, as well as any node properties that were defined (such as number of DHT connections, online/offline status). Visualisation is very helpful in debugging a simulation, however it is advised to limit the number of nodes to around 50 due to performance issues.

5 Results

Multiple results were obtained for all metrics by applying them to scenarios with various load distributions, each with several variants (several sets of parameters).

All measurements were performed in a model with crash-stop failures of 40% of the nodes during all of the simulation. More precisely, the number of online hosts decreases linearly starting at the half-point of the simulation. The simulation was run for 100 nodes. In the simulation, the processing time of matchmaking request is equal to 0. That is, the service node which received the request replies instantaneously. In practice the processing could often take more than several tens of seconds. At any given time window, a number of requests is generated according to a given distribution. They are uniformly distributed among the non-ring nodes which then initiate the matchmaking procedure.

Most of the metrics have one figure with a plot for each specific set of parameters of a load distribution. The considered load distributions are, as described in Sect. 3: uniform, step, and peak. For uniform distribution there are 3 parameter sets: low (1 request per second), medium (5 requests per second), high (15 requests per second). These quantities of requests are comparable to the following sizes of populations of players (assuming short games of average 5 min): 1 req/s = 300 players, 10 req/s = 3 000 players, 15 req/s = 4 500 players. The same set of parameters applies to peak distribution, where the load starts from

0 and reaches the parameter value at the 60% of the simulation time. The step distribution has 2 parameter sets: from low to medium, from medium to high. The load changes at 60% of simulation time. Plots showing a value for each peer separately are usually sorted based on the value (from biggest to smallest).

Figures 1, 2 and 3 present the value of Overload Time Ratio for uniform, step, and peak request distributions. Overload Time Ratio is relevant only for peers, which at some point provided service (in the sense of SelfAid network) to other nodes. The number of nodes providing service increases with the demand for the service (as all nodes have the same, quite restrictive, limitations to bandwidth). In Fig. 1, number of the nodes affected for the OTR changes from around 3% (low load) through more than 20% (medium load), up to about 45% (high load). The OTR value varies significantly for the affected nodes. A similar situation occurs for step and peak distributions. By comparing all distributions, one can see that the OTR of the most burdened peers in the step distribution is larger than in the uniform distribution. Also, the plot shape is similar for peak and step distributions, however the values for the most burdened nodes differ. That is because, the step distribution results is the one with the most abrupt change in load quantity and so it gives the least amount of time to adapt.

Total network traffic sent and total network traffic received is displayed in Figs. 4, 5, 6, 7, 8 and 9, respectively. It is important to note, that the amounts of data are shown on a logarithmic scale. It can be observed that there is a disparity in the amount of bandwidth consumed at the nodes, particularly in the case of low load. The plots for data sent and data received are similar in shape (for a given distribution). Both of the previous observations are tied to the fact that most of the traffic is attributed to the application level messages (request for an opponent, and the response to that request), which were assigned the same size in the simulation. Requests are sent uniformly (at any given point in time) by all nodes and responses are generated by the ring nodes. Thus, naturally the ring nodes will be subject to a substantially higher amount of traffic. This can be observed by comparing, for a given distribution, data traffic plot with the OTR plot. For example, for peak distribution (Figs. 6, 9 and 3) in all plots medium and high are relatively close to each other, while low is clearly separated. Also, the points at which the plots become flat coincide with the number of nodes with positive OTR values. The traffic required to sustain the structure of the SelfAid network, excluding DHT-specific messages which were not considered in measurements, is not demanding.

System responsiveness is shown in Figs. 10, 11 and 12. The overall level of system responsiveness is good from user's satisfaction point of view. The worst result was the average of close to 5 s for one peer, which is negligible when compared to the time usually required to process a matchmaking request. The value of system responsiveness increases with the load. As more requests are generated, a bigger amount is dropped when nodes are overloaded. The difference between low and medium distributions is significant. The medium and high distributions present similar results (medium is only slightly worse). The plots are not ideally flat for any distribution due to a number of reasons. First, while overloaded ring nodes drop all incoming application messages which means that a player may be treated unfairly and wait for a long time. As visible in all plots, there are a few nodes with significantly higher and lower system responsiveness values than average. It seems that the deviation from the average value roughly follows a normal distribution. Second, nodes fail during the simulation. This is especially visible in step distribution plots, because the load increases with time. As we know from analysing OTR plots, OTR value rises with load, which means that more requests will be dropped. Therefore, the nodes that failed before the load increased were serviced by ring nodes with lower average OTR value which ultimately resulted in lower system responsiveness.

The values of Service Node Response Faults are presented in Figs. 13, 14 and 15. SNRF shows how many requests were ignored by a given node due to load on that node being too big. The more requests there are in general, the more nodes provide the service and have a chance to be overloaded and ignore requests. So these figures are closely related to the Overload Time Ratio figures. However, the relation is not linear. For example, for uniform distribution and high load (Figs. 1 and 13) the most overloaded node has 30% OTR and 2100 SNRF, while the 25th most overloaded node has 15% OTR and 500 SNRF.

The metric deviation was tested for the uniform distribution (Figs. 16, 17 and 18). The OTR deviation (Fig. 16) is mostly negative for the first 20 peers, and positive for the rest. At the most significant point reaching 12% points difference (for high load). This could be caused by some of the longer functioning ring members (in the first 20) failing, which caused an increase in burden for the remaining ones. A similar trend is present in the SNRF deviation (Fig. 18). System responsiveness is consistently affected by the presence of failures (Fig. 17). The average time required for completing the request to find an opponent is increased by up to 2 s. It is a significant increase in terms of percentage, but negligible from the pragmatic point of view as it is common for the whole process to take more than tens of seconds in practice.

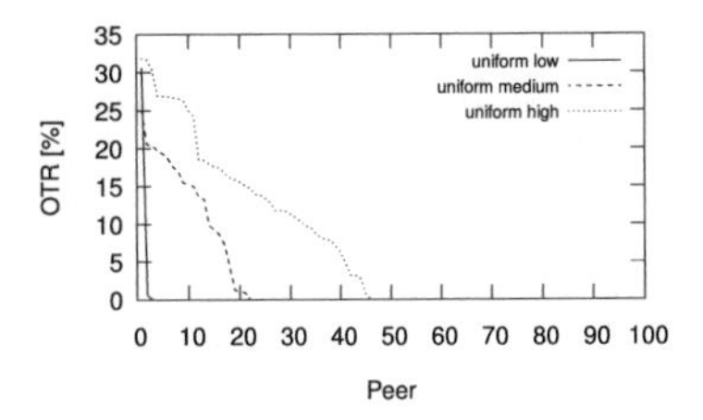

Fig. 1. Overload Time Ratio for uniform request distributions

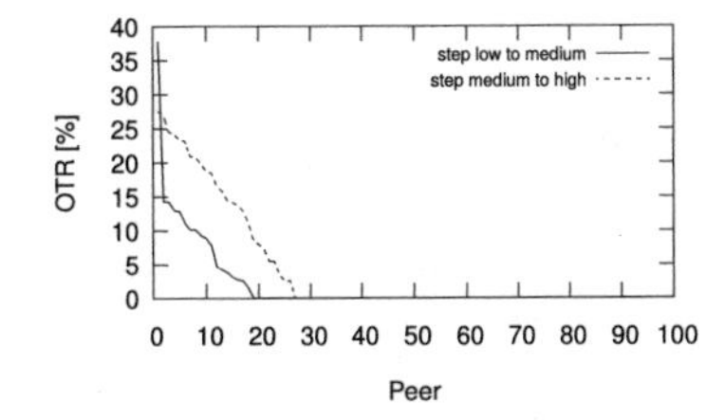

Fig. 2. Overload Time Ratio for step request distributions

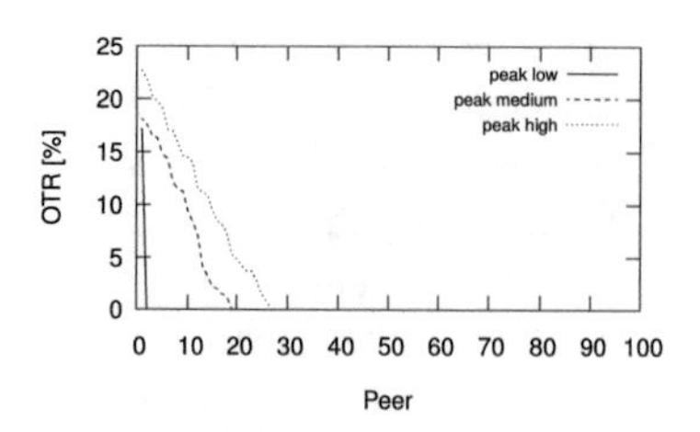

Fig. 3. Overload Time Ratio for peak request distributions

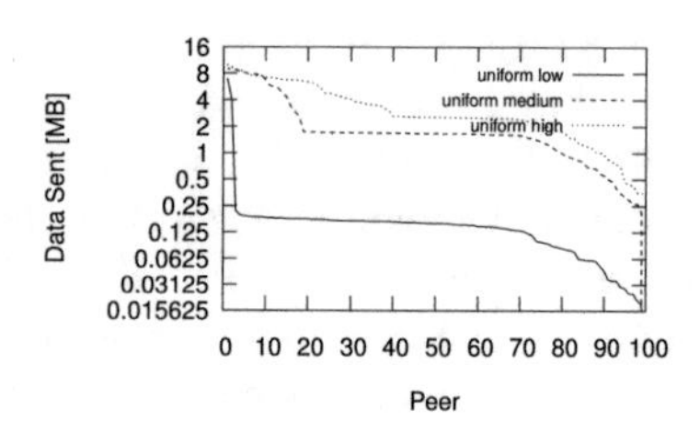

Fig. 4. Data Sent for uniform request distributions

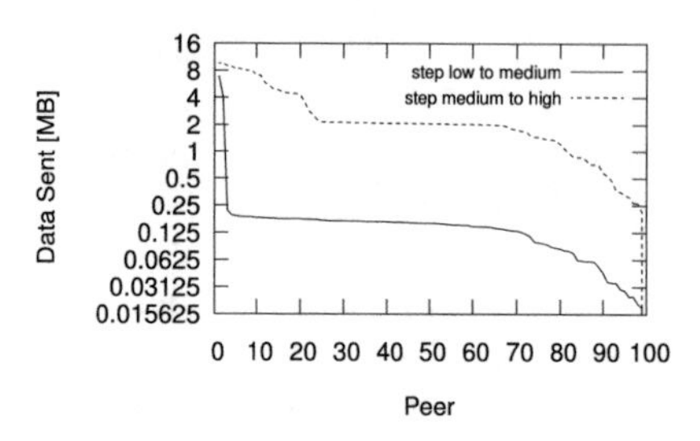

Fig. 5. Data Sent for step request distributions

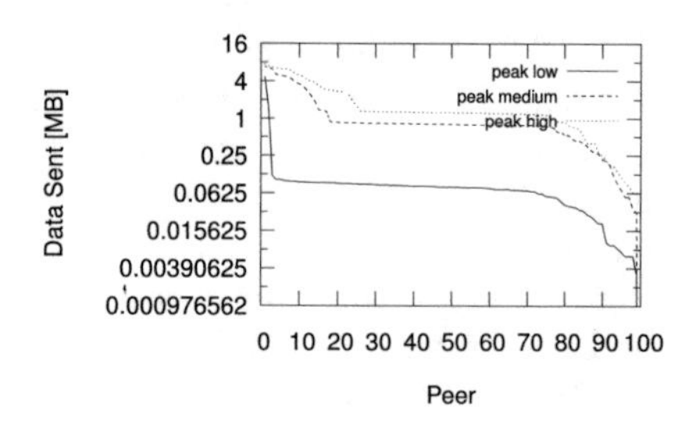

Fig. 6. Data Sent for peak request distributions

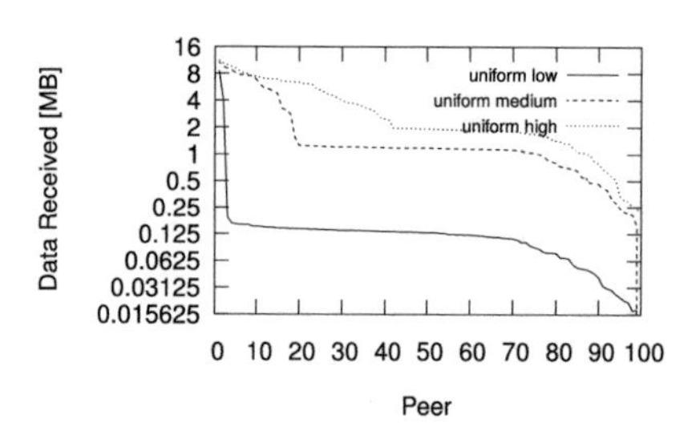

Fig. 7. Data Received for uniform request distributions

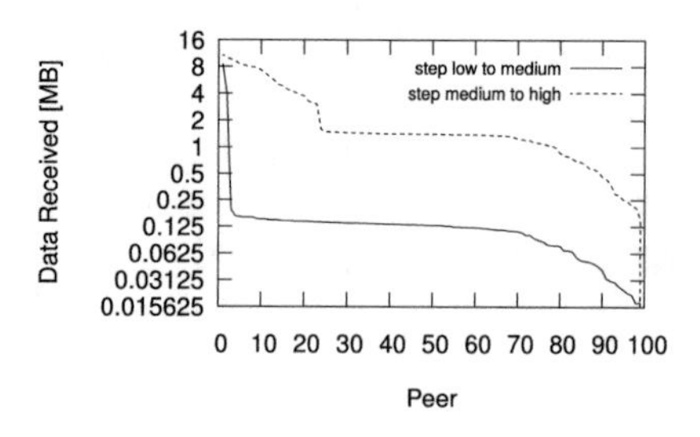

Fig. 8. Data Received for step request distributions

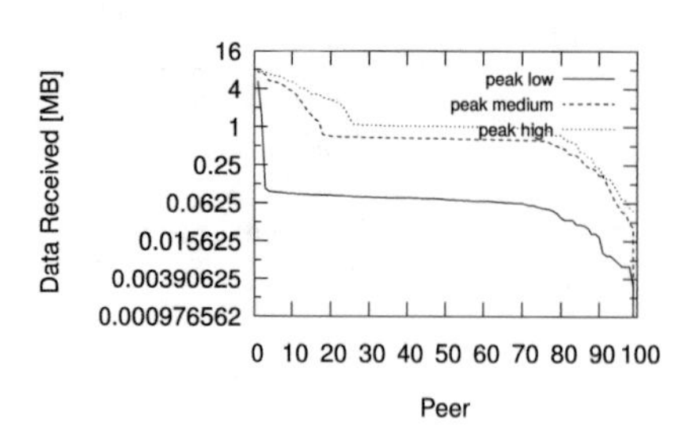

Fig. 9. Data Received for peak request distributions

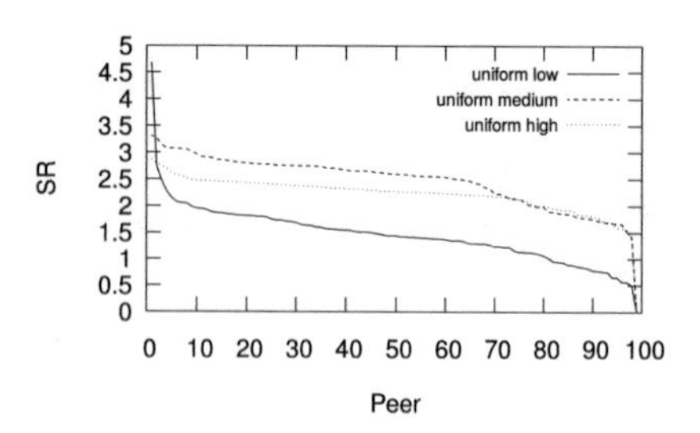

Fig. 10. System Responsiveness for uniform request distributions

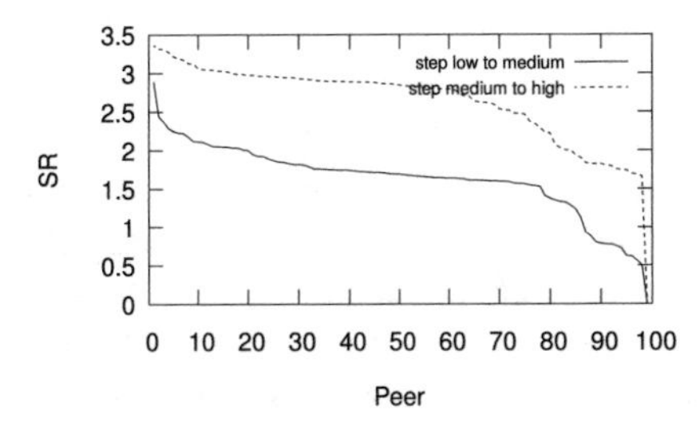

Fig. 11. System Responsiveness for step request distributions

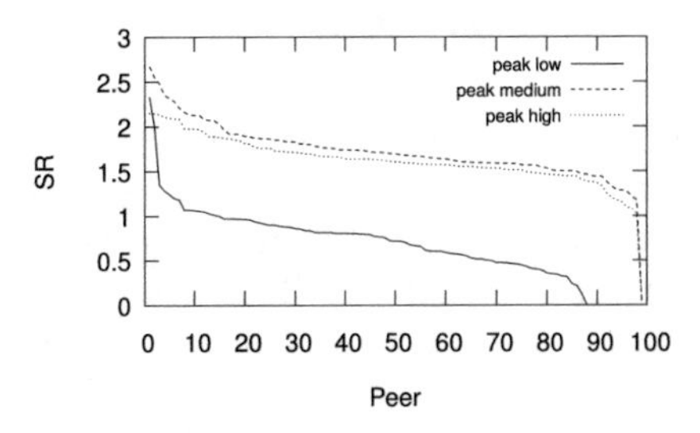

Fig. 12. System Responsiveness for peak request distributions

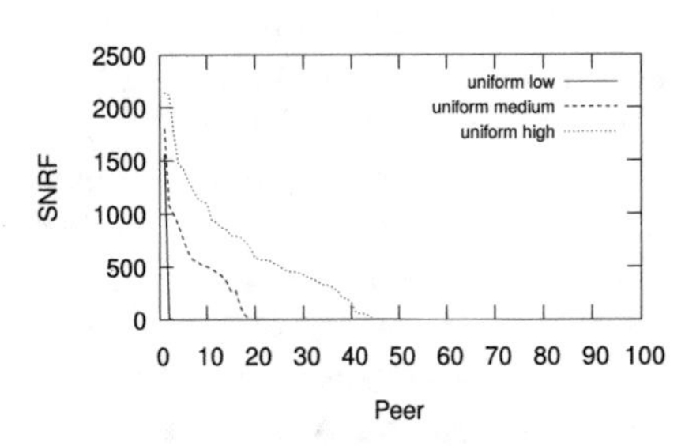

Fig. 13. Service Node Response Faults for uniform request distributions

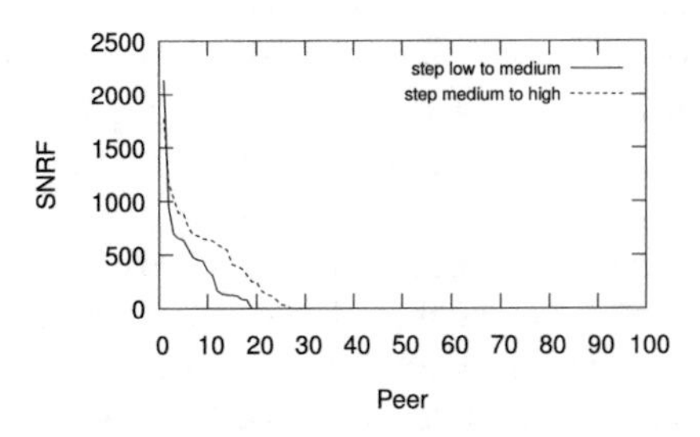

Fig. 14. Service Node Response Faults for step request distributions

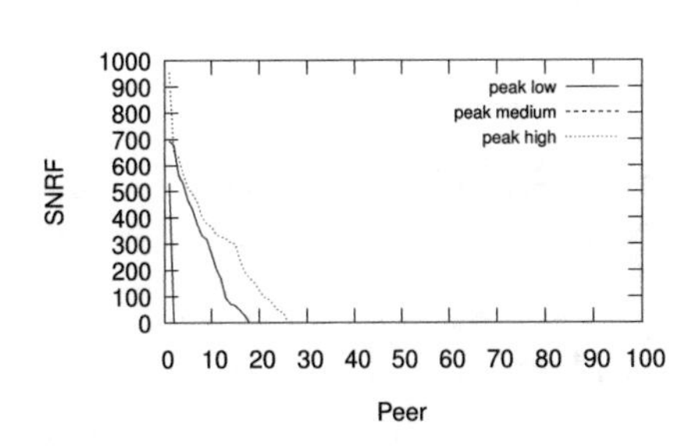

Fig. 15. Service Node Response Faults for peak request distributions

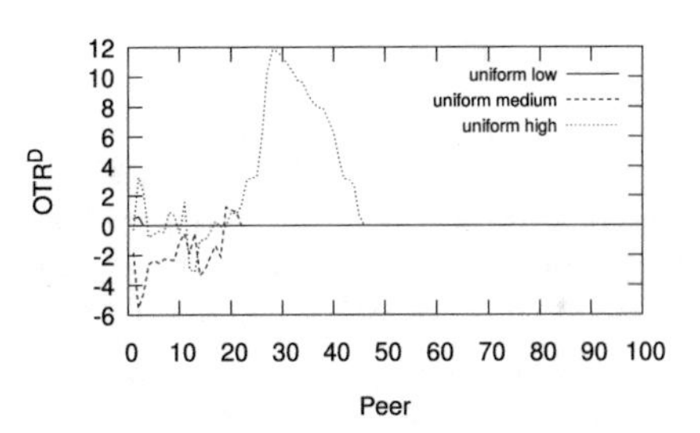

Fig. 16. Overload Time Ratio metric deviation

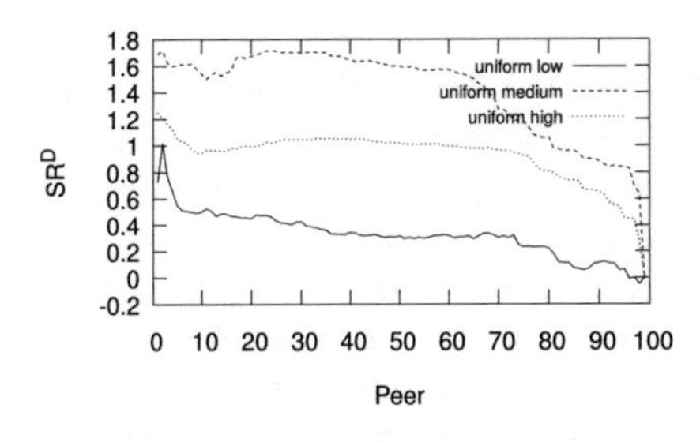

Fig. 17. System Responsiveness metric deviation

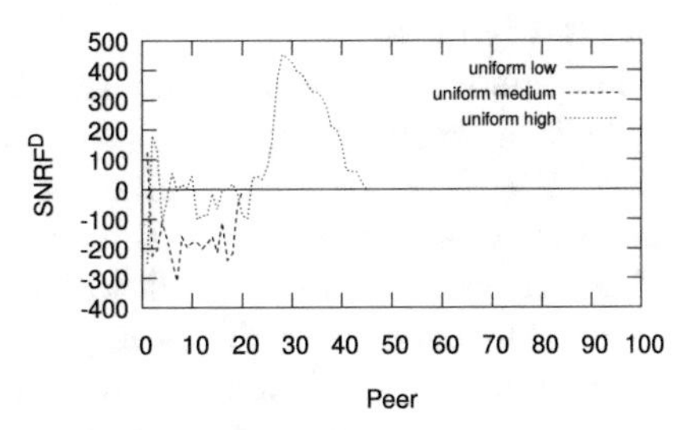

Fig. 18. Service Node Response Faults metric deviation

6 Related Work

Along with the proliferation of P2P networking, several evaluation studies have been carried out to evaluate various characteristics of different P2P systems. Among the evaluated features, the network performance and traffic analysis are often considered. To assess those attributes, papers focus on the latency, and bandwidth measurements, as well as on the payload information and flow dynamics [9,11,14]. In addition, the packet size distribution for packets belonging to P2P data flows are examined [18,19]. Those papers analyze also the overlay structures, content delivery methods, topology formation strategies for content delivery and scheduling strategies within P2P systems [16,17]. Although until recently, most papers have been interested in the traffic evaluation and identification issues, in the current research the efficiency and quality of service of P2P services are often analyzed [8,12,13]. Other studies are concerned with the user behavior issues [10,15,20]. Also, the characteristics of peers and resources participating in P2P systems are examined [11]. The performed studies show that there is significant heterogeneity in peers bandwidth, availability, and transfer rates. They also analyses how host dynamics within P2P networks affect performance and reliability. However, according to the best knowledge of the authors, none of existing papers tackles the problem of evaluation of the cost of running the application, user satisfaction, and resistance to failures, which are discussed in this paper.

7 Conclusions

This paper proposed effectiveness metrics for the SelfAid Network. Formulas and rationale for application were provided for all of them. The propositions consisted of: Overload Time Ratio (OTR), total network traffic sent, total network traffic received, System Responsiveness (SR), Service Node Response Faults (SNRF), and metric deviation. The importance of testing with various load distributions in the domain of online computer games was discussed, and a selection of adequate distributions was listed. The metrics were applied to the simulation of the SelfAid Network, implemented in PeerfacSim simulator. Performance of the SelfAid simulation viewed by the lens of proposed metrics was shown for multiple load distributions.

The simulation showed that the SelfAid behaved well with respect to user satisfaction (system responsiveness). Furthermore, it turned out that the shape of SNRF and OTR plots is often very similar, however there exist significant differences due to the average amount of burden changing over time. The OTR and SNRF values varied significantly for the ring nodes across all distributions. The SelfAid network could be improved to distribute load more evenly among the ring nodes. In addition, it was clear that the traffic required for maintaining the operation of the SelfAid network is not a significant burden. Therefore, it is possible to introduce new, bandwidth-consuming features to the protocols of SelfAid network. Moreover, the analysis of metric deviation showed not only how much failures impact performance, but also provided additional insight into the events which occurred during the simulation.

Future work includes investigating how to construct multiple failure scenarios in order to capture borderline cases. It would also be interesting to propose improvements to the SelfAid Network using the presented experimental data.

References

1. Boroń, M., Brzeziński, J., Kobusińska, A.: SelfAid network–a P2P matchmaking service. In: Multimedia and Network Information Systems, pp. 183–191. Springer, Cham (2017)
2. Stingl, D., Gross, C., Rückert, J., Nobach, L., Kovacevic, A., Steinmetz, R.: PeerfactSim.kom: a simulation framework for peer-to-peer systems. In: 2011 International Conference on High Performance Computing and Simulation (HPCS), pp. 577–584. IEEE, July 2011
3. Stoica, I., Morris, R., Liben-Nowell, D., Karger, D.R., Kaashoek, M.F., Dabek, F., Balakrishnan, H.: Chord: a scalable peer-to-peer lookup protocol for internet applications. IEEE/ACM Trans. Netw. (TON) **11**(1), 17–32 (2003)
4. Maymounkov, P., Mazieres, D.: Kademlia: a peer-to-peer information system based on the XOR metric. In: International Workshop on Peer-to-Peer Systems, pp. 53–65. Springer, Heidelberg, March 2002
5. Ripeanu, M.: Peer-to-peer architecture case study: Gnutella network. In: 2001 Proceedings ofFirst International Conference on Peer-to-Peer Computing, pp. 99–100. IEEE, August 2001
6. Ratnasamy, S., Francis, P., Handley, M., Karp, R., Shenker, S.: A scalable content-addressable network, vol. 31, no. 4, pp. 161–172. ACM (2001)
7. Rowstron, A., Druschel, P.: Pastry: scalable, decentralized object location, and routing for large-scale peer-to-peer systems. In: IFIP/ACM International Conference on Distributed Systems Platforms and Open Distributed Processing, pp. 329–350. Springer, Heidelberg, November 2001
8. Fan, H., Sun, X.: A multi-state reliability evaluation model for P2P networks. Reliab. Eng. Syst. saf. **95**(4), 402–411 (2010)
9. Gummadi, K.P., Dunn, R.J., Saroiu, S., Gribble, S.D., Levy, H.M., Zahorjan, J.: Measurement, modeling, and analysis of a peer-to-peer file-sharing workload. ACM SIGOPS Operating Syst. Rev. **37**(5), 314–329 (2003)
10. Goswami, A., Gupta, R.: Evolutionary Stability of Reputation Management System in Peer to Peer Networks. arXiv preprint arXiv:1703.08286 (2017)
11. Gummadi, P.K., Saroiu, S., Gribble, S.D.: A measurement study of napster and gnutella as examples of peer-to-peer file sharing systems. ACM SIGCOMM Comput. Commun. Rev. **32**(1), 82–82 (2002)
12. Hughes, D., Warren, I., Coulson, G.: Improving QoS for peer-to-peer applications through adaptation. In: 2004 Proceedings of 10th IEEE International Workshop on Future Trends of Distributed Computing Systems, FTDCS, pp. 178–183. IEEE, May 2004
13. Iguchi, M., Terada, M., Fujimura, K.: Managing resource and servent reputation in P2P networks. In: 2004 Proceedings of the 37th Annual Hawaii International Conference on System Sciences, pp. 9-pp. IEEE, January 2004
14. Karagiannis, T., Rodriguez, P., Papagiannaki, K.: Should internet service providers fear peer-assisted content distribution? In: Proceedings of the 5th ACM SIGCOMM Conference on Internet Measurement, p. 6. USENIX Association, October 2005

15. Li, Q., Qin, T., Guan, X., Zheng, Q.: Analysis of user's behavior and resource characteristics for private trackers. Peer-to-Peer Netw. Appl. **8**(3), 432–446 (2015)
16. Perényi, M., Dang, T.D., Gefferth, A., Molnár, S.: Identification and analysis of peer-to-peer traffic. J. Commun. **1**(7), 36–46 (2006)
17. Pouwelse, J., Garbacki, P., Epema, D., Sips, H.: The BiTtorrent P2P file-sharing system: measurements and analysis. In: International Workshop on Peer-to-Peer Systems, pp. 205–216. Springer, Heidelberg, February 2005
18. Sen, S., Wang, J.: Analyzing peer-to-peer traffic across large networks. In: Proceedings of the 2nd ACM SIGCOMM Workshop on Internet Measurement, pp. 137–150. ACM, November 2002
19. Tutschku, K.: A measurement-based traffic profile of the eDonkey filesharing service. In: International Workshop on Passive and Active Network Measurement, pp. 12–21. Springer, Heidelberg, April 2004
20. Ullah, I., Doyen, G., Bonnet, G., Gaiti, D.: A survey and synthesis of user behavior measurements in P2P streaming systems. IEEE Commun. Surv. Tutorials **14**(3), 734–749 (2012)
21. Steam Charts - An ongoing analysis of Steam's concurrent players. https://steamcharts.com

Cloud Computing and High Performance Computing

Two-Layer Cloud-Based Web System

Krzysztof Zatwarnicki(✉) and Anna Zatwarnicka

Department of Electro Engineering, Automatic Control and Computer Science, Opole University of Technology, ul. Prószkowska 76, 45-758 Opole, Poland
k.zatwarnicki@gmail.com, anna.zatwarnicka@gmail.com

Abstract. Providing high quality of Web services is now very important for most of the services. The application of cloud computing in Web systems enables lower costs and increases efficiency and reliability. In the paper we present a design of two-layer decision-making method in a cloud-based Web system TLCWS (Two-Layer Cloud-based Web System). It uses the neuro-fuzzy approach to enable the distribution of HTTP requests. The proposed method is design to minimize the response time of each request. The simulation experiments show that the proposed approach is more than just adequate and that it is possible to shorten the response time required for requests significantly.

Keywords: Cloud computing · HTTP request distribution
Neuro-Fuzzy modeling

1 Introduction

The Internet accompanies humans in almost every aspect of modern life. This has resulted in the rapid and constant development of Web-based system technologies. Web systems are not only becoming more stable, but also reliable and can handle large number of upcoming HTTP requests.

In the past, in order to achieve both high reliability and efficiency, companies have had to build private data centers. The costs of accomplishing and maintaining such a center are very high. Suffice to say, not every company could afford to have its own center. Nowadays, in order to reduce operating costs, cloud computing is used instead. Cloud computing is a data processing model based on extensive infrastructure. It is provided as a service by external center [8]. All aspects of the cloud system architecture depend on the service provider. The provider specifies the architecture of available cloud system. There are three main models of architecture used in practice [5]:

1. Software as a Service (SaaS). Consumer is able to use the provider's applications running on a cloud infrastructure. The consumer does not control the cloud infrastructure and software.
2. Platform as a Service (PaaS). The consumer does not control the underlying cloud infrastructure but has control over the deployed applications.
3. Infrastructure as a Service (IaaS). The consumer has control over operating systems, storage, and deployed applications.

L. Borzemski et al. (Eds.): ISAT 2018, AISC 852, pp. 125–134, 2019.
https://doi.org/10.1007/978-3-319-99981-4_12

Most Web systems belong to the SaaS model of architecture. Each Web service in the cloud is completely separate from others. It is crucial in such systems to adapt them to customer requirements in terms of bandwidth and service time. Systems architecture and the way HTTP requests are processed should also help achieve the highest performance and availability.

The way in which HTTP requests are processed in cloud systems can be similar to the one known from cluster-based Web systems. Such systems have been in use for over 20 years, and during this time, have been explored and testified further by extensive research carried out in this field.

In this paper, the method used to distribute HTTP request in Cloud-based system with the SaaS architecture will be proposed. The method uses an algorithm that minimizes the HTTP request response time based on neuro-fuzzy model. This algorithm have already been presented in our previous papers and was used to minimize request response time in cluster-based Web systems [13, 14].

This paper is divided into four more sections apart from this one. Section 2 contains related works. In Sect. 3, we describe the proposed system distributing HTTP requests in cloud systems. Section 4 presents the conducted experiments with the results, and Sect. 5 contains concluding remarks.

2 Related Work

Several distribution algorithms for cloud computing systems have been proposed over the last decade. As a result, research was carried out on cloud systems with various architectures, including SaaS, PaaS and IaaS. All of the algorithms can be classified into three main categories: static [7, 10, 11], dynamic [12] and adaptive [13, 14].

In most cases, tasks are the units that are managed in the data distribution systems. For example Coninck et al. [4] proposed a dynamic auto-scaling algorithm that reduces the execution time of tasks in a manner that shows consideration for the deadline as a constraint. M. Kumar and Sharma [6] proposed an algorithm that not only reduced the span time of tasks but also increased the utilization ratio considering the priority of tasks as quality of services parameter.

A separate group of issues relate to the distribution of HTTP requests in Cloud-based systems with SaaS architectures. The request distribution algorithms known from cluster-based Web systems are most often used in this case. One of the most frequently used algorithms is Round Robin (RR) that dispatches each request to the subsequent server. Another and one of the most efficient algorithms is Least Loaded (LL) that chooses the server with the lowest value of load factor [3]. An effective and subsequent algorithm is the Client-Aware Policy algorithm (CAP) [2]. It divides the requests into classes and then distributes requests according to RR algorithm for each class separately.

The research results indicate that the most efficient algorithm for dealing with the requests distribution in cluster-based Web systems is the Fuzzy-Neural Request Distribution algorithm (FNRD). This dispatches the HTTP requests to those Web servers that offer the shortest response time [13]. The response time is estimated with the use of the neuro-fuzzy model approach.

In this article we propose a new HTTP requests distribution method in a Cloud-base Web systems. New method will again use FNRD distribution algorithm in the Two-Layer Cloud-based Web System called TLCWS.

3 Two-Layer Cloud-Based Web System

Cloud computing systems are often hosted world-wide in multiple locations. These locations are composed of regions and zones. Each region is a separate geographic area. Each region has multiple, isolated locations known as zones (Fig. 1). Each zone is isolated, but the zones in a region are connected through low-latency links. This kind of organization of cloud computing system is the most common one [1, 8].

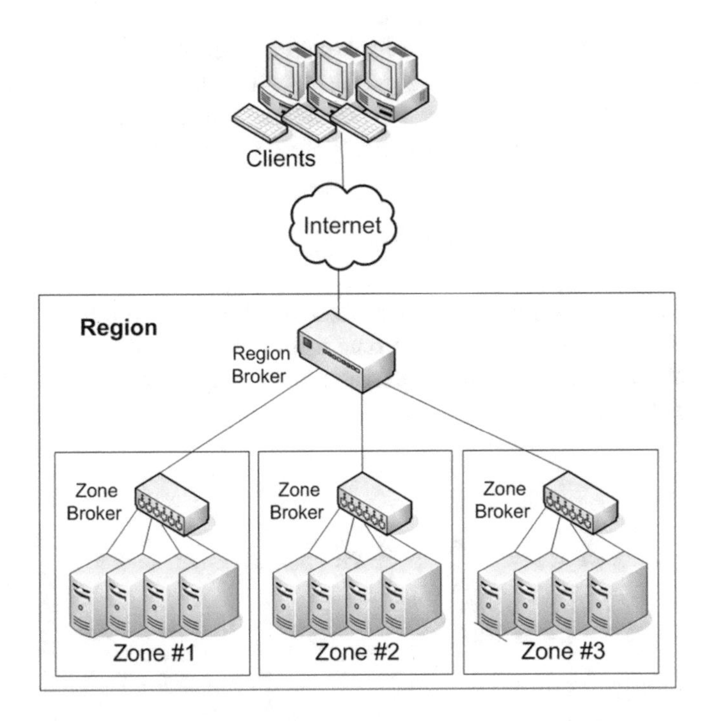

Fig. 1. Architecture of one region in cloud computing Web system.

In the proposed cloud-based Web system, clients (web browsers) send HTTP requests to the region, region broker receives them inside the region. Then, inside the region, the region broker chooses the zone to provide the service request. According to our method, the web zone offering the shortest estimated response time is chosen. Therefore, requests are redirected to the zone brokers, then zone brokers choose Web servers with the shortest response times for requests. We are assuming that all of the zones and Web servers that are inside can service exactly the same HTTP requests, and the resources are available from every Web server.

In conclusion, decisions about requests redirecting are taken separately on two layers: in region and in zones brokers. The main purpose of controlling the redirection of HTTP requests in the system is to minimize the requests response time.

The brokers in the region and zone layers are logically constructed in the same way and contain elements that perform the same functions.

In our approach, the estimation of the request response time is implemented using the neuro-fuzzy models constructed in a similar way to the one we proposed for FNRD system in cluster-based approach [13].

Broker Design

The design of region and zone brokers are similar. The difference lies in where HTTP requests are sent. In the first case, requests are sent to the zones while in the second case to Web servers. In both cases brokers treat them in the same way, therefore, zones and web servers will be called executors.

The main task is to propose such a Broker design which will determine - for each incoming HTTP request r_i, $i = 1, \ldots, I$, on the base of the knowledge on the load of the executor $M_i^1, \ldots, M_i^w, \ldots, M_i^W$ and the awareness on the past response times $\tilde{s}_1, \tilde{s}_2, \ldots, \tilde{s}_{i-1}$ to requests – the executor z_i out of W, which will be chosen according to formula:

$$z_i = \min_w \{ \hat{s}_i^w : w \in \{1, 2, \ldots, W\} \}, \quad (1)$$

where $\hat{s}_i^w$ is an estimated response time to request r_i for the w th server.

The request response time s_i^w is the time that elapses from the moment when the decision to allocate the request to the executor is made by the broker, until the HTTP response is obtained.

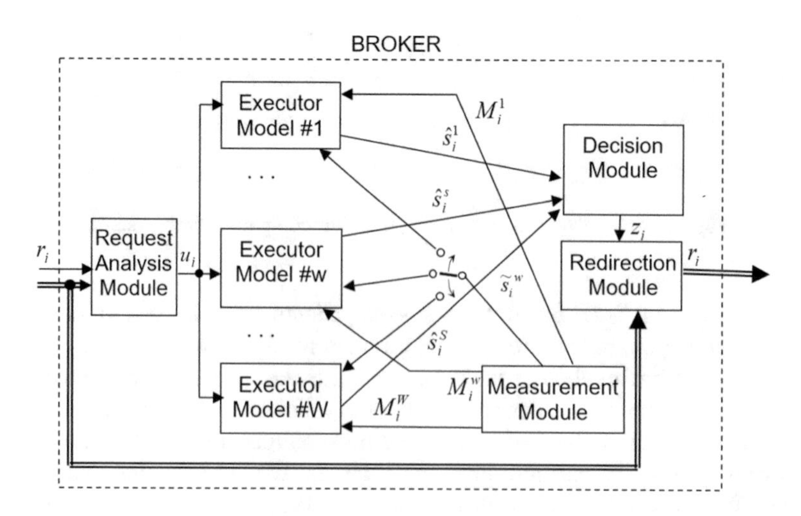

Fig. 2. Broker design.

The broker consists of the following main elements (Fig. 2): request analysis module, executor models, decision module, redirection module and a measurement module.

The request analysis module in the broker analyses incoming request and retrieves the address of requested object u_i. The broker contains W models of the servers. Each model corresponds to one executor and estimates its response time $\hat{s}_i^w$, taking into account the workload M_i^w of executor w at the moment of arrival of i-th request r_i. When the request execution is completed, the measured response time $\tilde{s}_i^w$ is used in the adaptation process, in order to change the parameters of the executor model. The decision-taking module chooses the executor according to formula (1). Redirection module sends the request r_i to chosen executor z_i.

The executor model is a main element of the Broker. It let to estimate the response time before the request is serviced. The adopted executor model consists of four functional modules (Fig. 3): classification module, estimation mechanism, adaptation mechanism and parameter database module.

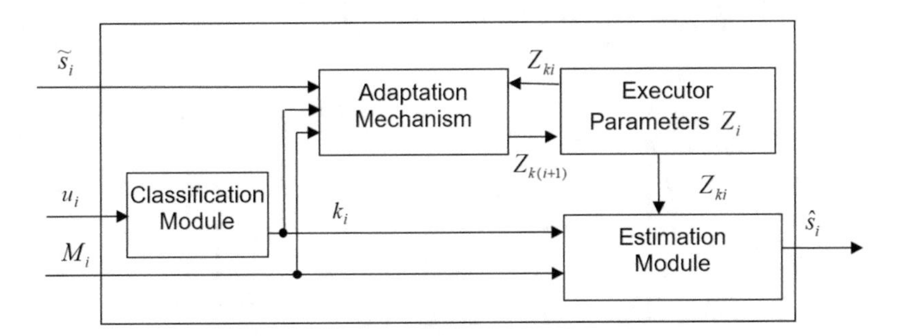

Fig. 3. Executor model.

The classification mechanism classifies incoming requested in a way that response times for requests belonging to the same class should be similar. The classification module possesses information of the size and type of HTTP objects served by executors. The output of the module is a class k_i of the requested object, where $k_i \in \{1, \ldots, K\}$, and K is the number of determined classes. The adaptation mechanism, executor parameters and estimation module form together neuro-fuzzy model estimating response time $\hat{s}_i$ for given w-th executor. The scheme of the neuro-fuzzy model is presented on Fig. 4.

The service time is calculated on the base of information of the requested object class k_i and the load $M_i^w = \left[e_i^w, f_i^w\right]$ of the executor, where e_i^w is the number of all requests being concurrently serviced by the w-th executor and f_i^w is the number of dynamic request.

The parameter database $Z_i = \left[Z_{1i}, \ldots, Z_{ki}, \ldots, Z_{Ki}\right]$ stores parameters for different classes, where $Z_{ki} = \left[C_{ki}, D_{ki}, S_{ki}\right]$, $C_{ki} = \left[c_{1\,ki}, \ldots, c_{l\,ki}, \ldots, c_{(L-1)\,ki}\right]$ and $D_{ki} = \left[d_{1\,ki}, \ldots, d_{m\,ki}, \ldots, d_{(M-1)\,ki}\right]$ are parameters of input and output $S_{ki} = \left[s_{1\,ki}, \ldots, s_{j\,ki}, \ldots, s_{J\,ki}\right]$ fuzzy set functions (Fig. 4a). Fuzzy set functions for input are

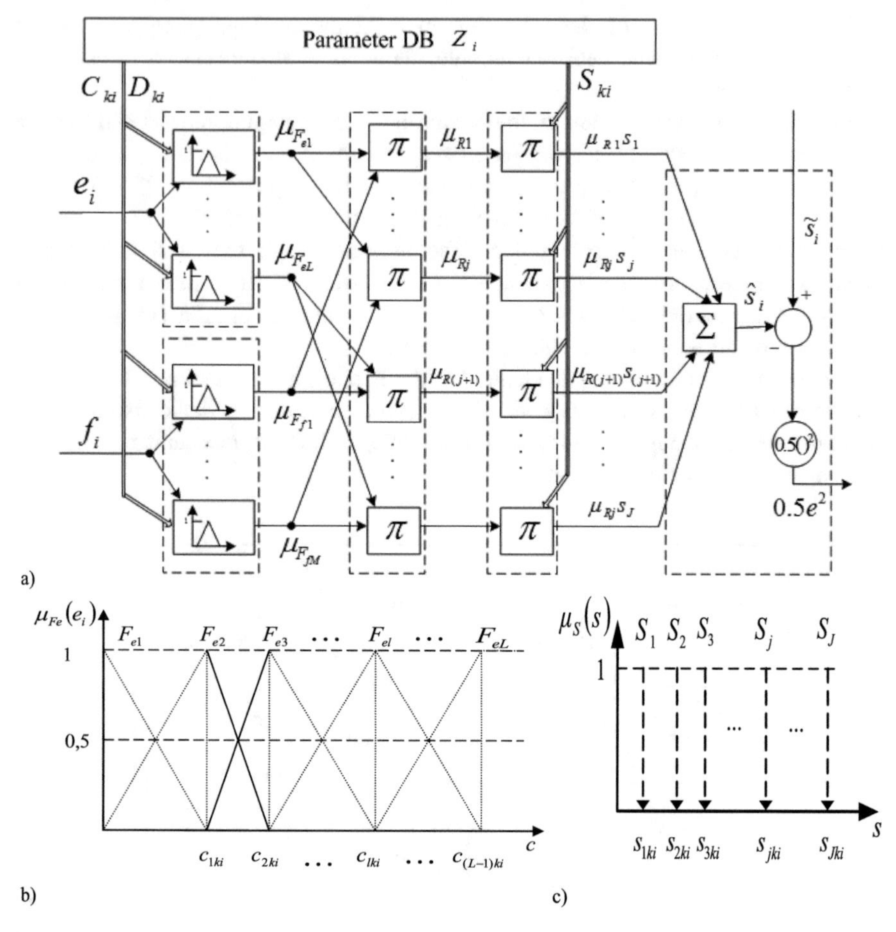

Fig. 4. Server model: (a) neuro-fuzzy model, (b) input fuzzy sets functions, (c) output fuzzy sets functions.

presented on Fig. 4b $\mu_{F_{el}}(e_i)$, $\mu_{F_{fm}}(f_i)$, $l = 1, \ldots, L$, $m = 1, \ldots, M$. Fuzzy sets functions for outputs $\mu_{S_j}(s)$ are singletons (Fig. 4c). The number of fuzzy sets for inputs were chosen experimentally $L = M = 10$, and $J = L \cdot M$.

The service time is calculated as follow: $\hat{s}_i = \sum_{j=1}^{J} s_{jki}\mu_{R_j}(e_i, f_i)$, where $\mu_{R_j}(e_i, f_i) = \mu_{F_{el}}(e_i) \cdot \mu_{F_{fm}}(f_i)$. The parameters C_{ki}, D_{ki}, S_{ki} are modified during the learning phase with use of the Back Propagation Method. The output fuzzy sets parameters are modified in following way: $s_{jk(i+1)} = s_{jki} + \eta_s \cdot (\tilde{s}_i - \hat{s}_i) \cdot \mu_{R_j}(e_i, f_i)$, while parameters of input fuzzy sets are calculated as follows $c_{\phi k(i+1)} = c_{\phi ki} + \eta_c(\tilde{s}_i - \hat{s}_i) \sum_{m=1}^{M} \left(\mu_{F_{fm}}(f_i) \sum_{l=1}^{L} \left(s_{((m-1)\cdot L + l)ki} \partial \mu_{F_{el}}(e_i) / \partial c_{\phi ki}\right)\right)$ and $d_{\gamma k(i+1)} = d_{\gamma ki} + \eta_d(\tilde{s}_i - \hat{s}_i) \sum_{l=1}^{L} \left(\mu_{F_{el}}(e_i) \sum_{m=1}^{M} \left(s_{((l-1)\cdot M + m)ki} \partial \mu_{F_{fm}}(f_i) \Big/ \partial d_{\gamma ki}\right)\right)$, where η_s, η_c, η_d are adaptation ratios, $\phi = 1, \ldots, L-1$, $\gamma = 1, \ldots, M-1$ [14].

4 Simulation Model and Experiment Results

In order to evaluate the proposed distribution method, simulation experiments were conducted with the use of the OMNeT++ discrete event simulator [9]. The simulation program consists of the following modules: the HTTP request generator, the region and zone brokers, Web servers and database servers (one for each zone) (Fig. 5).

The request generator in the simulator contains clients that behave in a similar way to the modern Web browser. Clients in the simulation were sending requests using up to 6 concurrent connections. The cloud-based Web service was serving simulated Web page of parameters (size and type of pages) exactly the same way as described in the case of Wordpress 4.4 from one of the academic websites.

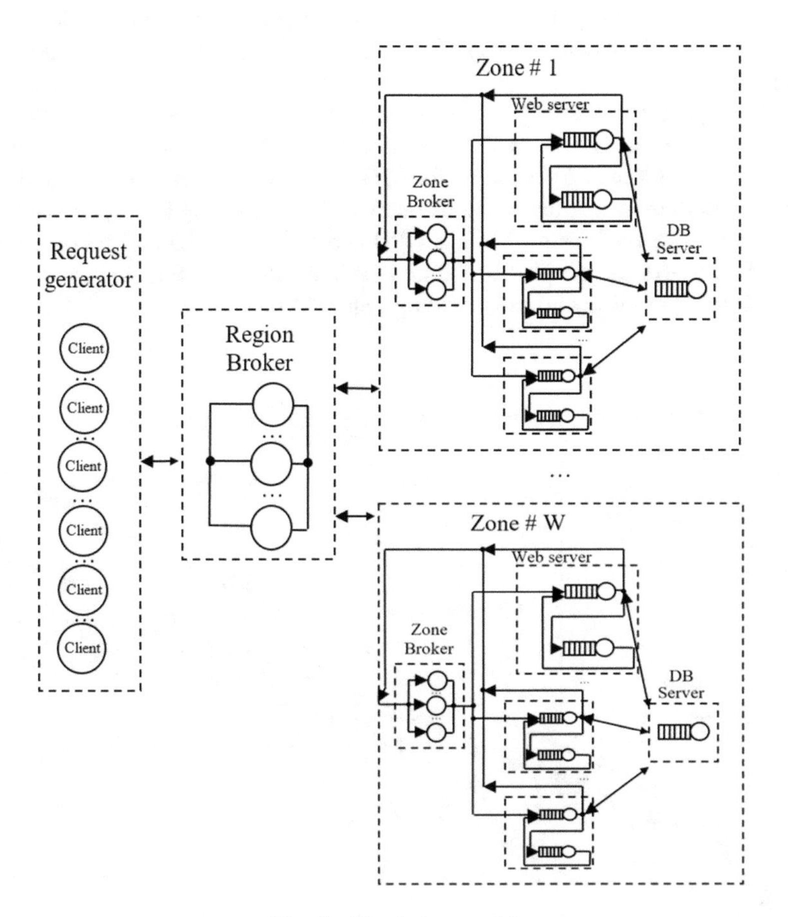

Fig. 5. Simulation model.

Both the region and zone brokers were acting in the same way and distributing requests with use of following strategies:

- Round Robin (RR),
- Client-Aware Policy algorithm (CAP),
- Least Loaded (LL) choosing the server with the lowest number of request served by executor at given moment,
- TLCWS strategy proposed in the article,
- TLCWS adaptive strategy in which the process of modifying model parameters is carried out adaptively.

TLCWS strategy was proposed in two variants. In the first one, fuzzy-neural models were learning the behavior of executors in learning phase and after that, the real experiment was conducted using constant parameters of the model. In TLCWS adaptive strategy, the model was adapting not only in the learning phase but also during the experiment.

In the experiments, we used three zones each with three web server. Every zone contains one database server. In the model of the Web server, we have implemented two queues, one for the CPU and one for SSD drive. We acquired the service times of HTTP requests serving for the Wordpress service by performing the experiments using the computer with Intel Core i7 7800X CPU and Samsung SSD 850 EVO driver.

The Fig. 6 presents diagram of the mean request response time in the load of the system (number of clients actively sending requests) function.

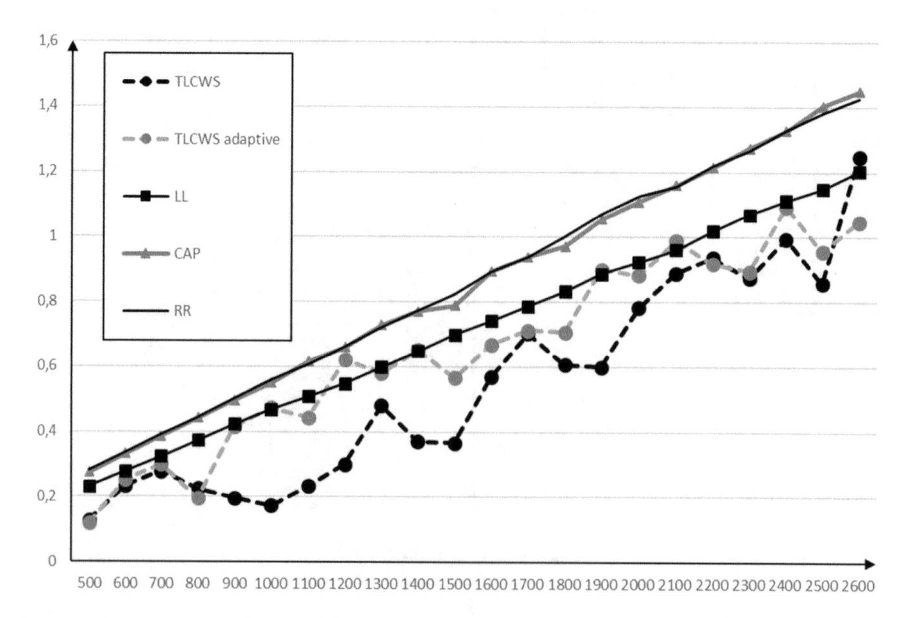

Fig. 6. Results of experiments. Mean request response time in function of number of clients.

As we can see, the obtained results are the best for TLCWS strategy. In some cases, the response time is twice times lower for TLCWS than for LL strategy. However, despite the good results, it is worrying that the results are not equally good in the whole range of the experiment.

Presented research is, for the moment, a preliminary to further in-depth research on the proposed method, however, the results are satisfactory and show that it is possible to achieve much better with our proposed strategy than using classical approaches.

5 Summary

The paper has presented a design of Two-layer decision-making method in cloud-based Web system using neuro-fuzzy approach. The application of the method can make it possible to improve the cloud-based Web system quality as compared with the other reference algorithms used in region and zone brokers. The simulation experiments have showed that the approach proposed is adequate and owing to the application of the method developed it is possible to shorten request response times significantly.

Research on the presented method should be continued in order to achieve greater stability in the test results.

References

1. AWS Documentation, Regions and Availability Zones. https://docs.aws.amazon.com/AWSEC2/latest/UserGuide/using-regions-availability-zones.html. Accessed 21 Apr 2018
2. Cardellini, V., Casalicchio, E., Colajanni, M., Mambelli, M.: Web switch support for differentiated services. ACM Perf. Eval. Rev. **29**(2), 14–19 (2001)
3. Cardellini, V., Casalicchio, E., Colajanni, M., Yu, P.S.: The state of the art in locally distributed Web-server systems. ACM Comput. Surv. **34**(2), 263–311 (2002)
4. Coninck, D.E., et al.: Dynamic auto-scaling and scheduling of deadline constrained service workloads on IaaS clouds. J. Syst. Softw. **118**, 101–114 (2016)
5. IaaS, PaaS and SaaS – IBM Cloud service models. https://www.ibm.com/cloud/learn/iaas-paas-saas. Accessed 21 Apr 2018
6. Kumar, M., Sharma, S.C.: Priority Aware Longest Job First (PA-LJF) algorithm for utilization of the resource in cloud environment. In: INDIACom, pp. 415–420 (2016)
7. Li, J.: An greedy-based job scheduling algorithm in cloud computing. J. Soft. **9**(4) 2014
8. Microsoft Cloud IT architecture resources. https://docs.microsoft.com/en-us/office365/enterprise/microsoft-cloud-it-architecture-resources. Accessed 21 Apr 2018
9. OMNeT++ Discrete Event Simulator. https://www.omnetpp.org/. Accessed 21 Apr 2018
10. Samal, P., Mishra, P.: Analysis of variants in round robin algorithms for load balancing in Cloud Computing. Int. J. Comput. Sci. Inf. Technol. **4**, 416–419 (2013)
11. Suresh, A., Vijayakarthick, P.: Improving scheduling of backfill algorithms using balanced spiral method for cloud metascheduler. In: International Conference on Recent Trends in Information Technology, Chennai, TN (2011)

12. Thomas, A.: Credit based scheduling algorithm in cloud computing environment. In: Proceedings of the International Conference on Information and Communication Technologies, ICICT 2014, Bolgatty (2014)
13. Zatwarnicki, K.: Adaptive control of cluster-based web systems using neuro-fuzzy models. Int. J. Appl. Math. Comput. Sci. (AMCS) **22**(2), 365–377 (2012)
14. Zatwarnicki, K., Płatek, M., Zatwarnicka, A.: A cluster-based quality aware web system. In: Information Systems Architecture and Technology, vol. 430, pp. 477–486. Springer, Cham (2016)

Contribution Title Enterprise Architecture Approach to Resilience of Government Data Centre Infrastructure

Wojciech Urbanczyk[1(✉)] and Jan Werewka[2]

[1] Faculty of Computer Science, University of Vienna, Vienna, Austria
w.urban@brz.gv.at
[2] Department of Applied Computer Science,
AGH University of Science and Technology, Kraków, Poland
werewka@agh.edu.pl

Abstract. We can observe a paradigm shift with respect to government data centres. Due to the complexity of such a data centre system, new types of quality attributes should be used, such as resilience, trustworthiness, and sustainability. This paper offers a general approach to the transformation of government data centres for different enterprise levels from a resilience point of view. It presents an enterprise architecture approach to the development of a federal data centre, considering resilience as a quality attribute. The approach is supported by using ArchiMate: an enterprise architecture modelling language. Resilience is considered from various perspectives: business, application, infrastructure and physical. An agnostic solution is proposed which is based on architecture and design patterns.

Keywords: Data-centre architecture · Enterprise architecture · Resilience

1 Introduction

Current structures in organizational and technical areas of DC (Data Centres), especially in the public sector, date back to the 1980s. The best example of such legacy systems is the mainframe, which are still present in many data centres and still perform their duties in the banking, government, insurance, healthcare and automotive sectors.

According to Gartner more than 90% of the world's top 100 banks still use mainframes. They are consistently accurate, efficient, reliable and fail-safe. They are part of every DC, due to their role as "cash cow". There are already more modern languages used, but Cobol is indispensable to large banks, corporations and parts of government institutions. Such DC structures need to be adapted to the expectations of the IT industry connected with energy efficiency and cost savings. Transformations are necessary because the environment in which the data centres are located is changing - both with respect to technology and with respect to legal and geopolitical conditions. Legal aspects have to be adapted e.g. to the EU requirements. The EU gives guidance about data retention, for instance: the OLAF project in the Schengen agreement is typical of the national jurisdiction of the pan-European area. In addition, technological

L. Borzemski et al. (Eds.): ISAT 2018, AISC 852, pp. 135–145, 2019.
https://doi.org/10.1007/978-3-319-99981-4_13

changes such as the open data government, cloud computing and Web 2 are already happening. These transformations will entail increasing the scope of the data centres with respect to cooperation with international bodies (such as ENISA, UNISA and CERT on jurisdiction and the prevention of crime, for example) or establishing new standards in the ICT (Information and Communication Technologies).

The need to create conditions adapted to rapid improvements made by the IT industry while providing standards based solutions in the public sector is becoming more and more urgent. Many companies and providers of IT solutions in the public sector are unable to or find it very difficult to adapt their structures to dynamic processes in the ICT. To achieve improvements, they are forced to look for measures that guarantee only short-term success.

The concept of possible enterprise architecture, which will be created by the new paradigm of the public data centre, should address the following topics: the growth of complexity in the authorities' networks, the design of high-quality security concepts and the efficient and cost-effective use of advanced network technologies. The main requirements of the IT-architecture of the data centres are: security and scalability done by design; IT-architecture design independent of the trends; reuse of software components; process automation; adaptive computing; part of autonomic computing; self-configuring; self-optimizing; self-healing; self-protecting.

The expected benefits of new DC Architecture are the following: communication processes are divided into small elements; creation of clear structures; easier to manage; flexible for changes in the environment; mutual communication of different interdisciplinary systems; flexible allocation of resources; continuous improvement; flexible self-defence mechanisms; replacement of traditional security systems (IPS, IDS); increased fault tolerance and self-organization; formation of a new structure; taking over routine actions; reduction of complexity within the system.

From a data-centre investment-protection point of view, it is very difficult to predict which technologies will be of interest in the future and which will stand the test of time and provide long-term profitability. The current data centre is characterized by a very complex and complicated structure of its IT systems. At least a dozen drivers control the development of DC and its infrastructure. Additionally, the following factors are gaining in importance: better control of critical systems, energy management, network connectivity, virtualization and cloud computing, service availability, application of resilience and reliability.

In general, organizations and their leaders should be able to manage unexpected events by using their financial, technical, and social resources in different decision making processes. Based on this, a resilience framework for organizational prosperity and survivability is proposed [1]. The NAS defines resilience [2] as "the ability to prepare and plan for, absorb, recover from, and more successfully adapt to adverse events".

In the paper, an enterprise architecture approach based on ArchiMate language is proposed for developing a government data centre (GovDC). After a motivation model is defined, a landscape of products and services should be determined for all enterprise architecture layers.

2 First Section

The discussed approach for the GovDC is analysed using the example of Federal Data Centre (FDC) [10]. As the federal ICT service provider and operator of one of the largest data centres in Austria, the FDC sees itself as an integrator of the processes of public administration, the services of the IT industry and the needs of users. As a vendor-independent provider of information technology services, the FDC bases all its activities on the interests and the mission of the Republic of Austria. Federal ministries and their subordinate agencies, social insurance institutions, universities and state-owned companies are among the target markets for the FDC. As the ICT service provider of the federal government, the FDC is able to establish itself as a market leading e-government partner through quality-assured and efficient application development.

The main transformation drivers are high maintenance costs and high demand on energy on the one hand, and complex, complicated structure and the availability of new ICT technologies on the other hand. The goals are sustainable transformation, optimal governance and the development of appropriate competencies to perform transformations and maintain future system configurations. Governance solution may be used starting from general approaches like TOGAF [4] and moving to more specific COBIT [5] and ITIL [6].

The main principles supply continuous data delivery and services for customers, maintain legacy systems, apply cloud computing, and use a software-defined system approach. Software defined systems (SDSys) reduce the overhead for data centre management by abstracting out the functionalities responsible for organizing and by controlling the main elements of the data centre such as network, storage, compute, and security [7].

To obtain a general view of the enterprise architecture of a government data centre, it is important to have a holistic view with which to evaluate the complexity of the architecture [8]. Enterprise modelling facilitates the analysis, understanding, and communication of proposed solutions in different domains, which requires combining different enterprise models [9]. To deal with the problems at the beginning, the motivation model for changes in FDC should be determined. A good way to accomplish this is to use ArchiMate, which is used to model enterprise architectures [3]. A simple version of a motivation layer for FDC is given in Fig. 1. One of the largest problems enterprises face is poor business alignment of increasingly complex and costly IT systems. This relation is investigated by using ArchiMate language providing a complete set of relation description concepts for relationships between architecture fields while producing a simple and uniform structure as a result.

The federal data centre should support e-governance by realizing e-government functions. There are many definitions of e-governance [11]. For the purpose of the paper the following definition is used: "E-Governance comprises the use of information and communication technologies (ICTs) to support public services, government administration, democratic processes, and relationships among citizens, civil society, the private sector and the state". The implementation of new network technologies in e-government data centres requires the use of on-demand solutions using Ethernet and

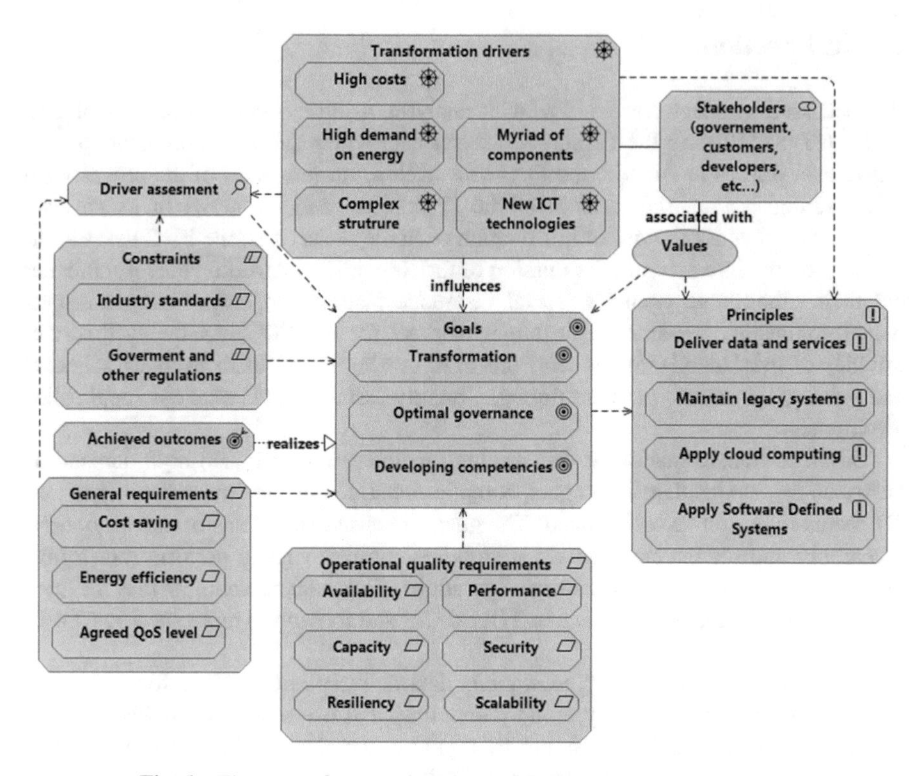

Fig. 1. Elements of the motivation model proposed for a data centre.

cloud computing [12], which leads to minimized operating costs and reduces the use of resources. An initial approach to e-government acceptance and use is given in [13]. The application of e-governance improves the efficiency of government functioning by removing redundancy at different levels. The four main categories of e-governance products in FDC [10] are given in Table 1.

Table 1 lists the main enterprise layers under consideration. The activities performed by the FDC are marked by X.

The corporate structure is divided into divisions that are responsible for specific tasks, projects and services. Let us consider the following as an example:

Standard Applications: This area includes planning, development and operation of the enterprise resources and planning systems of the federal government and outsourced companies. The range of services and operating solutions are proposed for budget management, personnel management and accounting based on standard business software and procedures for the federal budget and payments.

Individual Applications: Here, one is responsible for the creation of individual IT applications for public administration. These include financial, customs and judicial procedures for school boards and universities.

Table 1. General enterprise architecture landscape for FDC

Product types	G2G (Government to government) products	G2B (Government to Business) products	G2C (Government to Consumer) products	G2E (Gov. to Employees) products
Motivation layer	X	X	X	–
Strategy layer	X	X	–	–
Business layer	X	X	X	–
Application layer	X	X	X	X
Technical layer	–	X	X	X
Physical layer	–	–	X	X
Implementation and migration layer	–	–	X	X

E-government Solutions: The first e-government applications were developed and implemented in the FDC. In addition, many internationally acclaimed applications, such as HELP.gv.at were conceived here and are still operating today.

Infrastructure: The FDC is one of the top three data centres in Austria. The 1,500 IT offices run by the Federal Computing Centre are home to 45,000 PCs, and more than 500 applications are supported. With the volume of data served by the Federal Computing Centre, all residents of Vienna, Salzburg, Graz and Linz could be provided together. As the ICT service provider of the Austrian federal administration, the FDC is at present the market-leading e-government partner in Austria. Table 2 shows some examples of important systems of the FDC and of the types of functionality supported.

An important element of the company's philosophy is the optimal security of the entrusted data. Deliberate handling of risks, ongoing maintenance of safety rules, sensitization of all employees to safety issues, auditing compliance with safety guidelines and a separate parallel computer centre guarantee the highest possible safety standards. This is also confirmed by the certification according to ÖNORM A 27001, which is awarded by an independent expert and renewed annually.

3 Strategy and Reference Architecture for Government Data Centres

The FDC, and most data centres, must meet new challenges that will dominate the shape of the CIT industry in the near and distant future. An important problem lies in the implementation of the latest technologies. Cloud computing, digitalization, blockchain, virtualization and big data force us to look for new resources.

The new kind of software systems that support cloud platforms is extremely complex, and new approaches for resilience and engineering dependency are needed.

Table 2. Federal data centre – a part of enterprise architecture landscape. (Abbr. Catg – Product category, Oprt – Operation, Serv- Services, Dsgn – Design, Devp- Developing, Mnt – Maintenance, Copl – Compliance, SSpp – System support, SSoft – System software (OS, DBMS), SyInf - System infrastructure, Perm – Permises (Buildings IoTd – IoT devices)

Functionality	Type	Oprt	Serv	Dsgn	Devp	Mnt	Copl	SSpp	SSoft	SyInf	Perm	IoTd
Data.gv.at	G2C				X		X		X			
E. Records in gov.	G2B		X									
E-voting	G2C			X		X		X	X	X	X	
Finanz online	G2C		X	X		X		X		X		
Help.gv.at	G2C	X						X	X	X		
Sap	G2C			X		X		X	X	X		X
Trust centre services	G2B		X		X	X	X	X	X	X	X	X
Business service portal	G2B	X	X	X	X	X		X	X	X		X
Automation of court	G2B		X		X	X		X	X	X	X	
Peppol	G2B											
Stork	G2G			X	X	X		X		X		
Spocs	G2G							X	X			
Epsos	G2G			X		X	X	X	X	X	X	X
Cross-border ehealth	G2G			X	X	X	X	X	X	X	X	X
E-CODEX	G2G	X	X	X	X	X		X	X	X		
Euritas	G2G	X			X		X	X				
Eur. order for payment	G2G		X		X	X		X	X	X		
Twinning	G2B	X	X	X	X	X	X	X	X	X	X	X
		Operation		Development				Deployment				

The paper discusses [14] key benefits and architectures of the government cloud as well as deployment models, service models and selection strategies. Reengineering of existing solutions (e.g. redundancy) could be insufficient. Cloud computing providers promise easy and stable access to IT resources (applications) in addition to needs-based use (on demand) and billing (pay per use). Access will be also quick and safe. This sounds fantastic on paper and usually works very well. The reality, however, causes the greatest pain to IaaS and SaaS providers. An additional parameter complicating the implementation of new resilience methods is the variety (non-standardization) of solutions proposed by leading companies in cloud computing technology.

Providers offering their services through cloud platforms must guarantee reliability of services through automatic system reconfiguration, system analysis and/or resilience methods. The following aspects characterize the transformation of a data centre infrastructure to cloud technology: (1) Applications that were standard provisioning have to be adapted to new platforms such as cloud computing. (2) Decisions concerning which platform to use are often at the gambling limit. (3) A provider that proposes IaaS must adapt its own infrastructure to new requirements. (4) Many solutions have their roots in solutions called stand-alone or legacy systems. (5) The complexity of the systems increases, which at the same time means that security is unsealed. (6) Given the number of systems and applications that are active in the data centre, it is possible to conclude that the probability of a complete failure, at least temporarily, of a service or application is relatively high. (7) Most solutions used in case of failures in the data centre are the result of "classic" solutions, such as redundancy.

The development of data centres may be based on three strategies: complete stacks with integrated software and hardware based on a selected vendor, custom stacks based on hardware and software solutions from different vendors or the open-source community, a hybrid solution which is a mix of the above two solutions. Each strategy has its own benefits and trade-offs. The first solution may be costly in the long term as it depends on vendor strategy and capacity. A custom solution requires additional resources with proper competencies, but it can involve flexible and cost-effective solutions.

Reference architecture for a government data centres should be defined on the basis of the strategy selected. Reference architectures accelerate work, reduce operating expenses and improve the quality of software system development, mainly due to reuse [15]. The purpose of reference architecture is to provide guidance on developing architectures for new versions of a system or extended families of systems and products. Paper [16] offers a systematic review of the existing studies in designing cloud based solutions and shows the importance of cloud reference architecture.

4 Reactive System and Patterns Approach to Resilience of a GovDC

The Government Data Centre can be considered from a different point of view. A typical approach is to choose a set of quality attributes with assigned thresholds. The ISO/IEC 25010 standard defines the systems and software quality models [17] entitled SQuaRE (Systems and Software Quality Requirements and Evaluation).

The proposal here is to consider GovDC as a reactive system, which means that the methods used for designing reactive systems may be applied to the centre. In the classic approach [18] reactive systems are described by their characteristics: highly interactive, non-terminating process, interrupt driven, state-dependent response, environment oriented response, parallel processes and stringent real-time requirements. The description of reactive systems changed because the systems consist of many different components with different complexity. The reactive manifesto [19] defines reactive systems by their features (Fig. 2). A message driven pattern is defined as the basis for all other traits. Responsiveness seems to be the main goal of reactive systems, because all other traits depend on it. Responsiveness is important in developing reactive systems based on quality attributes such as resilience, elasticity and message driven pattern. For the predefined set of quality attributes, architectural tactics may be proposed. An example

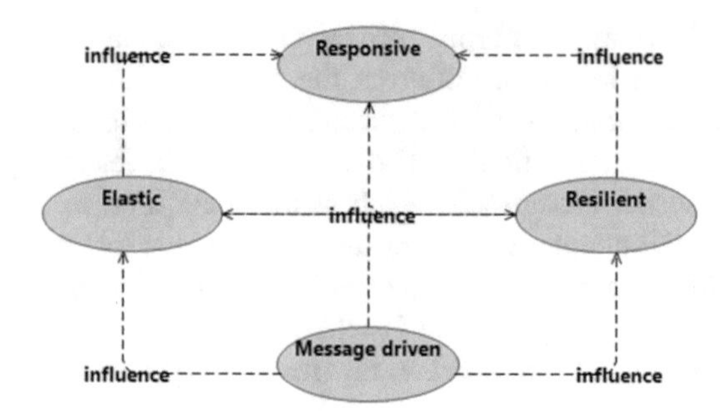

Fig. 2. Reactive systems dependences according to the reactive manifesto.

can be found in the quality-attributes approach used for defining reactive systems solutions applied to cloud sensors [20].

The paper [21] introduces a unified definition of reliability, survivability, and resilience. *Resilience* is defined as the time taken by a system to return to equilibrium (or acceptable resistance or reliability after a failure). In other words, this is the period starting from the moment when the system's resistance level drops below the persistence level and subsequently rises (due to built-in recovery or repair mechanisms) above the persistence level. Persistence is the minimum level of resistance that a system must maintain to be operational. For example, a persistent system may be specified as $S(t) > ST$, where $S(t)$ is the resistance (reliability) measured with the survivor function and ST is the persistence threshold the system must maintain to avoid being "destroyed" by catastrophic failures.

In the traditional approach availability is increased by extending the non-interrupted system working time:

$$Availability = \max_{MTTF \to \infty} \frac{MTTF}{MTTF + MTTR},$$

where MTTF (mean time to failure) is the average time from the start of the system's proper operation to the time a failure occurs, and MTTR (mean time to recovery) is the average time from the occurrence of a failure until the proper operation of the system is restored. MTTF should be maximized by taking active measures to reduce the probability of error occurrence. If MTTR is relatively small compared to MTTF, operation of the system is acceptable. This approach to increasing MTTF usually requires a lot of infrastructure overhead: using redundant hardware, operating HA (high availability) clusters, using multiple network connections (hopefully across different switches), and so on.

Another approach to availability is based on resilience:

$$Availability = \max_{MTTR \to 0} \frac{MTTF}{MTTF + MTTR}.$$

A resilient design does not try to maximize MTTF, as the basic assumption is that errors are generally neither preventable nor predictable but simply happen - that is,

MTTF is considered uncontrollable. Intensive research is now being carried out aimed at defining a formal metric for resilience. A general resilience metric (GR) is proposed by Nan and Sansavini in [24] as follows:

$$GR = R \times (RAPI_{RP}/RAPI_{DP}) \times (1/TAPL) \times RA.$$

The robustness R (which means the lowest performance level reached) and recovery ability RA are directly proportional to resilience. The time-averaged performance loss, $TAPL$, is in inverse proportion. The recovery speed $RAPI_{\mathrm{RP}}$ has a positive and loss speed *RAPIDP* has a negative effect on resilience.

The main problem is to develop resilience solutions for data centres in a holistic and agnostic way. To obtain required resilience, design patterns which are suitable to the resilience quality attribute may be used. Many software systems support resilience by applying paradigms. A clear introduction to resilient pattern and software design is given in [22, 23]. In these works resilience is based on isolation (shed load, bulkheads, complete parameter checking), supervision (monitor, error handler, escalation), latency control (circuit breaker, bounded queues, fail fast, timeouts, fan out & quickest reply) and loose coupling (asynchronous communication, idempotency, relaxed temporary constraints, self-containment, location transparency, event driven, stateless).

Microsoft proposes cloud design patterns [25] for Azure cloud. The following support resilience (for PaaS): (1) Bulkhead, isolation of elements; (2) Circuit breaker; (3) Compensating transaction; (4) Health endpoint monitoring; (4) Leader election; (5) Queue-based load levelling; (6) Retry; (7) Scheduler agent supervisor.

The PasS resilience is not of primary interest to a data centre, but it is at least important to know how resilience is protected and which resilience patterns are used. The main interest of the data centre is resilience at the IaaS level. Many resilience solutions are proposed for infrastructure which are of a general nature. Thomas Erl et al. [26] propose a set of patterns for reliability, resilience and recovery.

In the design of GovDC enterprise architecture, the building blocks should be established on the basis of patterns. This means that the resilience pattern should be adapted for all enterprise (business, application, infrastructure, physical) levels.

5 Conclusions

By analysing the methods and state of the art for data centre solutions, a kind of platform should be created to "coordinate" all the resilience measures. One disadvantage is growing complexity. As complexity increases, so does costs. There are still questions to be answered concerning measuring resilience, multilevel resilience, providing a small test cloud to examine the impact of changes, preventing potential failures or embracing them.

The proposed approach is holistic and is based on the GovDC enterprise architecture. The approach consists of developing the enterprise motivation model, defining architecture landscape, strategy and reference architecture on the basis of reactive systems, then developing solutions for business, application, infrastructure and physical layers.

References

1. Tengblad, S., Oudhuis, M.: The Resilience Framework Organizing for Sustained Viability. Springer, Singapore (2018)
2. Linkov, I., Palma-Oliveira, J.M. (eds.): Resilience and risk: methods and application in environment, cyber and social domains. In: Proceedings of the NATO Advanced Research Workshop on Resilience-Based Approaches to Critical Infrastructure Safeguarding, Azores, Portugal, 26–29 June 2016. Springer, Dordrecht (2017)
3. The Open Group, ArchiMate® 3.0 Specification (2016). http://pubs.opengroup.org/architecture/archimate3-doc/toc.html. Accessed 27 Jan 2017
4. The Open Group, TOGAF Version 9.1, 10th New edn. Van Haren Publishing, Zaltbommel (2011)
5. ISACA, COBIT 5: A Business Framework for the Governance and Management of Enterprise IT (2012). http://www.isaca.org/Cobit/pages/default.aspx. Accessed 13 May 2017
6. Vicente, M., Gama, N., de Silva, M.M.: The value of ITIL in enterprise architecture. In: 2013 17th IEEE International Enterprise Distributed Object Computing Conference, pp. 147–152 (2013)
7. Darabseh, A., Al-Ayyoub, M., Jararweh, Y., Benkhelifa, E., Vouk, M., Rindos, A.: SDDC: a software defined datacentre experimental framework, pp. 189–194 (2015)
8. Lakhrouit, J., Baina, K.: Evaluating complexity of enterprise architecture components landscapes, pp. 1–5 (2015)
9. Caetano, A., Antunes, G., Bakhshandeh, M., Borbinha, J., da Silva, M.M.: Analysis of federated business models: an application to the business model canvas, ArchiMate, and e3value, pp. 1–8 (2015)
10. Sharma, R., Kanungo, P.: An intelligent cloud computing architecture supporting e-Governance. In: The 17th International Conference on Automation and Computing, pp. 1–5 (2011)
11. Bannister, F., Connolly, R.: New problems for old? defining e-governance, pp. 1–10 (2011)
12. Urbanczyk, W.: How the implementation of new network technologies influenced the changes of e-government federal data centres, pp. 53–57 (2013)
13. Nunes, S., Martins, J., Branco, F., Goncalves, R., Au-Yong-Oliveira, M.: An initial approach to e-government acceptance and use: a literature analysis of e-Government acceptance determinants, pp. 1–7 (2017)
14. Liang, J.: Government cloud: enhancing efficiency of e-government and providing better public services, pp. 261–265 (2012)
15. Turek, M., Werewka, J., Sztandera, K., Rogus, G.: Assessment of software system presentation layers based on an ECORAM reference architecture model (2015)
16. Breivold, H.P., Crnkovic, I., Radosevic, I., Balatinac, I.: Architecting for the cloud: a systematic review, pp. 312–318 (2014)
17. ISO/IEC/ IEEE: Systems and software engineering—Architecture description, International Standard. ISO/IEC/IEEE 42010 (2011)
18. Wieringa, R.: Design Methods for Reactive Systems: Yourdan, *Statemate*, and the UML. Morgan Kaufmann Publishers, Boston (2003)
19. The Reactive Manifesto. https://www.reactivemanifesto.org/. Accessed 06 Jan 2018
20. Skowroński, A., Werewka, J.: A quality attributes approach to defining reactive systems solution applied to cloud of sensors, pp. 789–795 (2015)
21. Ma, Z.: Towards a unified definition for reliability, survivability and resilience (I): the conceptual framework inspired by the handicap principle and ecological stability. In: 2010 IEEE Aerospace Conference, pp. 1–12 (2010)

22. Uwe Friedrichsen, “Patterns of resilience.” 15:37:10 UTC
23. Friedrichsen, U.: Resilient software design in a nutshell: Software Architecture Conference | Microservices training | O’Reilly. https://conferences.oreilly.com/software-architecture/sa-eu/public/schedule/detail/61746. Accessed 08 Jan 2018
24. Nan, C., Sansavini, G.: A quantitative method for assessing resilience of interdependent infrastructures. Reliab. Eng. Syst. Saf. **157**, 35–53 (2017)
25. Microsoft Azure, “Cloud Design Patterns.” https://docs.microsoft.com/en-us/azure/architecture/patterns/. Accessed 10 Jan 2018
26. Erl, T., Cope, R., Naserpour, A.: Cloud Computing Design Patterns. Prentice Hall, New York (2015)

The Research of Grover's Quantum Search Algorithm with Use of Quantum Circuits QX2 and QX4

Arkadiusz Liber(✉) and Laurentiu Nita

Faculty of Computer Science and Management, Wroclaw University of Science and Technology, Wybrzeze Wyspianskiego 27, 50-370 Wroclaw, Poland
arkadiusz.liber@pwr.wroc.pl

Abstract. We researched the implementation of Grover's Algorithm on two qubits and carried out a series of experiments on the Quantum Processors QX2 and QX4 [1] to see how it can be optimized. We found that by adjusting the Oracle function of the algorithm, we can extend the solution base, from 2^n to 5^n, where n is the total number of available qubits that are used to build the algorithm. The method presented in this paper makes possible to execute Grover's Algorithm with 25 different Oracle functions and obtain all unique solutions by using only 2 entangled qubits, if we choose to expand the solution base from $\mu \in \{0, 1\}^n$ to μ belonging to a set of 25 unique output values, where μ is the item being searched using Grover's method.

Keywords: Quantum computing · Grover's algorithm
IBM quantum processor

1 Grover's Algorithm on Two Qubits

1.1 Purpose and Achievement of the Algorithm

The subject of the research conducted by the authors is the Grover algorithm and its implementation in real quantum hardware. The idea of this algorithm was presented in the 1996 paper titled "A fast quantum mechanical algorithm for database search" [2]. Grover proved that the computational complexity of the quantum search problem is $O(\sqrt{N})$. The result was so promising that it triggered intensive research into the possibility of using Grover's algorithm in practice. Until recently, the study of quantum algorithms was mainly of a theoretical nature or in the form of implementations of simulators on classical computers. Work in this area was also the subject of previous research by the authors [3]. The use of superconducting quantum processors is of particular importance here. The physical basis of operation of such systems is the subject of many works in the field of solid-state physics. An important example is a paper published by Yuo and Nori [4]. Using the technology of superconducting quantum circuits, IBM constructed and made available QX series quantum processors for testing [1]. This allowed the authors to conduct a series of studies related to the quantum algorithms optimization and information dissipation.

L. Borzemski et al. (Eds.): ISAT 2018, AISC 852, pp. 146–155, 2019.
https://doi.org/10.1007/978-3-319-99981-4_14

This article contains results related to the actual implementation, operation and optimization of the Grover algorithm in the QX2 and QX4 systems. Both the research on the process and computational complexity of information loss were carried out and their comparison with the best results achieved by Bennet et all at work [5].

1.2 Publications Discussing Algorithm Implementations

There are multiple ways to implement this algorithm [6, 7]. Implementations of Grover's Algorithm have already been done on IBM's Quantum Processors QX2 and QX4. Investigations done by a research group from Los Alamos National Laboratory in the United States of America showed that Grover could be implemented using an ancillary qubit, requiring three qubits in total to find a solution based on two qubits within $\{0, 1\}^2$ [8].

In 2016 IBM launched the Quantum Experience program, which allows open access to anyone interested in trying out algorithms on their QX2 and QX4 quantum processors [9]. Both machines contain five fixed-frequency superconducting transmon qubits [10] that are made possible to be entangled for the creation of a circuit. The User Guide IBM published to describe how to design a quantum circuit ready to run on one of the IBM machines contains a full description of Grover's Algorithm implementation suitable for execution on only two qubits [11]. The manual describes in detail how the algorithm is built and provides ready to execute circuits to obtain solutions in the $\{0, 1\}^2$ solution space.

1.3 Grover's Algorithm Circuit Design for QX2 and QX4 Machines

We take N as the total number of items in an unsorted list. Among these N items, there is an item μ we want to search. To find the μ item, given classical computation it would take us in the worst-case scenario N searches to perform if the item is on the last entry on the list. With Grover's and a quantum machine, we can reduce the number of searchers needed to $\sqrt{N}$ by building an oracle function that looks for the searched item and by doing an amplitude amplification to return the searched item. All our qubits start in the $|0\rangle$ state.

Description of the Grover's Oracle Function

Oracle function is a black box type of function we will refer to as U_f that we will use to encode the set of items we are searching for. The solution the algorithm is seeking will be encoded inside but the algorithm does not have a way to look at the choice of Oracle function, hence the solution is not known to the algorithm. All the algorithm will do is supply a value x and ask for the function that we will refer to as $f(x)$ to be executed. We define outcome as follows:

$f(\mathrm{x}) = 0$ for items that are not what we seek.
$f(\mu) = 1$ if we found the right item, where $x, \mu \in \{0, 1\}^n$ so that $N = 2^n$.

Next, we encode this Oracle function $F(x)$ in a unitary matrix U_f that can act on any of the standard basis states $|x\rangle$.

$$U_f|\mathrm{x}\rangle = (-1)^{f(x)}|x\rangle \tag{1}$$

The outcome we will look for with U_f is to find item μ. If the state being processed is x, then oracle function does nothing to the state. If the state looked at is $|\mu\rangle$, we will map:

$$U_f|\mu\rangle = -|\mu\rangle \tag{2}$$

Description of Grover's Amplitude Amplification
The second aspect of Grover's Algorithm is the amplitude amplification. Before this part is executed, Hadamard gates are used to place both qubits in superposition. The execution of the algorithm will start with this normal superposition state of all qubits, referred to as state $|s\rangle$ (3) using the bra-ket notation.

$$|s\rangle \frac{1}{\sqrt{N}} \sum_{x=0}^{N-1} |x\rangle \tag{3}$$

Upon measuring, the superposition states would collapse therefore our results would consist of up to $N = 2^n$ tries to guess the right items, where n is the number of available qubits to store information. The amplitude amplification trick is a well-known quantum procedure to amplify the right searched item, in our case μ, while at the same time shrink all other possible states referred to as x. What this (4) does is to inverse about the average the amplitudes of all states.

$$U_s = 2|s\rangle\langle s| - 1 \tag{4}$$

1.4 Order of Events to Execute Grover's Algorithm on a Quantum Chip

Step 1. Uniform superposition is set on both qubits.
We define t as being the number of steps required to execute Grover's Algorithm. At $t = 0$, we place all qubits in a normal superposition state (5).

$$|s\rangle = H^{\otimes n}|0\rangle^n \tag{5}$$

To refer to the state of our qubits, we use $|\psi_t\rangle$, at $t = 0$ we have

$$|\psi_t\rangle = |s\rangle \tag{6}$$

Step 2. Oracle function is applied to the steady superposition state.
We apply the Oracle matrix (4) to the state of our qubits (6). The μ value we search for is encoded in the definition of the Oracle matrix.

$$U_f|\psi_t\rangle = |\psi_{t'}\rangle \tag{7}$$

Step 3. The qubit state closer to the Oracle function is amplified.
We apply the matrix defined using bra-ket in formula (4) to do a reflection about the average amplitude. The negative magnitude (2) of the state we seek $|\mu\rangle$ will be boosted, and upon repeating this procedure, we will obtain the searched state. After t steps, we

get (8) which is the final representation in bra-ket of Grover's Algorithm. The total number of execution steps t necessary to perform the reflection amplification of the searched amplitude is $\sqrt{N}$.

$$|\psi_t\rangle = \left(U_s U_f\right)^t |\psi_0\rangle \tag{8}$$

Step 4. Execution of the circuit.

Figure 1 shows a circuit prepared to find the solution we identify with $q[2] = 0$ and $q[1] = 1$. All we require to build it are a set of unitary matrixes: H, X, and CNOT gates [11–13].

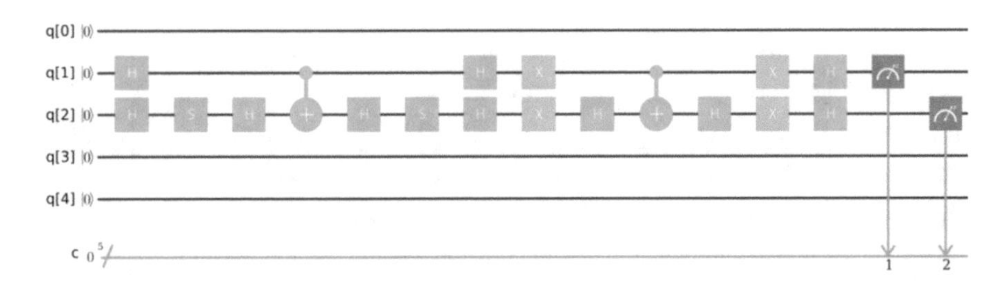

Fig. 1. IBM quantum composer with the full Grover's Algorithm circuit ready for execution. The circuit must be read from left to right.

The H Gate

The Hadamard gate maps $X \to Z$, and $Z \to X$. This act of mapping the state vector to X-axis allows for superposition to occur. This is the only logical gate from the set that provides for superpositions to be formed.

$$H = \frac{1}{\sqrt{2}}\begin{pmatrix} 1 & 1 \\ 1 & -1 \end{pmatrix} \tag{9}$$

The X Gate

It is the quantum version of the NOT classical gate, also known as a bit-flip, turns a 0 into a 1. Besides moving the qubit state vector, the X gate affects the axis on which measurement will occur by mapping X axis to $X \to X$ and Z axis to $Z \to -Z$.

$$X = \begin{pmatrix} 0 & 1 \\ 1 & 0 \end{pmatrix} \tag{10}$$

The S Gate

This Phase gate has the property that it maps $X \to Y$ and $Z \to Z$. This gate extends H to make complex superposition.

$$S = \begin{pmatrix} 1 & 0 \\ 0 & i \end{pmatrix} \tag{11}$$

The CNOT Gate
It applies an X gate to a target qubit if the control qubit is in $|1\rangle$ state. If the control qubit is in $|0\rangle$, it leaves the target qubit as is.

Step 5. Analyzing the output of the circuit to determine success criteria.
The circuit shown in Fig. 1 outputs the solution {0,1} from the possible solution set $\{0, 1\}^2$. By applying different Oracle functions, we obtain different results.

In Fig. 2 we see the circuit necessary to obtain solution {1,0}, which was achieved by changing the Oracle function to have S gates on $q[1]$ instead of on $q[2]$ as seen in Fig. 1. We can obtain solution {0,0} by adding S gates on both qubits and {1,1} by removing all S gates inside the Oracle function.

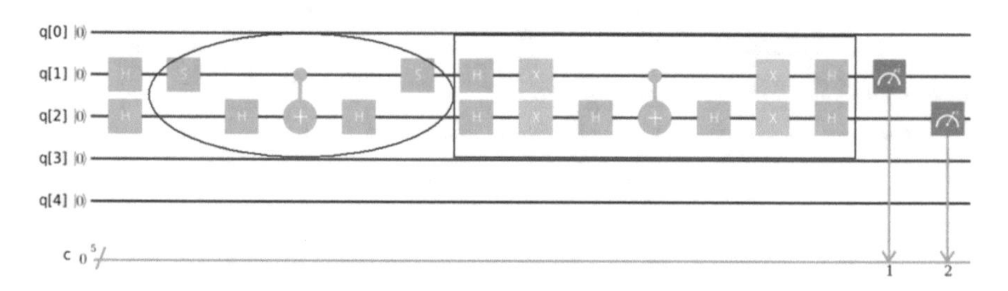

Fig. 2. The set of gates highlighted with the red ellipse determine the Oracle Function. The set of gates from the red rectangle perform the amplitude amplification. The circuit designed using IBM Quantum Composer.

1.5 Summarizing the Workings of Grover's Algorithm on Two Qubits

We initialize the algorithm with two H gates to create a uniform superposition of all states; then we add the Oracle function, U_f in which we encode the value μ we seek. The algorithm does not know this value since the Oracle function acts as a black box. To obtain searched value μ, where $\mu \in \{0, 1\}^n$, the amplitudes of all possible solutions are inverted about the average by the logic described for U_s (4).

Since the Oracle function acts as a black box, this opens the question if it would be possible to encode inside a value that would be a superposition state instead of a full state.

To formulate an answer for this question first, we consider the results of the Copenhagen Interpretation and its postulates that describe a model that could be used for computing. Basing on the following main principles: the probabilistic interpretation of a particle's spin created the concept of quantum superposition – the state of a particle is not defined before measurement but instead lives in a universe of probabilities and only when this probability universe collapses, which is triggered by the simple act of measurement, there will be a precisely measured state. This principle of superposition can be explained as "The laws of nature formulated in mathematical terms no longer determine the phenomena themselves but the probability that something will happen" [13].

An Oracle function defined within the superposition states of both qubit spins could potentially be used to expand the solution base since the act of measurement in our algorithm occurs once and only after the qubits have passed through the circuit.

2 Grover's Algorithm with Expanded Solution Base on Two Qubits

2.1 The Use of Custom Gates on QX2 and QX4

To be able to determine and build an Oracle function that does not involve full rotations of the two qubits, from $|0\rangle$ to $|1\rangle$ we need a gate that we could potentially use to rotate the qubit around its axis with less than a half of a full rotation. For this purpose, we have available in the advanced section of the IBM Quantum Composer UI the ability to configure our physical gate.

The best fit for the job is the U_3 gate since we will measure the spin only on one axis. We will only change the value in radians of the θ angle and execute the U_3 on the chip. The U_3 provides additional customization options that are given the value 0 since they are not necessary to achieve the purpose of this experiment. By simplifying the U_3 gate, we obtain (12) as its full description.

$$U_3(\theta) = \begin{pmatrix} \cos\left(\frac{\theta}{2}\right) & -\sin\left(\frac{\theta}{2}\right) \\ \sin\left(\frac{\theta}{2}\right) & \cos\left(\frac{\theta}{2}\right) \end{pmatrix} \tag{12}$$

2.2 Definition of the Expanded Solution Table for the Oracle Function

The representation of what makes a valid solution to Grover's Algorithm is arbitrarily chosen in advance. The standard interpretation of the result of a quantum circuit in literature is always chosen to be $\{0, 1\}^n$, hence 2^n possible outcomes.

Our experiments show that this is not necessary. We can choose to change the interpretation of the result to take into account many additional superposition states between $|0\rangle$ and $|1\rangle$ as long as these conditions are followed:

1. We do not perform any measurement that could lead to the collapse of the entangled superposition states of both qubits until the full execution of the algorithm is concluded
2. The quantum processor can produce results on the final measurement of each full execution stream that is between an error boundary big enough to ensure that the superposition states we chose as solution base do not overlap

For the experiments we performed, we decided to build Oracle functions that rotate the spin of both qubits only on the measurement axis.

2.3 The Expanded Solution Table for the Oracle Function

We started with qubit $q[1]$, where we applied a U_3 the gate that consists of θ values equivalent in degrees of 0, 60, 90, 120 and 180, converted to radians with up to 6 decimals to improve precision. The 2nd qubit we use, $q[2]$ receives U_3 gates in the Oracle function for each θ value of the U_3 the gate used in $q[1]$. By following through with this method, we obtain through experimental executions of the circuit with 8192 shots a table of 25 possible solutions, all unique.

We modified the Oracle function as seen in Fig. 3 to manipulate the result being searched for by using a U_3 on each qubit.

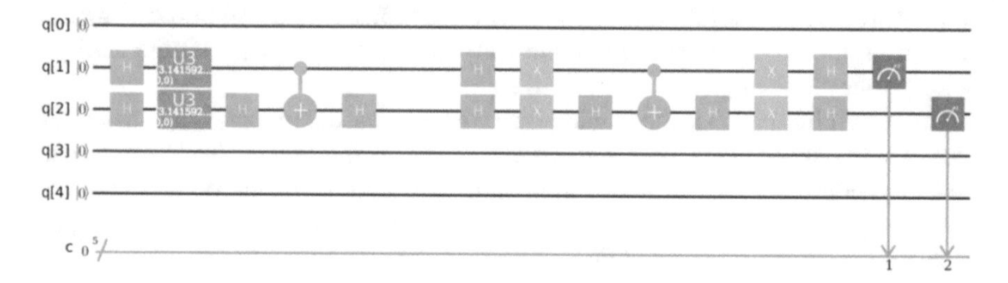

Fig. 3. A circuit example of Grover's Algorithm with U_3 gates on $q[1]$ and $q[2]$. The circuit designed using IBM Quantum Composer.

To obtain the solution basis, we executed 8192 repetitions of the circuit for each experiment and placed the results in Table 2, starting with θ of 0 for both $q[1]$ and $q[2]$. The probability distributions we obtain at the end of the 8192 repetitions are our result.

We consider this table as the full solution base of Grover's Algorithm on two qubits with an arbitrary choice for solution base that consists of using U_3 gates with θ values of 0, 60, 90, 120 and 180. The values behind the table columns $P_{(00110)}, P_{(00010)}, P_{(00100)}, P_{(00100)}, P_{(00000)}$ are the probability distributions from 0 to 1 of seeing that specific qubit configuration upon the collapse of the probability wave that must happen at the end of all repetitions of an execution stream. If we were to run the algorithm only once, our results would be only one of these four possible configurations of the system at the end of a single execution. By design, any quantum computation does require many repetitions due to the nature of the quantum world and the possibilities of error. The more significant the number of repetitions, the better our chance is to obtain a correct result. We chose 8192 shots since this was the most significant number of repetitions that are offered in the Quantum Composer UI to run for a circuit.

We can use the same approach with U_3 gates to construct Table 1, which is the solution set identical to Grover's Algorithm we showed in Sect. 1. Applying this change to the Grover's Algorithm, allows us to extend the solution space from Tables 1 and 2. Although we implement it with U_3 gates, the outcome is identical to that of the S gates that are used and discussed in Sect. 1. Both methods satisfy the conditions required to build the Oracle function.

Table 1. Results table of Grover's Algorithm on two qubits from Sect. 1.

$q[1]$ θ in rad	$q[2]$ θ in rad	$P_{(00110)}$	$P_{(00010)}$	$P_{(00100)}$	$P_{(00000)}$
0.000	0.000	1.000	0.000	0.000	0.000
0.000	3.142	0.000	0.000	1.000	0.000
3.142	0.000	0.000	0.000	0.000	1.000
3.142	3.142	0.000	0.000	0.000	1.000

Table 2. Results of the experiments containing all unique solutions obtainable which create the solution space we propose for Grover's Algorithm on two qubits.

$q[1]$ θ in rad	$q[2]$ θ in rad	$P_{(00110)}$	$P_{(00010)}$	$P_{(00100)}$	$P_{(00000)}$
0.000	0.000	1.000	0.000	0.000	0.000
0.000	1.047	0.750	0.000	0.250	0.000
0.000	1.571	0.504	0.000	0.496	0.000
0.000	2.094	0.250	0.000	0.750	0.000
0.000	3.142	0.000	0.000	1.000	0.000
1.047	0.000	0.750	0.250	0.000	0.000
1.047	1.047	0.569	0.184	0.181	0.066
1.047	1.571	0.370	0.127	0.381	0.122
1.047	2.094	0.191	0.062	0.555	0.192
1.047	3.142	0.000	0.000	0.753	0.247
1.571	0.000	0.509	0.491	0.000	0.000
1.571	1.047	0.375	0.374	0.125	0.126
1.571	1.571	0.246	0.258	0.251	0.246
1.571	2.094	0.128	0.375	0.369	0.128
1.571	3.142	0.000	0.000	0.495	0.505
2.094	0.000	0.258	0.742	0.000	0.000
2.094	1.047	0.193	0.559	0.064	0.184
2.094	1.571	0.126	0.371	0.126	0.377
2.094	2.094	0.063	0.188	0.190	0.559
2.094	3.142	0.000	0.000	0.242	0.758
3.142	0.000	0.000	1.000	0.000	0.000
3.142	1.047	0.000	0.755	0.000	0.245
3.142	1.571	0.000	0.497	0.000	0.503
3.142	2.094	0.000	0.258	0.000	0.742
3.142	3.142	0.000	0.000	0.000	1.000

3 Conclusions

We showed that if we consider the conditions put forward in Sect. 2.2, we can encode in the Oracle function of Grover's Algorithm information that is less than a full spin.

By doing so, we can compute solutions between 0 and 1 states of the entangled qubits, thus allowing additional flexibility for defining the solution space.

Because of the probabilistic nature of the atomic and nuclear scale of computation, to satisfy condition two from Sect. 2.2 of this paper, we placed in Fig. 4 the maximum allowed error rates of the final measurement for each circuit from Table 2, for all the sets of Oracle functions with θ values defined in the table.

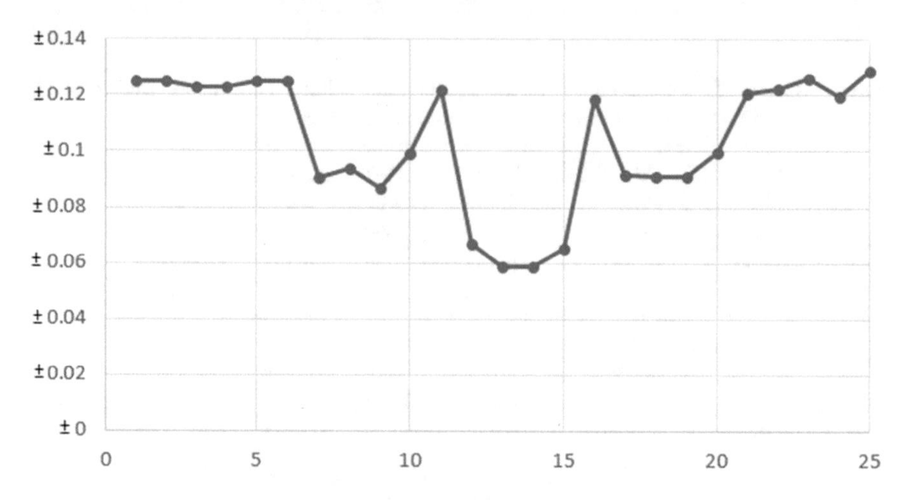

Fig. 4. A graph shows maximum allowed error rates for each execution streams of the circuit on the QX2 or QX4 quantum processors. The horizontal axis represents each output row from Table 2. The vertical axis is the maximum error rate allowed to obtain reproducible solutions

Using this method, we conclude the following changes to the Oracle function of Grover's algorithm are usable and provide unique solutions if the computation of the executions stream does not exceed the following error rates:

(a) ±0.12, we can use 12 Oracle functions: rows 1 to 6, 11, 21 to 25,
(b) ±0.08, we can use 21 Oracle functions: rows 1 to 11, 16 to 25,
(c) ±0.05, we can use all 25 Oracle functions.

We showed that given we have an error in the execution of the circuit within the range of values from Fig. 4, it is possible to have an expanded solution base from $\{0, 1\}^2$ up to a set of 25 unique solutions on the implementation of Grover's Algorithm on two qubits. Quantum supremacy [14] over classical computing is achievable if we improve the technology used to build the qubits and implement error correction techniques [15, 16] to take full advantage of superposition states as we discussed in this paper.

References

1. IBM, IBM Q experience Device. https://quantumexperience.ng.bluemix.net/qx/devices. Accessed 21 May 2018
2. Grover, L.K.: A fast quantum mechanical algorithm for database search. In: 28th Annual ACM Symposium on the Theory of Computing, pp. 1–8, 19 November 1996
3. Liber, A., Rusek, R.: Quantum digital signatures for unconditional safe authenticity protection of medical documentation. Higher School's Pulse **4**, 34–39 (2015)
4. You, J.Q., Nori, F.: Superconducting Circuits and Quantum Information. Physics Today, pp. 42–47 (2005)
5. Bennett, Ch.H., Bernstein, E. J., Brassard, G., Vazirani, U.: Strengths and weaknesses of quantum computing. SIAM J. Comput., 10 (1997)
6. Nielsen, M. A., Chuang, I.L.: Quantum Computation and Quantum Information, Cambridge University Press (2010)
7. Viamontes, G.F., Markov, I.L., Hayes, J.P.: More efficient gate-level simulation of quantum circuits. Quantum Inf. Process. **2**, 347 (2003)
8. Coles, P.J., Eidenbenz, S., Pakin, S.: Quantum Algorithm Implementations for Beginners. Los Alamos National Laboratory Publication, p. 10 (2018)
9. IBM Quantum Experience Homepage. https://www.research.ibm.com/ibm-q/. Accessed 17 May 2018
10. Koch, J., Yu, T.M.: Charge-insensitive qubit design derived from the Cooper pair box. Phys. Rev. A, (2007)
11. IBM Grover's Algorithm User Guide. https://quantumexperience.ng.bluemix.net/qx/tutorial?sectionId=full-user-guide&page=004-Quantum_Algorithms~2F070-Grover%27s_Algorithm. Accessed 17 May 2018
12. Gubaidullina, K.: Stability of Grover's algorithm in respect to perturbations in quantum circuit, Nanosystems: Physics, Chemistry, Mathematics, pp. 243–246 (2017)
13. Heisenberg, W., Born, M., Schrodinger, E., Auger, P.: On Modern Physics, Collier Books, p. 17 (1962)
14. Preskill, J.: Quantum computing and the entanglement frontier, pp. 10–18 (2012)
15. Padma Priya, R., Baradeswaran, A.: An efficient simulation of quantum error correction codes. Alexandria Eng. J., 8 (2017)
16. Quantum Computing Report. https://quantumcomputingreport.com/our-take/better-qubits-versus-more-efficient-error-correction/. Accessed 22 May 2018

From Sequential to Parallel Implementation of NLP Using the Actor Model

Michał Zielonka(✉), Jarosław Kuchta, and Paweł Czarnul

Faculty of Electronics, Telecommunications and Informatics,
Gdansk University of Technology, Narutowicza 11/12, 80-233 Gdańsk, Poland
miczielo@student.pg.edu.pl,
{qhta, pczarnul}@eti.pg.edu.pl

Abstract. The article focuses on presenting methods allowing easy parallelization of an existing, sequential Natural Language Processing (NLP) application within a multi-core system. The actor-based solution implemented with the Akka framework has been applied and compared to an application based on Task Parallel Library (TPL) and to the original sequential application. Architectures, data and control flows are described along with execution times for an application analyzing an online dictionary of foreign words and phrases.

Keywords: Parallelization · Actor model · Akka · TPL · NLP

1 Introduction

Nowadays, we deal with multi-core computers every day. Even a personal computer is usually equipped with a two- or quad-core processor with hyperthreading. Despite this, as our experience in cooperation with CI TASK[1] has shown, scientists who are not IT professionals, often write sequential applications and do not fully use the capabilities of multi-core hardware. They run their sequential applications in a multiprocessor environment and are uncomfortably surprised that they do not get acceleration of calculations. Proper parallelization is usually done by rewriting the program code from scratch or designing it to be parallel upfront especially for a multi-node cluster environment, so that the most time-consuming parts can be executed simultaneously by many processors. Unfortunately, such parallelization is generally very complicated and time-consuming.

To avoid rewriting the code of our NLP application from scratch, we can use an alternative approach within a single computational node. One solution proposed by Microsoft is TPL (Task Parallel Library) [1]. If we define several tasks in the program that are loosely coupled, we can attach these tasks to separate threads. TPL helps us to define tasks, run them asynchronously, and return the results. Having a computer with e.g. a quad-core processor with hyperthreading, we can easily use eight logical processors in our application. Unfortunately, the TPL solution has its limitations. Firstly, it is useful in a pipeline or divide-and-conquer model of parallel application, when there is no need for inter-task communication. TPL fails with more complex communication due to lack of support for sending messages between tasks. The second problem is that

[1] Centre of Informatics - Tricity Academic Supercomputer & networK

L. Borzemski et al. (Eds.): ISAT 2018, AISC 852, pp. 156–165, 2019.
https://doi.org/10.1007/978-3-319-99981-4_15

tasks must be run explicitly, one per thread. Finally, TPL supports multi-core calculations, so parallelization is limited to multicore processors.

All these problems are solved by the Akka [2] toolkit implementing the actor model [3]. In this model, application parallelization is based on the concept of actors - classes that perform specialized tasks. Objects that are instances of actor classes send each other data through messages that are placed in queues. Each queue is associated with an actor class that creates one or more instances depending on the load recognized by the queue length. After the data has been processed from the message, the actor's instance sends a message with the results to the queue of another actor's class.

Therefore, we decided to examine parallelization of sequential applications using the aforementioned example, and to compare the performance of parallel applications using both TPL and Akka solutions.

2 Related Work

Parallelized NLP is considered in the literature. For instance, a parallel parser with the work stealing approach is analyzed in [4]. Fine-grained parallelism is needed to obtain high speedup for this purpose according to [5] and on GPUs [6]. Our goal in the paper was parallelization of sequential elements of our application in a way that does not penetrate too much in the current architecture of the code and make it a scalable solution within a single multi-core node. There are many APIs to parallelize sequential applications [7]. OpenMP [8] is one of the popular APIs for this purpose. This is a well-known approach suitable for applications running on shared memory systems. OpenACC [9], similarly to OpenMP, allows extending sequential codes with directives and parallelization using GPUs. OpenCL [10] is an API allowing writing parallel programs for both multicore CPUs and GPUs with multiple threads running NDRange kernels. CUDA [11], focuses on NVIDIA GPUs Programming productivity, performance, and energy consumption when using OpenCL, OpenACC, OpenMP, and CUDA are discussed in [12]. Message Passing Interface (MPI) [13] allows to parallelize sequential applications for cluster environments and provides API for interprocess communication. For applications written in the .Net environment, we can often observe the use of Task Parallel Library to add parallelism and concurrency to the solution [1]. This approach will also be discussed and compared to the actor-model described later with the Akka toolkit [2].

3 Original Sequential Application

The original sequential application is an NLP (Natural Language Processing) application for analyzing an online dictionary of foreign words and phrases[2]. The dictionary consists of over 14,000 HTML files. Each file consists of one or several entries. There are over 30,000 entries in this dictionary.

[2] https://www.slownik-online.pl/.

In order to avoid frequent Internet access, the files have been downloaded once and stored locally. After processing by the HTML parser, the files have been divided into entries, which have been saved in the relational database for further analysis. Both the results of the entire analysis and the results of individual stages are recorded in the same database.

The NLP component of the application became the subject of our interest. The NLP algorithm used is well known [14] and can be described as a processing pipeline [15] (Fig. 1). First, the text is divided into paragraphs and sentences (corresponding to our dictionary entries). This has already been done while storing entries to the database. Then entries go to the process of tokenization, during which token sequences are generated. Tokens correspond to words and punctuation marks. Usually, in the next stage, tokens are subject to the process of parts of speech (POS) recognition, and then - parsing. In our application, the POS tagging step has been omitted, since dictionary entries have their own formal syntax. Instead, the correction step was added, during which syntax errors made by the authors of the dictionary are removed.

However, not all syntax errors can be identified and removed at this early stage. Some of them are revealed later - at the stage of tokenization, and some others - at the stage of parsing. Therefore, the process requires relapses. Errors detected at the parsing stage are returned to the tokenization step, which has to produce a different sequence of tokens. Sometimes it is required to return to the error correction stage.

The algorithm of the original application has been implemented in C# using the Visual Studio IDE and the local MS SQL Server database.

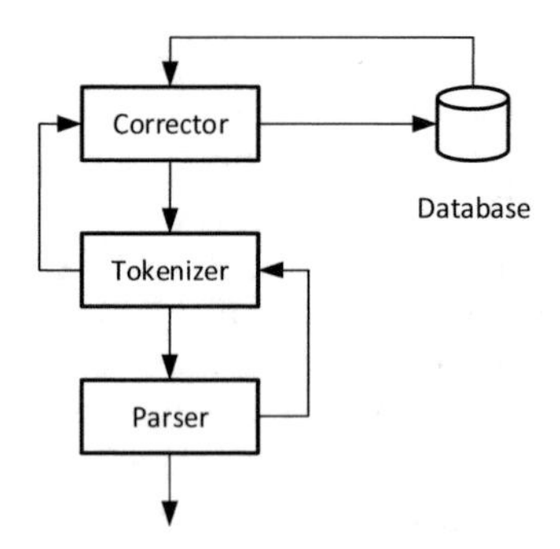

Fig. 1. Dataflow for NLP in the original sequential application

4 Parallelizing the Sequential Application

The most-seemingly natural way of parallelizing the application for a parallel environment would be to process all dictionary entries in parallel independently - in separate processes run on separate cores and machines. However, we parallelized our application to run in a single node environment available for an average scientist - such as a laptop with an Intel i7 processor – where the number of threads possible for parallel processing is rather small when using typical multi-core CPUs.

4.1 TPL Approach

In the first approach, we used the solution based on the *Task* class, which represents the operation performed asynchronously (although not necessarily in parallel). Each task is associated with a separate thread, which is placed in the *thread pool*. The thread pool is equipped with a core load balancing mechanism that matches the number of threads to run to the processor's capabilities.

TPL does not support synchronization between threads, so it is suitable rather for pipelining and the divide-and-conquer model. In both models, it is enough to take the result after the asynchronous operation has ended.

Since our application includes feedback loops, we introduced an additional revision step to simplify the algorithm. in which we grouped backward tokenization and backward correction operations. We defined four tasks (correction, tokenization, parsing, revision) in the processing pipeline (as shown in Fig. 2). We put buffer variables between the tasks to pass the results of one task to be processed by the second task.

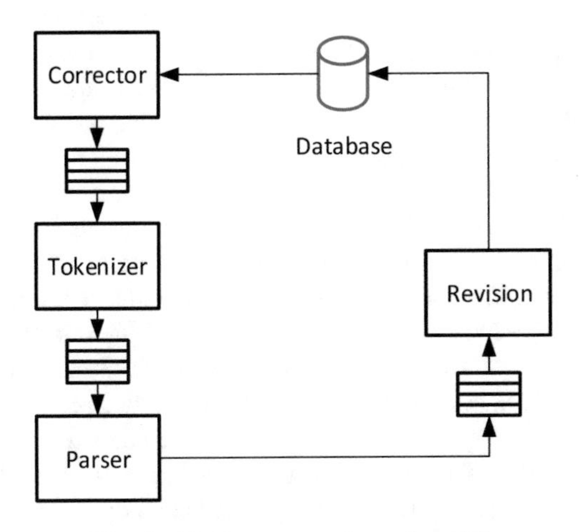

Fig. 2. Pipeline processing using TPL

We implemented the TPL algorithm using a common TaskFactory class instance with a StartNew() operation. Each StartNew() invokes a specific operation (Corrector, Tokenizer, Parser, Revision). These operations take the results of the previous stages which are stored in buffers. The whole algorithm was implemented without using "async" or "await" keyword, as we used a Task. WaitAll() operation to suspend the main thread until all the tasks finalize their processing (see below code sample).

```
var taskFactory = new TaskFactory();
var correctorStage = taskFactory.StartNew(
  ()=>Corrector(fileEntries.Count,correctorBuffer));
var tokenizerStage = taskFactory.StartNew(
  ()=>Tokenizer(correctorBuffer,tokenizerBuffer));
var parserStage= taskFactory.StartNew(
  ()=>Parser(tokenizerBuffer,parserBuffer));
var revisionStage= taskFactory.StartNew(
  ()=>Revision(parserBuffer));
Task.WaitAll(correctorStage, tokenizerStage,
             parserStage, revisionStage);
```

4.2 Approach Using the Actor Model and Akka Framework

The actor model was created to facilitate synchronization between the separate threads. Programmers writing parallel applications in C# usually use the classic *lock* instruction to describe the critical section. In many situations this is the best and easiest way. Unfortunately for large and complex systems, maintaining such code is very difficult, laborious and susceptible to deadlocks.

In the actor model we do not need to use any locks as the code in each actor class is executed only by one thread. Critical sections are therefore replaced by asynchronous messages. If one thread wants to read or change the state of other thread, it sends an asynchronous message. The most difficult part of the multi-threading is to modify the shared state. Using actor model, there is no need for such modification as a shared state does not exist. Each actor works independently of each other. If one actor's data is needed by another, the data is sent as non-modifiable asynchronous message.

The actor itself is an object that has an inbox, state and behavior. It communicates with other actors within the system using asynchronous, non-blocking messages (Fig. 3). Different types of messages are queued in the inbox (for remote actors the data types must be serializable) and they are processed in series.

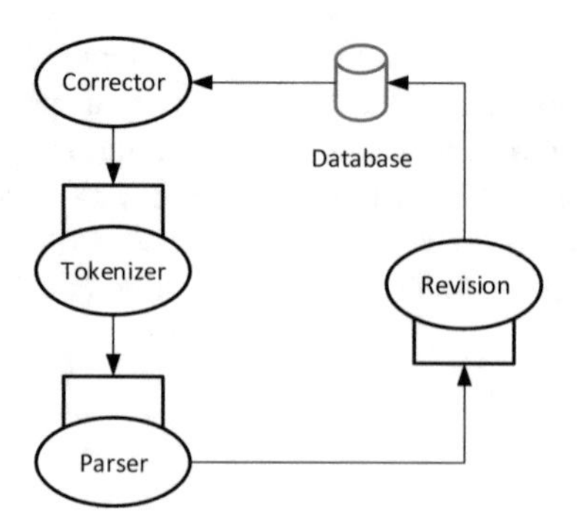

Fig. 3. Using Akka actors model

Akka is an open-source, free toolkit implementing an actor model. It was originally written in Scala on Java Virtual Machine, but a .NET implementation is also available.

Using Akka in our application, we created appropriate actor classes for our tasks. Actors among themselves submit indexes of data records ready to be processed. Below we can see the sample implementation of one of the actors.

```
public class CorrectorActor: ReceiveActor
{
  public IActorRef Tokenizer;
  public CorrectorActor(IActorReftokenizer)
  {
    Tokenizer = tokenizer
    Receive<FileEntryIndex>(index =>
    {
      Corrector.TryCorrectEntry(fileEntry[index]);
      Tokenizer.Tell(index);
    });
  }
}
```

FileEntryIndex is a class representing the message which the actor receives. If someone sends a specific message to this actor, it starts his work. Corrector is an object derived from the sequential version of the application and contains the logic responsible for dictionary entry correction. The delivered message contains an index of this entry. After appropriate actions, finally a new message is sent to the next module. Thanks to this, the actor responsible for Tokenizer stage will receive information about work for itself and Corrector will handle the next message at that time. In this simple way we create actors responsible for each module. Then we move on to our main methods and initialize the Akka system and actors.

```
ActorSystem = ActorSystem.Create();
IActorRefcorrectorActor=ActorSystem.
  ActorOf(props,"correctorActor");
for(var i=0; i<fileEntry.Count; i++)
{
  correctorActor.Tell(i);
}
```

Here we send the data record index to the actor instead of immediately referring to the method responsible for the stage (as we did in the sequential version).

In addition, we are able to define how many instances of each actor we want to create. This means that we can, for example, create several instances of actors responsible for the first stage, the second and so on.

At this point, the system responsible for the actors will prepare as many Corrector actor instances as we define in the numberOfActors variable. The user also has the option to apply the algorithm to load-balance which will be used by the system. For load-balance, we used the round-robin algorithm shown in the code below.

```
var props = Props.Create(
  ()=>new CorrectorActor(tokenizerActor))
  .WithRouter(newRoundRobinPool(numberOfActors));
var correctorActor = ActorSystem.ActorOf
  (props, "correctorActor");
```

Akka guarantees that operations inside an actor algorithm are executed sequentially, so transferring the logic of each module of the existing sequential version of the application does not pose any major problem.

5 Performance Tests

This chapter presents performance results of selected solutions tested in two environments that are described in the next section. It is worth noting that we measured only include the duration of calculations. The time used for reading and writing operations to the database has been omitted.

5.1 Test Environment

Tests have been carried out on two different machines. The first is a typical laptop with a quad-core processor. Its parameters were as follows:

- Operating system: Windows 10 Pro 64
- Processor: Intel Core i7-6820HQ (2.70 GHz, 8 MB L3 cache, 4 cores)
- Memory: 16 GB DDR4-2133
- Frameworks: .NET Core 2.0, Akka 1.3.2.0

In order to verify the scalability of the prepared solutions, the applications were also launched on a cluster node with many cores. Calculations were carried out at the TASK Academic Computer Center in Gdansk. The node parameters are as follows:

- Operating system: Ubuntu 16.04 64
- Processors: Intel Xeon Processor E5 v3 @ 2,3 GHz, 12-core (Haswell)
- Memory: 256 GB RAM DDR4
- Frameworks: .NET Core 2.0, Akka 1.3.2.0

5.2 Test Results

Figure 4 presents a comparison of application execution times using various numbers of actors to the versions using TPL and the sequential approach.

As we can see from the results, the best time for a quad-core machine is 29 s when using 28 actors. From the results on the cluster node, we see that for the sequential version and using fewer threads or actors, the results are slightly worse than on the first

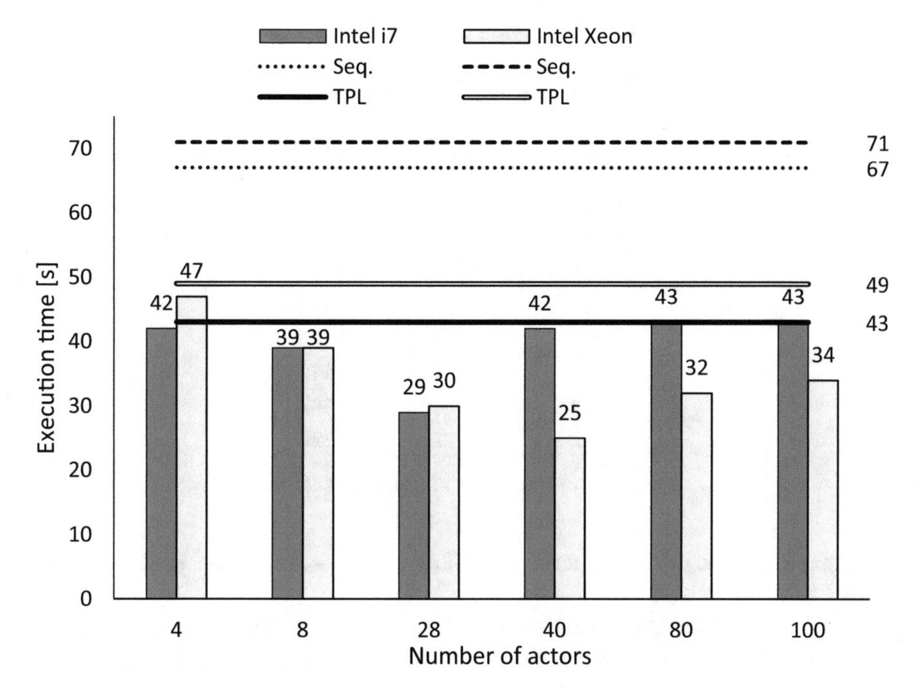

Fig. 4. Performance comparison

machine. This is due to the fact that the Intel Xeon E5 is a processor with a lower clock frequency the Intel Core i7.

With the use of higher number of cores (for 40 instances of actors), the application was executed in 25 s. This observation shows that the parallel approach is faster than sequential roughly 2.3x for Intel i7 and 2.8x for Intel Xeon. When performing parallelization, the code architecture was not interfered with, but only new modules were added using the existing sequential code elements. This makes the results satisfactory considering that we could easily convert a sequential application to its parallel version.

In Fig. 5, we can observe the CPU (Intel Core i7) load at a given time. In total, there are 8 logical processors (4 physical cores and 4 virtual cores due to Hyper

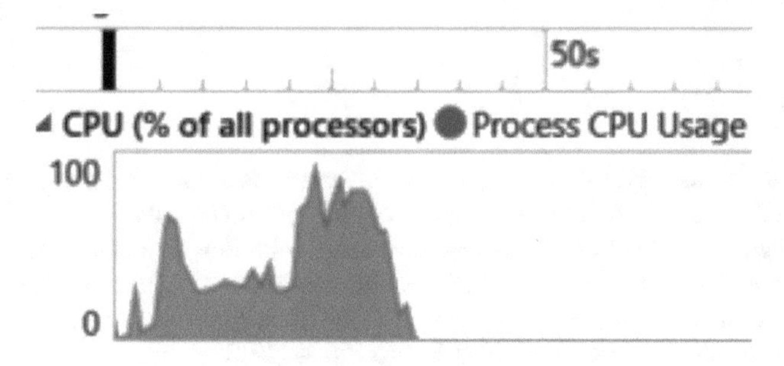

Fig. 5. CPU usage on Intel Core i7

Table 1. Intel Xeon CPU usage performance

CPU index	Avg usage [%]	CPU index	Avg usage [%]
0	86.4	12	73.4
1	85.6	13	73.1
2	73.4	14	50.6
3	73.1	15	49
4	72.3	16	50.5
5	68.5	17	49.1
6	71.4	18	44.2
7	70.5	19	42.5
8	71.1	20	45.5
9	67.4	21	38.4
10	68.2	22	32.7
11	68.5	23	34.6

Threading Technology). Akka used all available cores as presented in the aforementioned figure.

The level of processor utilization was also measured on the cluster node. Table 1 shows the average load of each logical processor during application execution.

6 Summary and Future Work

In the paper we showed that it is possible to parallelize an existing NLP application in an easy way using the actor model. The results showed that the obtained speedup is roughly 2.3 to 2.8 times using the actor based approach within a multi-core machine.

Programming with actor model separates the programmer from using threads. This simplifies writing efficient programs using the available processor cores. This is an important issue, because the computing power of new computers relies mainly on the number of cores. In the next step we plan to profile execution at a finer level to investigate bottlenecks. This will allow to model overheads and scalability of the actor model and the Akka implementation into our modeling and simulation environment called MERPSYS and perform performance tests on various CPUs from the MERPSYS database [16].

References

1. Leijen, D., Schulte, W., Burckhardt, S.: The design of a task parallel library. SIGPLAN Not. **44**(10), 227–242 (2009). https://doi.org/10.1145/1639949.1640106
2. de Castilho, R.E., Gurevych, I.: A broad-coverage collection of portable NLP components for building shareable analysis pipelines. In: Proceedings of the Workshop on Open Infrastructures and Analysis Frameworks for HLT (2014)
3. Wyatt, D.: Akka concurrency. Artima Incorporation (2013)

4. van Lohuizen, M.P.: Parallel processing of natural language parsers. In: Parallel Computing: Fundamentals and Applications (200)
5. Van Lohuizen, M.P.: Effective Exploitation of Parallelism in NLP (1999)
6. Lai, C.Y.: Efficient parallelization of natural language applications using GPUs. vol. Technical Report No. UCB/EECS-2012-54. University of California at Berkeley, Electrical Engineering and Computer Sciences (2012)
7. Czarnul, P.: Parallel Programming for Modern High Performance Computing Systems. CRC Press (2018). ISBN 9781138305953
8. Chandra, R., Dagum, L., Kohr, D., Maydan, D., Menon, R., McDonald, J.: Parallel programming in OpenMP. Morgan kaufmann (2001). ISBN 1-55860-671-8, 9781558606715
9. Wienke, S., Springer, P., Terboven, C., an Mey, D.: OpenACC—first experiences with real-world applications. In: European Conference on Parallel Processing, pp. 859–870. Springer, Heidelberg, August 2012. https://doi.org/10.1007/978-3-64232820-6_85
10. Stone, J.E., Gohara, D., Shi, G.: OpenCL: a parallel programming standard for heterogeneous computing systems. Comput. Sci. Eng. **12**(3), 66–73 (2010). https://doi.org/10.1109/MCSE.2010.69
11. Nickolls, J., Buck, I., Garland, M., Skadron, K.: Scalable parallel programming with CUDA. In: ACM SIGGRAPH 2008 Classes, p. 16. ACM, August 2008. https://doi.org/10.1145/1365490.1365500
12. Memeti, S., Li, L., Pllana, S., Kołodziej, J., Kessler, C.: Benchmarking OpenCL, OpenACC, OpenMP, and CUDA: programming productivity, performance, and energy consumption. In: Proceedings of the 2017 Workshop on Adaptive Resource Management and Scheduling for Cloud Computing. ACM, July 2017. https://doi.org/10.1145/3110355.3110356
13. Gropp, W.D., Gropp, W., Lusk, E., Skjellum, A.: Using MPI: portable parallel programming with the message-passing interface, vol. 1. MIT Press (1999). ISBN 0262527391, 9780262527392
14. Collobert, R., Weston, J., Bottou, L., Karlen, M., Kavukcuoglu, K., Kuksa, P.: Natural language processing (almost) from scratch. J. Mach. Learn. Res. **12**(Aug), 2493–2537 (2011)
15. de Castilho, R.E., Gurevych, I.: A broad-coverage collection of portable NLP components for building shareable analysis pipelines. In: Proceedings of the Workshop on Open Infrastructures and Analysis Frameworks for HLT, pp. 1–11 (2014)
16. Czarnul, P., Kuchta, J., Matuszek, M., Proficz, J., Rościszewski, P., Wójcik, M., Szymański, J.: MERPSYS: an environment for simulation of parallel application execution on large scale HPC systems. Simul. Model. Pract. Theory **77**, 124–140 (2017)

The Registration of Digital Images for the Truss Towers Diagnostics

Rafał Gasz[1(✉)], Bogdan Ruszczak[1], Michał Tomaszewski[1], and Sławomir Zator[2]

[1] Faculty of Electrical Engineering, Automatic Control and Computer Science, Opole University of Technology, ul. Prószkowska 76, 45-758 Opole, Poland
r.gasz@po.opole.pl

[2] Faculty of Production Engineering and Logistics, Opole University of Technology, ul. Sosnkowskiego 31, 45-272 Opole, Poland

Abstract. The paper presents a semi-automatic image registration method. It includes a discussion of the most frequently used image registration methods and an analysis of the effectiveness of the registration of images of the tested object. The tested object was a power line supporting structure. For the purposes of the study, a series of photographs of this object were taken and acquired with images generated on the basis of a 3D model of the scene in a CAD environment. The main objective of the image registration performed was to allow the comparison of the actual condition of the structure (e.g. resulting from vision inspection performed by means of flying machines) with the design documentation.

Keywords: Image registration · Registration evaluation · Power line diagnostics
Truss structures

1 Introduction

Image registration, commonly called overlaying of images, is the process of transferring several sets of image data to a single coordinate system. Image registration is often used in medical imaging, computer vision, automatic detection in military systems or processing and analysis of data and images from satellites [1–3]. Image registration allows you to compare or combine the data obtained in different ways, e.g.: a few photographs taken with different acquisition parameters, data from multiple sensors, data obtained at different time points or data received from several points of observation, etc. Image registration uses control points generated automatically or entered manually. The image registration application performs transformations on a single image, so that the main features are aligned with another image or multiple images a single image, so that the main features are aligned with another image or multiple images.

Studies concerning the diagnostics of high voltage power line supporting structures require an assessment of their condition. One of methods used is vision inspection, which requires the precise adjustment of the possessed documentation. Such a comparison of images allows one to identify the actual condition of the structure with model images generated on the basis of the design documentation in order to detect various types of

L. Borzemski et al. (Eds.): ISAT 2018, AISC 852, pp. 166–177, 2019.
https://doi.org/10.1007/978-3-319-99981-4_16

abnormalities, such as missing elements, deformation, displacement or inclination of the structure.

Image registration, even though it is the most common [1, 3] method used for the purposes of visualization and comparison of images from multiple acquisition devices, especially in medical diagnostics, is also widely used in other areas of science, including remote sensing and computer vision. Despite the wide range of applications in which image registration is used, a difficult task is to develop a general optimized method for all applications. For example, in astrophotography, the alignment and registration of images is applied to increase the signal to noise ratio for faint objects. Image registration is used to draw up a time log of events, such as the rotation of planets around the parent star. This technique can also be used for images with different sizes to allow merging, e.g. of photos taken with different telescopes or lenses. In turn, the registration of medical images to obtain data of a patient at different time points to detect changes or to monitor cancer often also involves the need to take into account deformations of the object caused by breathing, anatomical changes etc. Image registration is also an important part of creating a panoramic image (stitching).

2 The Methods of Image Registration Digital Images

Image registration methods can be classified according to various criteria, depending on the method of taking images that will be subjected to registration and the type of transformations in the process [1, 2, 9].

Depending on the method of image registration, the methods can be divided into 4 groups.

In the first group, images of the same scene are taken from different positions. The aim is to obtain the widest possible 2D view or 3D representation of the observed scene model. An example of this method may be remote sensing-mosaic, which consists in producing one picture of the entire test area with the use of several images (pieces of area).

The second group includes methods of performing operations on images taken at different times (multilateral analysis) in the same or similar conditions. The objective of the methods of this group is to find and evaluate changes in the scene, which appear between successive moments of acquisition. Examples of their use may be the monitoring of global land-use or landscape planning. The main advantage of their use is the ability to automatically detect changes in images, which is used in CCTV systems for monitoring, tracking movement, or in medicine when observing the progress of neoplastic patients.

The third group of methods uses multimodal analysis. It is the use of various sensors or cameras to gather information about one object [1, 9]. The aim is to integrate information from different sources in order to obtain a more complex and detailed representation of the scene. These types of images are taken for the purpose of remote sensing, panchromatic imaging and medical imaging [1, 2, 9].

The fourth group includes methods combining information about the model obtained by conventional methods (photographs) with computer models, e.g. satellite images with computer maps. This type of image registration is used in remote sensing for placing aerial or satellite data on the map or layers of GIS spatial databases, matching templates in computer visualizations in real-time, automated quality control of products and medical imaging to compare the image of the patient with his/her digital anatomy (classification of patterns).

Due to the diversity of recorded images, it is impossible to design a universal method that can be used for all image registration purposes [2]. The registration of each image should take into account, not only the established type of geometric deformation between images, but also radiometric deformation and distortion, noise, the accuracy of the image registration and data dependent on the features of the application.

Another criterion might be the division of image registration methods according to the type of geometric transformation used [17]. Depending on the particular image analysis [17], there are two groups of image registration methods [1]:

- global – retaining the position of individual pixels relative to each other allowing linear transformations, also known as rigid transformations, which include operations of rotation, scaling, and the like.
- local – allowing deformation of the image overlaid in order to better align individual objects, but without moving pixels relative to each other, making the methods faster and recommended where the position of the individual objects to each other has not changed, but only the observation point of objects changes.

Global algorithms are used in relation to images acquired with the use of various devices or where the object was observed from various locations or the imaging angle has changed. In contrast, local algorithms work best if additionally the shape of the object being imaged has changed. Regardless of the manner of the overlaying of images, the use of a global approach to image registration requires appropriate implementation so as to rely on the predetermined characteristics of the image or the all pixel intensity values. Among the characteristics helpful when overlaying images, one can distinguish points and lines marked by the user or generated by a dedicated algorithm for this purpose. The result of the edge detection or pre-detection of objects in both images may also be useful. However, in the approach based on pixel intensity, the full information about the image is used. In this approach, it is possible to impose a more accurate overlaying of images, but it simultaneously requires more computing power [2].

Image registration methods can also be classified based on the level of automation which they provide. In relation to this criterion, one can distinguish interactive, semi-automatic and automatic methods. Interactive methods provide tools for the manual setting of images reducing the subjectivity of users by performing some key operations automatically, while giving the user the ability to control the image registration process. The use of the semi-automatic method requires more steps, but the user has the possibility to verify the correctness of image registration. Automatic methods do not require user interaction – all the stages of image registration are performed unattended.

3 The Analysis of the Supporting Structure on the Basis of the Method of Virtual Images

The method of virtual images, as discussed in detail in [4, 5], is to check the completeness of the power line supporting structure with the use of the methods of analysis of digital images. It allows for an automated preliminary diagnosis and selection of structures for further detailed analysis. The proposed method uses images of supporting structures and virtual images, i.e. obtained in a virtual environment as a result of rendering the structural 3D model of the same type with a specific position of the observer. It is based on a comparative analysis of the current condition of the structure, as reflected in the images, with the pattern obtained from the model generated in the CAD environment.

The comparison of an image of the real object with a virtual image generated in the CAD environment provides information about all the missing, redundant or deformed structural elements. A comparative analysis of the two images must be preceded by the process of their image registration. Unfortunately, the object shown in the real photo will not be identical with the object shown in the virtual image due to: errors, lens distortion and perspective, wrong location of the center of the matrix, inaccurate calculation of the camera position in space, or for other reasons. Potential errors have a big influence on the final effect of the proposed method. Its usefulness is determined by the registration of a real image of the structure with its virtual photo with the highest accuracy. The article presents the implementation of selected methods of digital image registration and analysis of the results obtained.

4 The Experimental Part

The studies described in the article were carried out on the basis of a photo of the model supporting truss structure type H52. The model was printed on a 3D printer according to as-built documentation in order to take laboratory photos allowing to check the operation of the method presented above. The supporting structure is made in a 1:32 scale. For the study, the real photos were taken in laboratory conditions using homogeneous light, on a white background. With this approach, the authors avoided the problems associated with heterogeneous background and its possible filtration. The virtual image used in the image registration process was generated in a CAD environment on the basis of the 3D model, the one used to print the model structure on a 3D printer.

Figure 1 shows the different stages of the experiment. At the beginning, two digital images are prepared. The first is a photograph of the model structure of the powerline pole produced in a laboratory. The second image is a digital image rendered from the 3D model of the same structure after defining the appropriate scene, being a reflection of the conditions in which the first image was taken (camera-object system, with the designated coordinates of the location of the object relative to the camera, lens focal length parameter and the center of the photo).

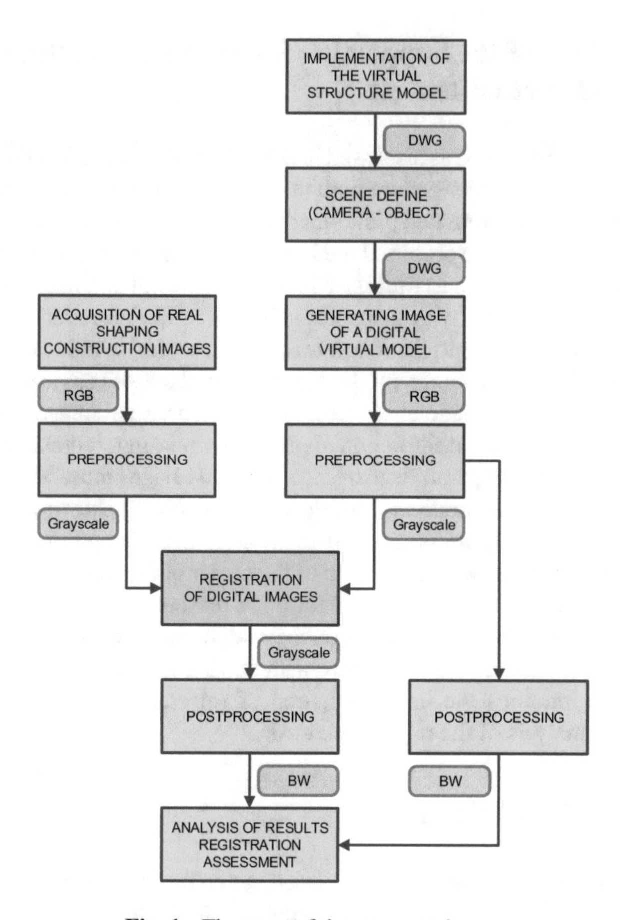

Fig. 1. The steps of the test experiment

Then the two images are subjected to preprocessing, whose main task is conversion to grayscale and contrast enhancement. The resulting pair of images is subjected to the image registration process using the selected methods. This results in a version of the image which after postprocessing (binarization + selected morphological operations) is compared (subtract) with the photo generated based on the 3D model (also subjected to similar postprocessing) to evaluate the effectiveness of the applied image registration.

To evaluate the effects of image registration, three measures of the study were selected: Jaccard Index, Accuracy and False Discovery Rate. Jaccard Index, often called Intersection over Union (IOU), is used to compare the similarity of sets [14] and is defined as the quotient of the power of the common part of the sets and the power of the sum of these sets. Accuracy (Acc), a measure determined on the basis of the confusion matrix, is used in assessing the performance of classifiers. The data labelled as positive and negative are subjected to classification, which assigns a predicated positive or negative class [16]. False Discovery Rate (FDR) is the expected proportion of errors for all incorrect assessments of the results [15].

5 The Test Results

The study analyzed semi-automatic image registration during which the operator specifies points indicating the pixels reflecting the same objects on pairs of images (select pairs of corresponding control points in both images) – the image presenting the supporting structure and the virtual image further referred to as Control Points (CP). On the basis of the coordinates of these points in accordance with the method of transformation, the image is subjected to image registration.

5.1 The Comparison of Different Transformation Methods Used in the Image Registration Process

In the experiments carried out, 3 methods of transformation used in the image registration process were compared: non-reflective similarity (non-reflected transformation of similarity), affine and projective.

Non-reflective similarity transformation can include rotation, scaling and mapping. After the transformation, the shapes and angles of objects are retained, and parallel lines remain parallel. Straight lines in the object are also preserved. The transformation is a subset of the transformation of similarity. It has four degrees of freedom and requires two pairs of points correlated in the images [13]. Affine transformation is a linear mapping method that retains the points, straight lines and planes. After transformation, sets of parallel lines remain parallel. The affine transformation technique is typically used for the correction of geometric distortions or deformations which occur at non-ideal angles of the camera. For example, satellite images use affine transformations to correct wide-angle lens distortion. Transforming and merging images with a large flat coordinate system are necessary to eliminate distortion [11, 12]. They allow for easier interaction and calculations that do not require taking into account image distortion. Projective transformation displays a three-dimensional geometric object on a 2D plane [11] and can be perceived as a general example of projective transformation. Strictly speaking, it provides transition from one plane to the other, but if one identifies the two

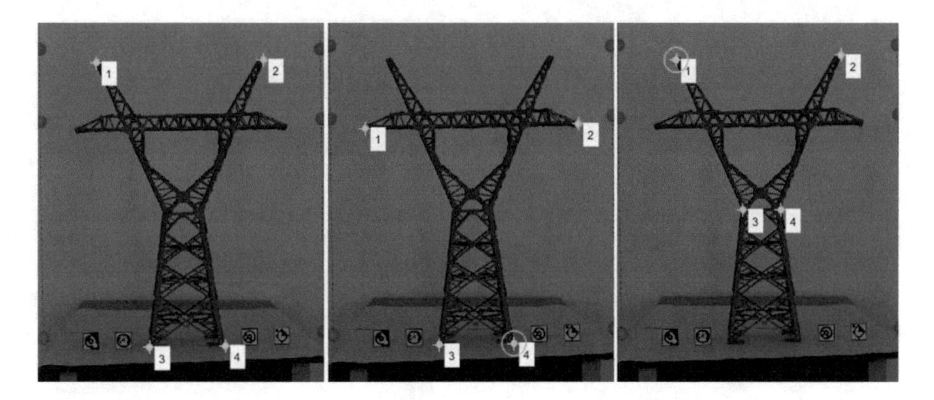

Fig. 2a. Example images prior to image registration – the same number of CPs in different places on the object

planes by e.g. establishing the Cartesian system in each of them, one can obtain the projective transformation from the plane to itself.

There were 15 sets of CP in images investigated for the purpose of this study. These sets were prepared by being divided into two groups. In the first group, incorporating sets 1–6, 4 CPs were used, and their location on the object was changed (Fig. 2a). In the second group (sets 6–15), the number of CPs was increased successively by one – 4 CPs in set 6 and 12 CPs in set 15, respectively (Fig. 2b).

Fig. 2b. Examples of images before image registration – increasing the number of CPs (respectively 4, 5 and 6 CPs)

As can be seen in Fig. 3, the best result for the analyzed sets of images in the whole experiment was recorded for the projective method in set 11, with accuracy of 0.987.

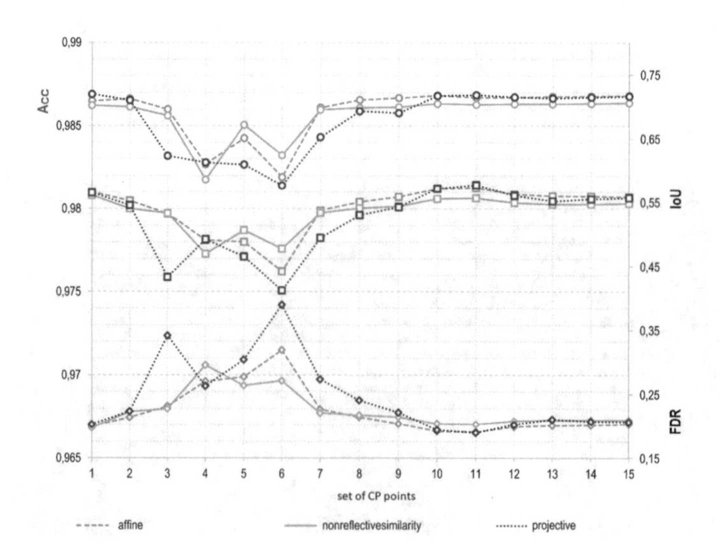

Fig. 3. Comparison of the average value of the ACC parameter for 15 CP variants using different methods of image registration

For the mentioned method the smallest IOU value of 0.189 was also achieved for this set.

For sets 3–6, in which the CPs are located in the central part of the object, the lowest image registration accuracy was obtained. For CPs located at small distances from each other (in the image) the results were the most disappointing. Increasing the number of points has a positive impact on the effectiveness of image registration performed to a certain threshold. The effectiveness of increasing the number of CPs is obtained (for all methods) with 8 points or more (set 11). The best result for the image registration operation was also achieved for 8 points.

The location of points has a large influence on image registration. The best results were obtained for object outermost CPs. Slightly better results, in comparison with other methods, were obtained using the affine method.

5.2 The Impact of the Location of the Image Registration Device on the Image Registration Quality

The location of the image registration device relative to the object has a large influence on the parameters determining the quality of image registration. In Fig. 4, there are two different cases of imaging: the front view and the side view, for which an image registration analysis was performed. The results are presented in Fig. 5.

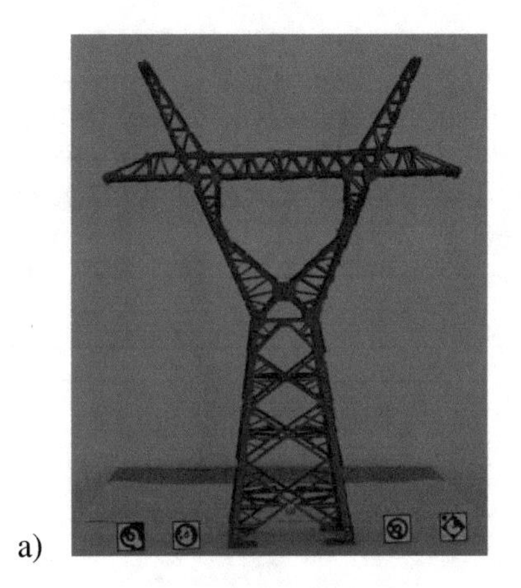

Fig. 4. Analyzed scenes: (a) front view, (b) side view

Better results were obtained when the camera was in front of the object (front view). The result may be influenced by the number of elements (angle of the truss) to be aligned. In the case of imaging from the side (side view), there is a greater number of elements covered by the camera. Worse results of image registration in this case are also caused by the number of shadows inside the object. In addition, the image registration error is intensified by the accuracy of the camera view setting in the virtual scene.

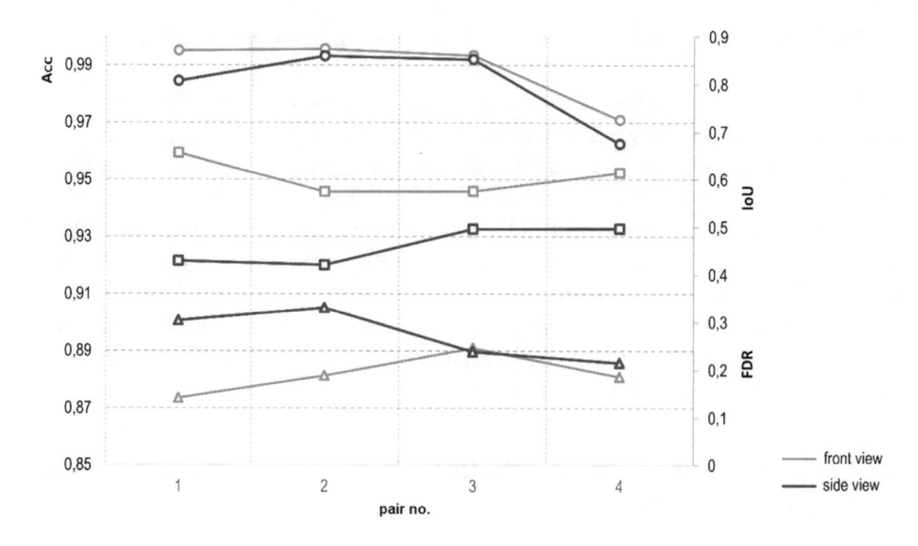

Fig. 5. Comparison of Acc for object views – front view, side view

5.3 The Impact of the Use of Correction on the Final Result of Image Registration of the Supporting Structures

This method described in Sect. 5.3 is to obtain a list of the missing elements of the power line supporting structure. The image obtained after image registration and comparing the photo of the real object and the image generated from the virtual 3D model contains a number of isolated pixels, which are among other things the effect of imperfections of the image registration process. In order to remove them (and leave only larger, dense blocks of pixels – gaps in the structure) the image formed after the comparison of input

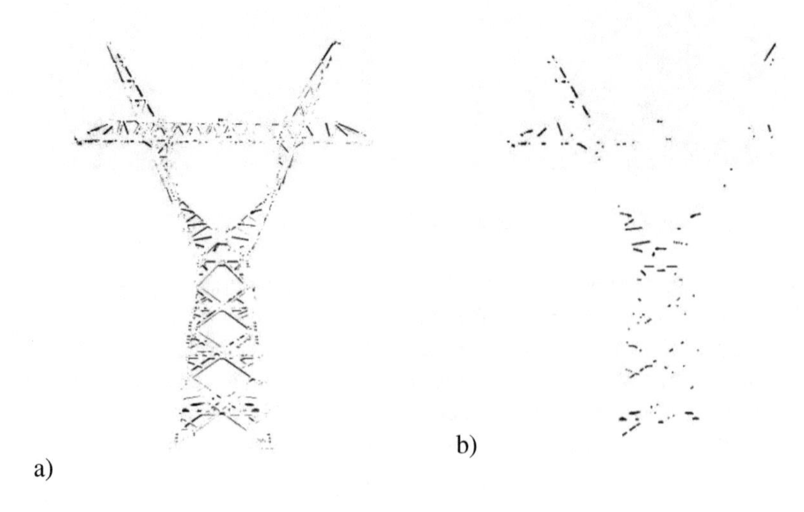

Fig. 6. The result is: (a) before the correction, (b) after the correction

images is corrected. It is made with the use of two alternating morphological operations – erosion and dilation – using a quadratic transformation matrix (Fig. 6).

The effects of the correction are shown in Table 1. It contains a comparison of the FDR values obtained and the corresponding number of pixels before and after the correction. After this operation, the images obtained contain less isolated pixels, which results in greater possibilities of classification of other larger groups of pixels (corresponding to the missing elements of truss).

Table 1. Impact of correction on the image registration quality

	Set	1	2	3	4	5	6	7	8
Before	fdr	0,141	0,304	0,188	0,33	0,246	0,238	0,185	0,214
	px	12942	43718	21837	40938	23990	20541	84958	83680
After	fdr	0,062	0,249	0,131	0,267	0,187	0,147	0,155	0,169
	px	5234	33328	14208	30256	16970	11325	68950	62168

As shown in Fig. 7, the best image registration results, also after the correction, were obtained with the affine method. On average, improvement of the IoU parameter was approximately 0.03.

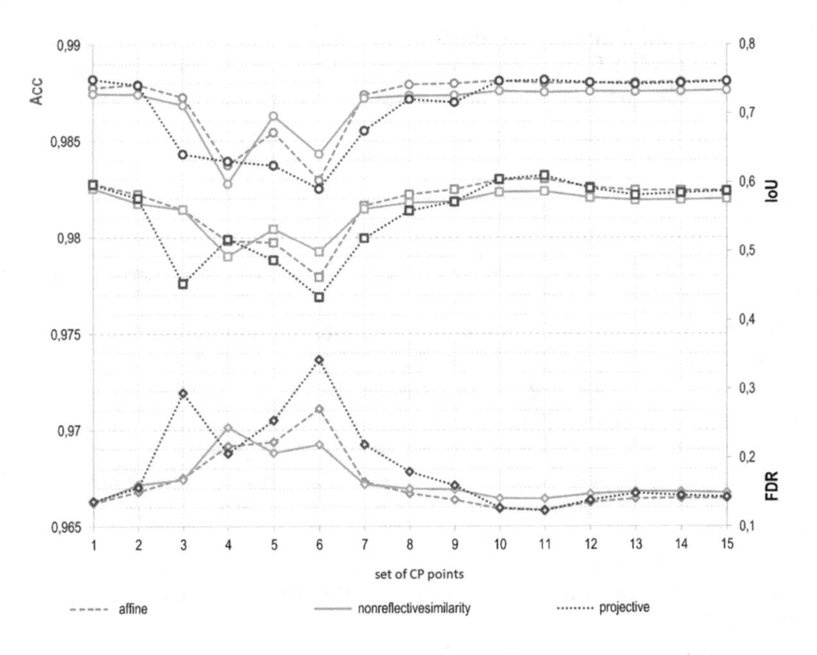

Fig. 7. Comparison of the average value of the accuracy parameter for 15 CP variants using different methods of image registration after correction

6 Summary

The aim of the experiment was to compare semi-automatic methods for image registration on the example of a truss structure. It may be noted that the designation of CPs has a significant effect on the performance of the process itself. The appropriate placement of the CPs is important – the most extreme elements of the recorded images, placing CPs close to each other or in the central part of object. This has allowed one to obtain results (IoU) at the level of 0.41–0.49. Large differences in results were obtained by increasing the number of CPs. The use of more than 8 CPs for image registration causes no further improvement of the efficiency of image registration. During manual image registration, satisfactory results were obtained with IoU of 0.56–0.57 (analysis without correction), and 0.57–0.60 (analysis following correction). Studies have shown that the choice of the appropriate image registration points has a significant impact on the final result of the process. Better results were obtained in the case of the front view.

The use of correction resulted in the expected effect of correcting isolated pixels (inaccuracy of image registration). The correction allowed one to improve the FDR results.

In the study, the possibility of using the automatic image registration method was also analyzed. Much better results were definitely obtained, however, for manual registration. During the testing of the automatic image registration mode, which among others used descriptors of SURF local characteristics, only in some cases were the correct results obtained. For most of the studied cases, we failed to properly perform image registration.

References

1. Zitová, B., Flusser, J.: Image registration methods: a survey. Image Vision Comput. **21**(11), 977–1000 (2003)
2. Vemuri, B., Ye, J., Chen, Y., Leonard, C.: Image registration via level-set motion: applications to atlas-based segmentation. Medical Image Analysis, March 2003
3. Ardeshir, A.: 2-D and 3-D Image Registration for Medical, Remote Sensing, and Industrial Applications. Wiley Press (2005)
4. Gasz, R., Zator, S.: Evaluation of selected elements of a power line with using CAD environment. In: Tomczuk, B., Waindok, A., Zimon, J., Wajnert, D. (eds.) Electrodynamic and Mechatronic Systems SELM 2013, pp. 47–48 (2013). https://doi.org/10.1109/SELM.2013.6562973
5. Zator, S., Gasz, R.: Ocena stanu konstrukcji wsporczych na podstawie zdjęć. Pomiary Automatyka Robotyka 4/2016, pp. 23–26 (2016). ISSN 1427-9126. https://doi.org/10.14313/PAR_222/23
6. Fischer, B., Modersitzki, J.: Ill-posed medicine – an introduction to image registration. Inverse Prob. **24**, 1–19 (2008)
7. Xie, J., Hsu, Y., Feris, R., Sun, M.: Fine registration of 3D point clouds with iterative closest point using an RGB-D camera. In: Proceedings of the 2013 IEEE International Symposium on Circuits and Systems, Melbourne, Australia (2013)
8. Huber, D., Hebert, M.: Fully automatic registration of multiple 3D data sets. Image Vis Comput (2001)

9. Takimoto, R., de Sales, M.: 3D reconstruction and multiple point cloud registration using a low precision RGB-D sensor. Mechatronics **35** (2016)
10. Chen, C., Stamos, I.: Semi-automatic range to range registration: a feature-based method. In: International Conference on 3-D Digital Imaging and Modeling (2005)
11. Goshtasby, A.: Image registration by local approximation methods. Image Vis. Comput. **6**, 255–261 (1988)
12. Karsli, F., Dihkan, M.: Determination of geometric deformations in image registration using geometric and radiometric measurements. Sci. Res. Essays **5**(3), 260–274 (2010)
13. Berman, M., Blaschko, M.: Optimization of the Jaccard index for image segmentation with the Lovász hinge. In: CVPR 2018 (2018)
14. Yoav, B.: Simultaneous and selective inference: current successes and future challenges. Biometrical J. **52**(6), 708–721 (2010). https://doi.org/10.1002/bimj.200900299
15. Alison, S.: Difference Between Accuracy And Precision. Englishtipsdaily.com. Accessed 5 Aug 2016

GAP - General Autonomous Parallelizer for CUDA Environment

Jan Kwiatkowski[1](✉), Dzanan Bajgoric[2], and Mariusz Fras[1]

[1] Department of Informatics, Faculty of Computer Science and Management, Wroclaw University of Science and Technology, Wybrzeze Wyspianskiego 27, 50-370 Wroclaw, Poland
{jan.kwiatkowski,mariusz.fras}@pwr.edu.pl

[2] ARM Norway, Olav Tryggvasons Gate 39-41, 7011 Trondheim, Norway
dzanan.bajgoric@arm.com

Abstract. Writing efficient general-purpose programs for Graphics Processing Units (GPU's) is a complex task. In order to be able to program these processors efficiently, one has to understand their intricate architecture, memory subsystem as well as the interaction with the Central Processing Unit (CPU). The paper presents the GAP - an automatic parallelizer designed to translate sequential ANSI C code to parallel CUDA C programs. The general processing architecture of GAP is presented. Developed and implemented compiler was tested on the series of ANSI C programs. The generated code performed very well, achieving significant speed-ups for the programs that expose high degree of data-parallelism. The results show that the idea of applying the automatic parallelization for generating the CUDA C code is feasible and realistic.

Keywords: Automatic parallelization · Loop transformations
Data-parallelism

1 Introduction

Even though today's processors are parallel in nature, most programs are still written in the traditional sequential manner. It is hard for people to think in parallel and the result of this is that very few programmers write good parallel code. It is even worse – many programmers don't write parallel code at all. This is why automatic parallelization could potentially come to rescue. With automatic parallelization, instead of writing a program for the given architecture, it is possible to use an automatic parallelization to transform sequential code into a parallel one for the given architecture.

Graphics processing units (GPU's) are massively parallel multiprocessors, where each processor consists of hundreds of cores. Such processors are designed to efficiently handle very large workloads in parallel manner. In order to utilize the great parallel power of the GPU, the same set of operations has to be executed over large number of dataset entries. Onwards, operations on different dataset entries have to be mutually independent so that they can run in parallel. Unfortunately, not all problems are data parallel, it means that when executing programs on the GPU we cannot expect that for all of them performance will be improved.

L. Borzemski et al. (Eds.): ISAT 2018, AISC 852, pp. 178–189, 2019.
https://doi.org/10.1007/978-3-319-99981-4_17

Automatic transformation of sequential programs into a parallel form has been a subject of research for computer scientists from several decades, although real parallelism is possible only with the new architectures, which give the opportunity to execute a number of threads at the same time, for example GPU.

One of the first publications which can be recognized as the starting point of this topic is Utpal Banerjees "trilogy of books" [1–3]. The content of them cover most of the problems concerned with automatic parallelization techniques, from introducing the mathematical apparatus through the presentation of different loop transformations and ways of its parallelization to data dependence analysis. The problem of loop transformation is also presented in [4, 5]. The next most important books dedicated to automatic parallelization that represent the overview of the problem are [6, 7]. Their authors focuses on overview of complier techniques concerning with automatic parallelization, vectorization and parallelization of numerical programs for scientific and engineering applications. In the last decade the research concerned with automatic parallelization has changed. Currently more and more often the so-called Polyhedral Model is used. The polyhedral model for compiler optimization is a powerful mathematical framework based on parametric linear algebra and linear integer programming [8, 9]. The paper [10] is dedicated to ANSI C to CUDA C automatic parallelization for affine programs whereas [11] is our own paper where the first experiments with the GAP are presented.

The paper presents the GAP, an automatic parallelizing compiler that accepts programs written in ANSI C and automatically generates CUDA C programs. The compiler takes on the following responsibilities: discovering data parallelism in input programs and finding the segments which are worth paralleling, discovering data dependence constraints, applying program transformations that will expose the data parallelism by reducing the effect of data dependencies and automatically generate CUDA C programs.

The paper is organized as follows. Section 2 presents the structure of the GAP as well as the way how it uses different parallelizing techniques. The next section focuses on the description how the CUDA code is generated and how the compiler is used. Section 4 illustrates the experimental results obtained during testing of the GAP compiler, which results are compared with results presented in [11] as well. Finally, Sect. 5 outlines the work and discusses ongoing work.

2 The GAP Compiler Design

The GAP compiler is written using C++ language with use of open-source Armadillo C++ linear algebra library (http://arma.sourceforge.net/). The advanced operations on matrices is critical processing step and Armadillo turned out to be an excellent solution for this purpose. The library was extended to support features required for algorithms used in the GAP, among others: unimodular matrices, diophantine equations solving, echelon reduction, and diagonalization.

The basic operation of the GAP is to analyze input program code to find loops and to parallelize them. The dependences in loop nests that make efficient parallel processing possible or not (i.e. it is possible that the operations will reference the same memory location throughout the loop nest execution) are discovered with use of the diophantine equations built on loop indexes variables and nest boundaries. On the basis of this analysis, the GAP performs loop transformation to automatically parallelize code.

In order to simplify data dependence analysis and parsing, the GAP assumes several restrictions on the input program listed below.

- Only perfect loop nests with positive unit strides can be used (all statements are contained within the innermost loop). The nests with more than four loops are not supported. Outermost loop must have a scalar integer lower and upper bound (its value should be known at compile time) and inner loops must have a single linear integer function of index variables as lower and upper bound.
- Array subscripts that appear in the nest body have to be linear functions of the index variables. Constant parts have to be variable whose values are known at compile time and that are defined within the function containing the loop nest. It is assumed that memory regions occupied by the two arrays of different names are always distinct. Breaking this assumption by using pointers will not be detected and may lead to unreliable dependence analysis results. The arrays size is required when allocating and populating the GPU memory.
- Branches are supported by assuming that every outcome of the branch will materialize, this can lead to discovering a dependence between statements that actually does not exist which is not critical as it will not lead to invalid transformations.
- Pointer analysis is another important factor that is not included in the current version of the GAP.

The compiler consists of four main modules: Compiler Frontend, Utility Module, Dependence Analyzer and Transformation Engine [11]. The modules provide various routines that are used as building blocks for algorithm in the compiler. The GAP general processing scheme with use of these routines is shown in Fig. 1.

The source code is parsed by Compiler Frontend with the help of Clang library which allows the GAP to support both C and C++ inputs. For each function definition the depth-first-search (DFS) algorithm is used to find all loops in source code representation. Next, each loop is analyzed and all data required for dependence analysis is collected. If the loop nest violates any of the mentioned constraints, processing for this loop is terminated. The loop nest representation (LNR) data includes index variables for each loop (loop nest vector), lower and upper bounds of loop, the list of assignments statements, output and input variables of each statement.

The LNR is used in dependence analysis performed by the Dependence Analyzer. Two different algorithms can be used depending on the type of the loop. For regular and rectangular loops (RR loops) and variables with identical coefficient matrices Uniform Linear Dependence Algorithm (ULDA) is used. This algorithm, with help of calculated unimodular dependence distances fully determines all the dependent iterations in loops without enumeration of all iterations (all iterations that are distant from each other by a given uniform distance are dependent). The General Linear Dependence Algorithm (GLDA) handles more general nests supported by the compiler (i.e. also non-RR loops). It finds dependencies in the loop nest and all pairs of iterations that are mutually dependent. It is performed with use of a system of linear diophantine equations solving

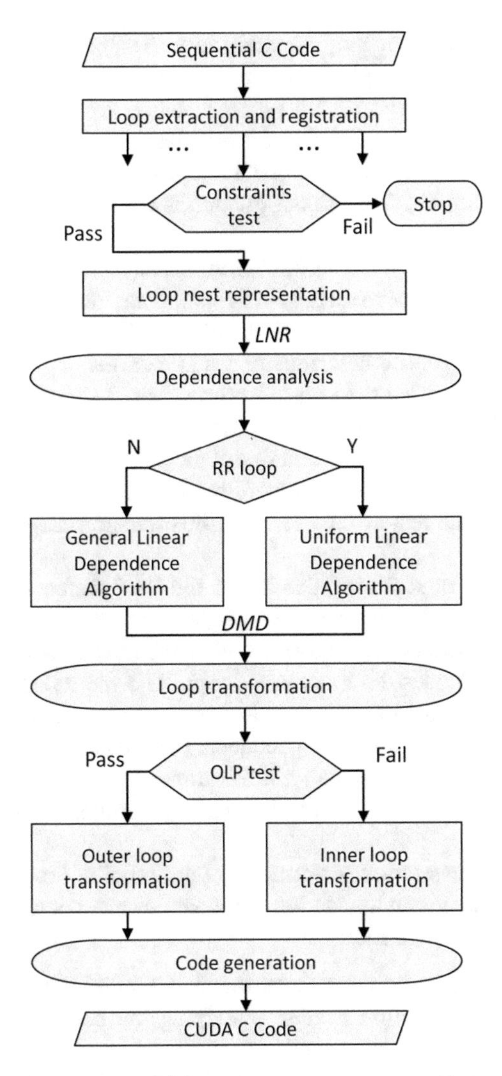

Fig. 1. General process diagram of the GAP automatic parallelizer

procedure based on the echelon reduction algorithm applied to input and output variables of statements in the loop nest. Additionally, the test considers lexical ordering of statements. This algorithm can be very time consuming as it enumerates all the pairs of dependent loop nest iterations.

On the basis of generated dependence model data (DMD), the Transformation Engine module performs loop transformation to parallelize code (if it is possible). First the outer loop parallelization test (OLP test) is performed. OLP is possible only when the rank of the distance matrix is smaller than the number of loops in the nest (refer to [2] for detailed discussion of these and other transformations). Then unimodular outer loop transformation is performed. If OLP is not possible or it produces a single dependence-free loop, the inner loop parallelization (ILP) is applied. The inner loop parallelization transforms the original loop nest by pushing all the dependences up to the outermost loop. The routine uses a set of distance vectors for the original nest (produced by the Dependence Analyser) and it calculates the unimodular transformation matrix that generates the transformed loop nest from the original one while respecting all its dependence constraints. This transformation is always possible for two or more levels deep nests.

In order to fully utilize the GPU, one has to run as many parallel operations per kernel invocation as possible. Thus, the main goal of the loop transformation stage is to produce a nest with as many independent iterations as possible. The other goal is to run as few kernels as possible for the entire nest, ideally a single kernel per entire nest. The GAP uses these goals when selecting the loop transformation. As inner loop parallelization pushes all the dependencies to the outermost loop, the number of CUDA kernels that must be run is equal to the number of iterations of the outermost loop of the transformed nest. If the number of iterations of the outermost loop is minuscule compared to the number of iterations in the rest of the nest, then this transformation may yield a good performance.

This is not the case in general, and this is why the GAP prefers the outer loop parallelization when possible. This transformation produces $m - rank$ dependence-free outer loops (where m is the depth of the nest and $rank$ is the rank of the distance matrix) as well as $rank - 1$ dependence-free innermost loops. The more dependence-free outer loops the better, as this means that all those iterations can be run in parallel. Likewise, the deeper the loop containing all the dependencies the better, as this means that fewer iterations will have to be run sequentially. If the number of iterations to be run sequentially is large, it might be better to perform inner loop parallelization and run multiple CUDA kernels.

The last step is code generation performed by the Compiler Frontend for each translation unit (assuming that given translation unit contains any parallelizable nests). The compiler produces three source files:

- Modified source file corresponding to the translation unit being compiled.
- The CUDA kernel declaration header containing the declaration of each kernel generated for the given translation unit.
- The kernel definition source containing the definition for each generated CUDA kernel.

3 Code Generation and Using the Compiler

The code generator as a part of the Compiler Frontend is in charge of traversing the loop nest's CFG (control flow graph) and generating the final CUDA C code. Information about the type of transformation is passed together with a CFG to the generator. Most of the real-world CUDA C programs consist of five main stages:

1. Allocating the memory on the GPU
2. Copying data from the CPU memory over to the GPU memory

3. Launching kernel that will process the data on the GPU
4. Copying data from the GPU back to the CPU memory
5. Releasing the GPU memory.

To illustrate the code generation assume that we have a transformed triple loop that has all the dependencies in the innermost or outermost loop. Furthermore, the loop nest body contains a statement whose input and output variables are elements of a two-dimensional array. The code generator will generate the CUDA C code in the following steps:

1. Generates the code that allocates the GPU memory and then if allocation fails because the GPU is out of resources, terminates the program.
2. Generates the code that copies a n-dimensional host array to a 1-dimensional GPU array. The data is first copied to the 1-dimensional host array that is then copied in one bulk to the GPU memory. It is much more efficient to transfer large bulk of data at once.
3. Generates the CUDA C kernel that runs all the doall loops of the nest in parallel. The problem is how to map the iterations of these doall loops that are nested together onto the CUDA threads. The GAP will flatten the original double loop (obtained by removing the loop that carries all the dependencies) in such a way that each iteration is handled by the separate CUDA thread. This step will differ depending on which transformation has been applied. If no transformation has been applied then the above describes the overall process of the CUDA C kernel generation [11].
4. Modifies the original nest's statements that are now part of the CUDA C kernel. Loop transformation also provided the lower and upper bounds for the transformed loop nest.
5. Setting up the launch configuration and launching the kernel.
6. Generates the code that copies the data from GPU back to a 1-dimensional host array that is then copied to a n-dimensional destination host array defined by the sequential program.

The rest of the function will remain the same as in the original C program. The only part that changes is the area around the loop nest that is being parallelized. The code before and after in the CUDA C program will still execute on the CPU and its logic will remain intact.

The following steps should be done when using the GAP to generate CUDA C code:

1. Invoke the GAP and pass all ".c files" that should be compiled "./gap source1.c source2.c… sourceN.c".
2. The GAP will analyze each of the translation units separately (".c file" and all included headers), locating functions and loop nests. For example, let's assume that translation unit "sourceK.c" has several loop nests that the GAP decided to parallelize. The GAP will create 3 files for this source:
(a) a single header file containing kernel declarations (each kernel implements a single loop nest),
(b) a single source file containing kernel definitions,

(c) a modified "sourceK.c" source file where loop nests have been replaced with the kernel invocation statements invoking CUDA C kernels implementing the original loop nests.

3. Once the GAP compilation completes, the user would use "nvcc" compiler to compile all the ".cu" files produced by the GAP and any ".c sources" that were not compiled with the GAP to produce a parallelized code.

4 Case Studies

To confirm the usefulness of designed and implemented compiler different experiments have been performed. All measurements were done with a server equipped with Intel(R) Core(TM) i7 CPU, X 980 @ 3.33 GHz processor with Tesla K20c graphics card running under Ubuntu 16.04.1 LTS (RAM ~23.5 GB). The presented results are the average from the series of 10 experiments. Many of the programs used for experiments come from the real world, and the only modifications that were made are to make them fulfill the restrictions of the compiler. The sequential programs were hand-optimized, and some of them were taken from the Parboil Benchmark suite (http://impact.crhc.illinois.edu/parboil/parboil.aspx).

During experiments, four different times were measured: execution time of sequential programs, which were then parallelized by the GAP, parallel execution time (including data transfer overhead), parallel execution time without data transfer overhead, and execution time of programs from the CuBlas library. Two types of experiments were performed-firstly, the same tests as presented in [11] to confirm that results in general are independent from used hardware platform (different GPU was used), and secondly tests that compared execution time of the native CUDA programs taken from the CuBlas library with the code generated by the GAP on the basis of the same function from the BLAS library to check how effective are the programs generated by the GAP.

In the first test, a relatively simple matrix addition algorithm (2-levels deep nest) was parallelized. There is no loop carried dependence in this case; therefore, we can expect that large speedup can be obtained. However, as it can be noticed based on the results presented in Fig. 2, speed-up is extremely small. The main reason of it is that the considered problem does not cause enough operations to keep GPU busy to amortize the main overhead, the data transfer. When the time of data transfer was omitted, the obtained speed-up was between 48 and 100 (for the largest matrix). Therefore, it can be concluded that, even when available, "processing units" can be used efficiently, the obtained speed-up does not have to be large. So, from the practical point of view, it does not make much sense to ignore this overhead. Comparing the obtained results with previous experiments [11], it can be noticed that the shape of charts is similar with the exception of the larger speed-up obtained using new hardware.

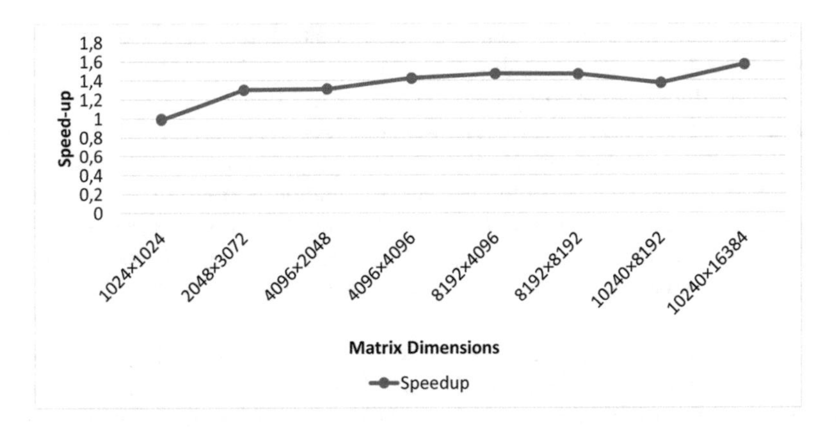

Fig. 2. Speedup obtained by parallelizing matrix addition with the GAP compiler

In the next test, the 3D Stencil Computation program, which is the part of Parboil benchmark suite, was parallelized. It is an iterative algorithm, the main code is a body of the Stencil3DHost function (3-levels deep nest), that is invoked a number of times in a loop. Therefore this function was parallelized by the GAP, moreover it was determined that there is no dependence in the nest, and therefore, all three loops can be recognized as "doall" loops.

Figure 3 presents results of this test. The obtained speed-up shows that the 3D stencil computation exposes high degree of data parallelism. The speed-up increasing with the size of the problem. The reason for this is that with the greater size, the ratio of the number of operations per byte increases. Moreover, speed-ups calculated when the data transfer is included and without it are close. Similarly to the first test, the shape of charts from the previous and current experiments are similar.

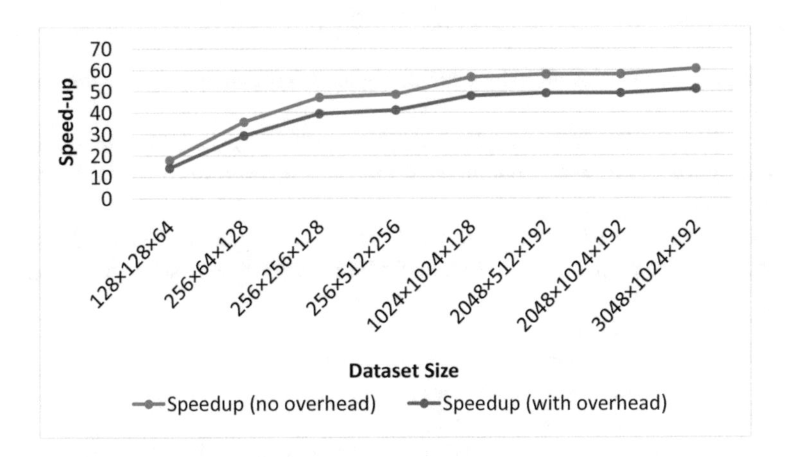

Fig. 3. Speedup obtained by parallelizing 3D stencil computation

Then, the GAP was used to parallelize the computation of matrix Q, representing the scanner configuration calibration that is used in 3D magnetic resonance image reconstruction algorithms in non-Cartesian space. Analyzing the procedure source code (available at the Parboil Benchmark suite website), it can be determined that there might be a dependence in the last two statements, as the input and output variables are the same. However, the innermost loop can run in parallel. Fortunately, as loop carried dependence exists only at level 1 (outermost loop), the GAP compiler was able to transform this loop nest into an equivalent loop nest, where the outermost loop is a "doall" loop and the innermost loop carries all the dependences using Loop Permutation transformation.

Figure 4 shows the obtained speed-up. It can be noticed that MRI-Q computation exposes a very high degree of data parallelism. As the overhead introduced by the data transfer to and from GPU memory is very small, only the speed-up calculated using wall-clock time is presented. A more detailed discussion related to results of all the above tests can be found in [11].

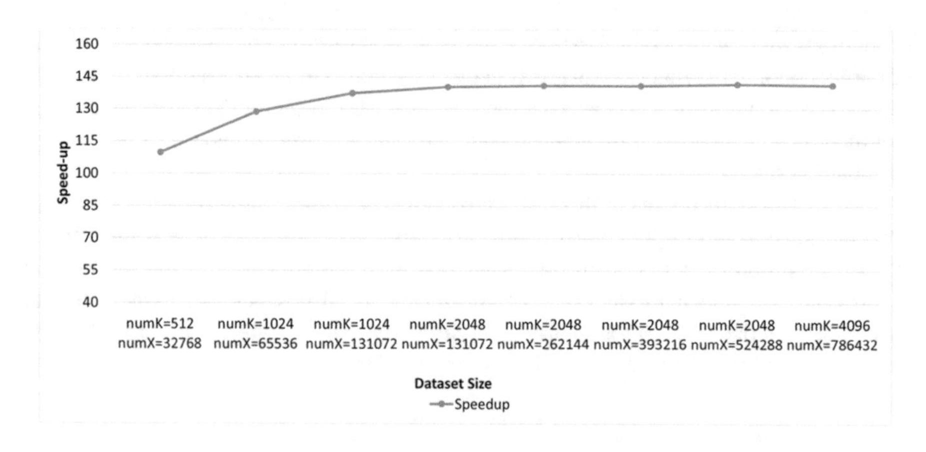

Fig. 4. Speedup obtained by parallelizing the MRI-Q computation algorithm

During the next 2 experiments, the execution time of CuBLAS routines with a code generated by the GAP, which parallelized the same routine from BLAS library, have been compared.

In the first test, the SSYR2 function that performs the symmetric rank-2 update was used. The source code parallelized by the GAP (2-levels deep nest) is presented below:

```
DO j = 1,n
  DO i = j,n
    temp1 = alpha*y(j)
    temp2 = alpha*x(j)
    a(i,j) = a(i,j) + x(i)*temp1 + y(i)*temp2
  CONTINUE
CONTINUE
```

In this case, the GAP dependence analyzer determined that there are no loop carried dependencies within this nest. Hence, the GAP generated a CUDA kernel where each CUDA thread handles one iteration of this nest. The execution times presented at Fig. 5 in both cases were very close, the difference is not visible in the figure. Sometimes the code generated by the GAP outperformed the routine from CuBLAS.

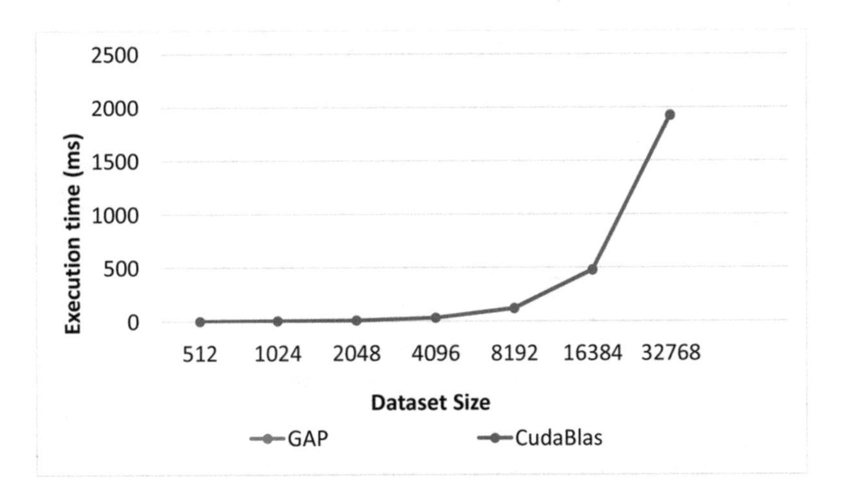

Fig. 5. Execution time of the SSYR2 routine

In the last test, execution times of the STRSM routine that solves the triangular system with the code generated by the GAP were compared. The part of the C code parallelized by the GAP (3-levels deep loop nest) is presented below.

```
DO j = 1,n
  DO k = 1,m
    DO i = k + 1,m
      b(i,j) = b(i,j) - b(k,j)*a(i,k)
    CONTINUE
  CONTINUE
CONTINUE
```

GAP's dependence analyzer will discover loop carried dependencies at level 2 and 3 of this test. In other words, the second and third loop carry dependencies and must run sequentially and only the outermost loop can run in parallel.

The GAP current loop transformation logic always prefers the outermost loop parallelization that pushes the dependencies down towards the innermost loop. As this loop nest has exactly such a structure (the outermost loop is dependence-free and all loop carried dependencies are within two inner loops) the GAP will not transform this nest at all. It will generate a CUDA kernel, where each CUDA thread will run the two inner loops sequentially, which is clearly very inefficient as the number of iterations that each of the threads will run can be enormous.

One simple addition to the compiler that would improve the performance is to choose innermost loop parallelization in case where there are more inner loops with carried dependencies than dependence-free loops after the outermost loop parallelization (Fig. 6).

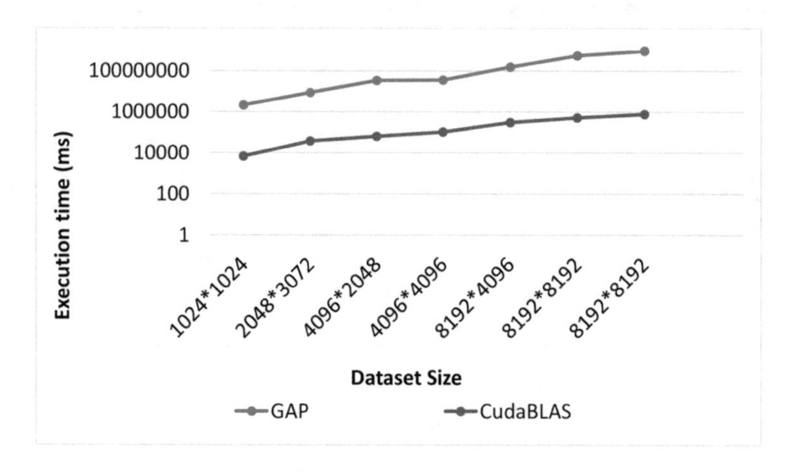

Fig. 6. Execution time of the STRSM routine (logarithmic scale)

In this case, the innermost loop parallelization would produce a nest where all dependencies are contained within the outermost loop and the 2 inner loops are dependence-free and can therefore run in parallel. The drawback of such approach would be that the generated CUDA kernel would have to be run as many times as there are iterations in the outermost loop, which can also lead to performance issues. However, as the data transfer is still done only once (to the GPU before the first kernel is run and from the GPU once the last kernel is completed), this approach would definitely dramatically improve the performance of the generated program. It will be implemented in the next version of the compiler.

5 Conclusions and Future Work

The first experiments with the GAP have shown that the compiler is capable of generating efficient parallel programs without any help from the programmer, however it needs a lot of improvements. Two kinds of improvements will be considered, the first related with the possible structure of the input programs, and the second with the dependence analyzer. The first relatively simple feature that is currently missing is the support for non-unit and negative loop strides. Secondly, the loop nest model will be extended to support non-perfect loop nests as many real-world programs contain nests that have statements outside the innermost loop. Moreover, the compiler has to be able to analyze more complex nest bodies that contain various types of statements, and more precise handling of the conditional statements must be introduced to the dependence model.

Pointer analysis is another important factor that is ignored in the current version of GAP compiler and will be included in its next version.

Currently, two types of unimodular transformations, - inner loop parallelization and outer loop parallelization, - as well as loop interchange and loop permutation are implicitly supported, as they are special case of unimodular transformations. The next transformation that will be supported is loop unrolling, which may reduce the overhead of managing the loop execution and it will reduce the number of branches.

The next planning improvements are much more complex. The GAP in the current version pays no attention to the CUDA memory model. The reason why CUDA exposes the global, shared and constant memory to programmers is to allow them to customize the memory access of their kernels to achieve the best performance. Onwards, compiler should apply a more elaborate heuristic when deciding which transformation to apply to the nest. And finally, the GAP currently misses any algorithm for estimating the potential benefits that could be obtained by the parallelization of a given loop nest. Sometimes parallelizing the code that is in its nature sequential makes no sense and will lead to terrible performance.

References

1. Banerjee, U.: Loop Transformations for Restructuring Compilers: The Foundations. Kluwer Academic Publishers, New York (1993)
2. Banerjee, U.: Loop Transformations for Restructuring Compilers: Loop Parallelization. Kluwer Academic Publishers, New York (1994)
3. Banerjee, U.: Loop Transformations for Restructuring Compilers: Dependence Analysis. Kluwer Academic Publishers, New York (1994)
4. Allen, R., Kennedy, K.: Automatic loop interchange. In: Proceedings of the SIGPLAN 84 Symposium on Compiler Construction, Montreal, pp. 233–246 (1984)
5. Wolfe, M.J.: Advanced loop interchange. In: Proceedings of the 1986 International Conference on Parallel Processing, St. Charles, Illinois, pp. 536–543 (1986)
6. Zima, H., Chapman, B.: Supercompilers for Parallel and Vector Computers. ACM Press, New York (1991)
7. Midki, S.M.: Automatic Parallelization: An Overview of Fundamental Compiler Techniques. Morgan Claypool Publishers, California (2012)
8. Quillere, F., Rajopadhye, S.V., Wilde, D.: Generation of efficient nested loops from polyhedra. Int. J. Parallel Program. **28**(5), 469–498 (2000)
9. Bondhugula, U.K.R.: Effective automatic parallelization and locality optimization using the polyhedral model. Ph.D. thesis, The Ohio State University, Ohio (2010)
10. Baskaran, M.M., Ramanujam, J., Sadayappan, P.: Automatic C-to-CUDA code generation for affine programs. In: Proceedings of the 19th International Conference CC2010, Paphos, Cyprus, pp. 244–263 (2010)
11. Kwiatkowski, J., Bajgoric, D.: Automatic parallelization of ANSI C to CUDA C programs. In: Wyrzykowski, R., Dongarra, J., Deelman, E., Karczewski, K. (eds.) PPAM 2017. LNCS, vol. 10777, pp. 459–470. Springer, Cham (2018). https://doi.org/10.1007/978-3-319-78024-5_40

Cloud Computing and Network Analysis

Cesar de la Torre(✉) and Juan Carlos Polo

Universidad de Fuerzas Armadas - ESPE, Sangolquí, Ecuador
cadelatorre@espe.edu.ec

Abstract. Today, the demand for security infrastructure is focused on perimeter defenses, for example, firewalls, proxies, IPS and content filtering. The idea associated with cloud computing has dramatically changed from the last-mentioned concepts, due to new ideas associated with the implementation of corporate IT services over the Internet. The virtualized applications running on leased servers and accepting network connections are denominated cloud services. Cloud services classify in publics and privates, although it exists another type known as hybrid clouds, which is mixed of the before mentioned ideas. On these days, a very important paradigm is how to delivery security services that satisfy the requirements of all corporate users. Therefore, the principal service analyzed in this paper is the security as a service known as SECaaS. The importance of the before mentioned concept is the possibility to analyze contents in a controlled environment, where policies forbid the execution of files with super user privileges and the contents are well verified. Finally, the importance of this paper is focused in the necessity of the optimal distribution of security infrastructure, deployed as cloud services, which simplify and accelerate the detection of compromised endpoints based on generated traffic logs analysis or user behaviors.

Keywords: Computer system security · Cloud services

1 Distribution of Software, Hardware and Security as an Infrastructure

Cloud computing services could be implemented in two versions private or public, by the way, it exists also a mixed version of both which is identified as a hybrid cloud (see Table 1). The basic requirements of enterprise and corporation cloud computing is the necessity of:

- Dynamic scale-in and scale-out of application by the provisioning and de-provisioning of resources, dependent of user requirements on demand. Usually this concept is associated with services virtualization.
- Monitoring of resource utilization to support dynamic load balancing and re-allocation of applications and resources.
- Implementation of security technologies which need to be optimally distributed in the cyberspace for a rapid development of new policies and measures in order to avoid cyberattacks and malware widespread [7].

L. Borzemski et al. (Eds.): ISAT 2018, AISC 852, pp. 190–198, 2019.
https://doi.org/10.1007/978-3-319-99981-4_18

Table 1. Deployment of security components

Administration	Location	Service requirement
Internal	User office/Desk	Antimalware, Data Leak Protection, client agents
Internal	Remote office	Security Management Services, local services.
Internal	Data centers (Private cloud)	Consolidate servers, virtualized applications. Central Security Management Consoles.
External	Third-party (Public cloud)	Recollection of enterprise events (Public Cloud), virtualized environments for event and data analysis. Infrastructure as a service (IaaS)
External	Third-party (Public cloud)	Distribution of analysis results and good practices, Platform as a Services
External	Third-party (Public cloud)	SECaaS (Security as a Services). Security instances and security technology on-demand.

1.1 Internal Security Services

These days, network administrators are interested in delivering high performance network cores and distribution levels, ensuring optimal visibility of applications and offered services; at the same time, it is crucial to reduce to the minimum the IT risk against network downtimes. These two-separate processes are simultaneously vulnerable to cyber-attacks against critical infrastructure originated in the inside network or behind the network perimeter, because it exists the human error paradigm and a group of unsatisfied users which are working with corporate network resources and advanced privileges in their daily normal activities. Therefore, it is very important the correct deployment of security components and security protocols to guarantee an optimal respond against network or system threats and eventually human fails.

1.2 External Security Services

The services distributed in external sites (third party locations) are optimal to guarantee the best time respond against threats and eliminate vulnerabilities. Because risk minimization involves analysis in controlled environments, generally virtualized and managed by very high level of experts in different areas, which is not possible to integrate in corporate staffs. Besides, the public cloud or cloud services integrated in Security as a Service are connected to corporate networks thought cypher channels, and well deployed policies which eliminate the possibility of being cracked or packet infected, and rather the most vulnerable points are hosts where information is stored or services are running.

2 Cloud Business Model

The Cloud Business Model Framework is analyzed in three layers or categories, which represent the principal environments to deploy cloud components: Infrastructure as a Service, Platform as a Service and Applications. The security services could be deployed in any of this category, but the goal is different in each one. The following Table 2 present an analysis of different services and instances developed for each Cloud Business Model [7].

Table 2. Business Model

Model	Service	Instance of
Infrastructure as a Service	Storage	Amazon
Infrastructure as a Service	Computing	Sun
Infrastructure as a Service	Security	Sophos Virtual NGFW and UTM
Platform as a service	Business	Sales force
Platform as a service	Development	Morph labs
Platform as a service	Security	CYREN web security gateways
Applications	Software as a Service	SAP
Applications	On demand WS	Xignite
Applications	Security	EXO5 + AVIRA, Antimalware and DLP

Several organizations have developed their own Cloud APIs (application-programing interfaces) to guarantee the cross-platform integration and to bring uniformity or standardization in protocols, and at the same time leverage cloud resources into cloud providers. All these goals allow to integrate Security as a Service in today's corporate networks.

3 Cybercrime and Malware as a Service

During the last five years the increase of exploiting security holes, commonly known as vulnerabilities, have increased dramatically. Financial institutions, governments and companies have implemented a lot of network services which are accessed from Internet, and private and public clouds dominate the information system technologies; simultaneously, organized groups of gangs have changed side to cybercrime. The criminal's main goal is to steal assets, services or financial information that can eventually be turned into hard currency.

There are attackers who are making money by infecting and linking up thousands of computers, often referred to as "botnets". These are then monetized in a variety of ways. For example, some criminal organizations rent out the computing power of these networks to send phishing spam. In other cases, criminals perpetrate extortion schemes by threatening to take a business' website down through distributed denial of service attacks. They look into the Cloud Computing trend with "Malware-as-a-Service" offerings, where cyber criminals sell capacity and services on a subscription basis to each other. Increasingly, cyber criminals are focused on stealing financial and personal information that can be directly monetized, even serious companies like Facebook or Google were accused of selling illegally personal information of their clients.

Criminals recently targeted RSA to get access to intellectual property based on their two-factor authentication (SecurID) implementation. Epsilon marketing provided access to email accounts information with business relationship details that could be used to allow phishing attacks to get through spam filters. The rampant Zeus Trojan, or Zbot, steals users' online banking credentials [6]. It does this by hijacking an online banking session and re-writing web pages on the victim's PC to look like the bank's website. This spoof tricks users into typing in their banking credentials, which the Zeus Trojan then uploads to a criminal server to be aggregated and sold to the highest bidder [5].

In response, financial institutions have increased layers of security around authentication and identity verification, including two-factor authentication for bank transactions. Virtually overnight, however, malware has evolved to circumvent these new safeguards. "Online thieves have adapted to this additional security by creating special programs, real-time Trojan horses, that can issue transactions to a bank while the account holder is online, turning the one-time password into a weak link in the financial security chain". Cyber gangs as response adopted a new and very ingenious form to introduce malware in network endpoints which steal credentials and transfer it to servers controlled by criminal cyber groups. Few years ago, one of most sophisticate groups of cybercriminals known as *RAMSOWARE* introduced a very dangerous malware denominated *Angler.* The entire process of workstation malware infections can occur completely invisible, requiring no user action and generating for cybercriminals enormous revenues (see Fig. 1) [2].

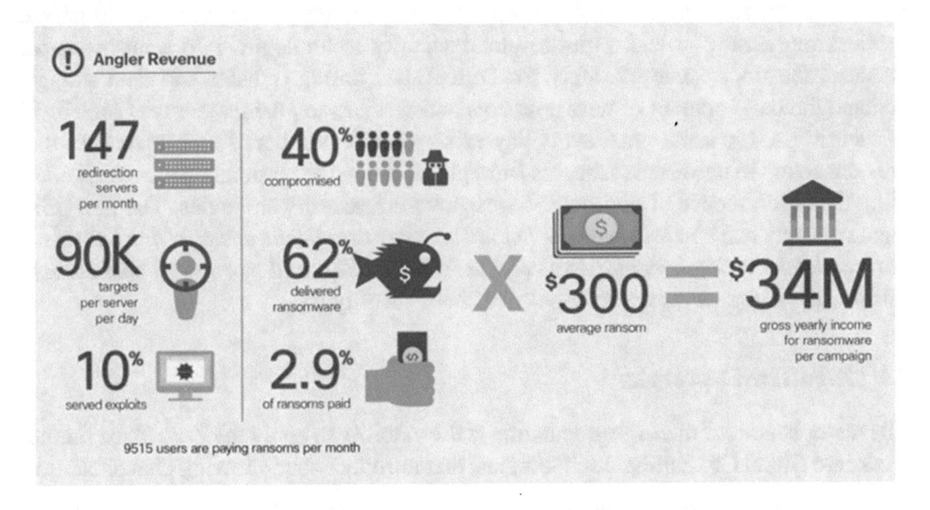

Fig. 1. Illustration of angler exploit revenues generated by users paying [2].

Another critical case to analyze is the *Wordpress site compromised by Criptowall,* which similar to Angler generated an enormous sum of money for cybercriminal group of gangs, as a consequence of malware introduction into corporate networks and put in evidence the necessity of new technologies and protocols to fight against cybercriminal gangs (see Fig. 2) [2].

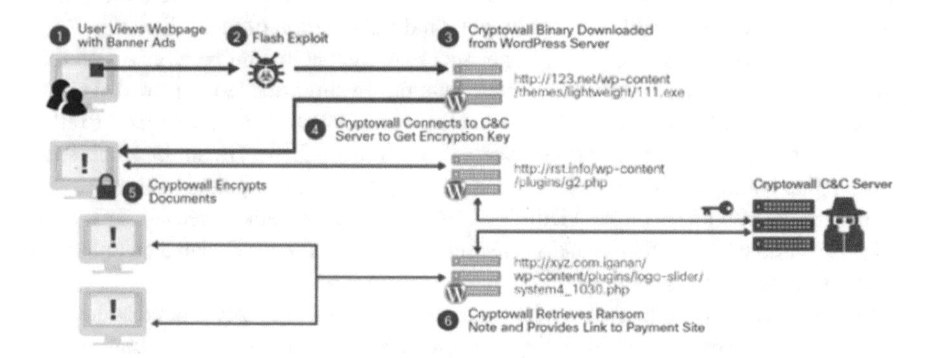

Fig. 2. Evolution of Wordpress site compromised by Criptowall [2].

4 Security Breaches and Trust

During the last year, they were detected many Network IPs which represent a significant threat for Internet users, as consequence of diverse installed software by administrators or as consequence of hacker's actions. Between the principal threats found on the compromise network IPs are mentioned Malicious Hosts, Scanning Hosts, Spamming Hosts, Phishing, Ransomware, Tor networks and others.

The evolution of networks in the last 10 years and the introduction of virtualized services represent a crucial point to cloud security technologies into a business; for instance, the importance of safety for critical data during transfer and data storage, demand the development of new protocols and services to guarantee a new high level of security. At the same time, every day exists a major number of mobile devices that are connecting to networks resources from places out of the corporate perimeter, generating the obsolescence of perimeter-centric network security strategies. The new principal concepts can be announced as *"cloud security architecture is effective only if the correct defensive implementation is divided into smaller, and more protected zones"*, this process is known as micro-segmentation of security [6].

4.1 Zero Trust Networks

The major benefit of micro-segmentation is the viability to apply the Zero Trust Framework into Cloud Computing Technologies, that introduce the following characteristics:

- Zero Trust is applicable across all industries and organizations.
- Zero Trust is not dependent on a specific technology or vendor.
- Zero trust is scalable.
- There is no chance of violating Civil Liberties.

The theory is that *"even if one small zone is compromised, the breach will be contained to a smaller fault domain, and finally will compromise critical enterprise data"*. However, the correct distribution of defensive devices must be correlated with the right recognition of security issues that will arise with security management [4].

4.2 Requirements for a Network Forensic Solution

Indeed, network security has been analyzed for two decades and recently they were introduced the principal concepts for *network forensic solutions;* which must provide three essential capabilities to be applied in network security zones:

- Capturing and recording data
- Discovering and analyzing data
- Obtaining network analyzes and visibility of network traffic (Zero Trust NAV).

The correct application of the before mentioned concepts require to separate traffic in two domains: Internet domain where is assumed that application traffic is free and uncontrolled and Intranet domain were application control criteria is applied based in three concepts: *visibility, control and reports, manage bandwidth* (see Table 3).

Table 3. Classification of control criteria

Criteria	Domain	Area of influence
Visibility	Intranet	All traffic in the internal network
Control/Logs & reports	By policy By user By time Blocking applications[a]	All traffic in Intranet
Manage bandwidth	Business critical	DB, Sales force applications, VoIP/Video
	Socio Business	Social networks, Instant messenger.
	Non-critical	Spotify, YouTube.
	Undesired	Anonymity services TBB, uTorrent, Kaza, etc.

[a]Include blocking infected application as consequence of avoid malware widespread.

To complete the classification, it is important to consider the necessity of a full evaluation and identification of traffic protocols, and considering the explicit concept of Zero Trust Networks, *any protocol that is not explicit authorized in the network is denied in the full segments* [4]. The principal criteria to identify an application and to classify its impact on the network are:

- The risk level that represent.
- Typical characteristics of the application.
- Implemented technologies to support.
- Category of application in Internet web and application public services.

4.3 Cloud Security Optimization Techniques

The principal goal to analyze is an efficient method that maximize network flows, in terms of association with domains business critical and socio business; allowing to identify and reduce threats in network traffic. At the same time, it is increased the average network security and optimize the throughput assigned to Cloud Services, in that way that the communication to different instance in selected Business Model will be optimums, this throughput can be calculated as follow.

$$\max_{flow} Benifit = \min_{networkThreats} f(givenNetworks, userRisk, matrixFlow, networkThreats) \quad (1)$$

5 Methodology and Hypothesis

In order to calculate the correct structure of small zones and small domains to remediate security breaches and to reduce attacks, it is required an appropriate identification of critical nodes, where the traffic could be commuted to other small zones [8]. Consequently, it is possible analyze this problem as graph network[1], (see Fig. 3) [1]. The principal steps of methodology to be analyzed are:

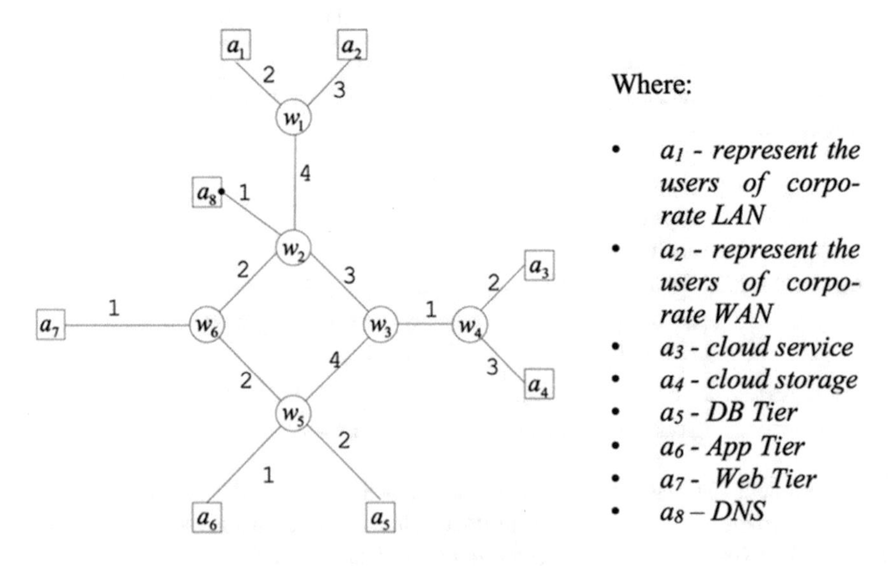

Fig. 3. Graph representation of application flow mapping.

- Design map dependencies among the elements in delivering services, which downtime and increase productivity.
- Better utilization of network resources, supported by adequate measures of traffic with respective reporting and planning.

[1] Typical graph network is formed by nodes and edges G = {N, E}.

- Classification of all traffic along each service delivery path, in that way to contribute for a faster classification, according to control criteria.

5.1 Hypothesis

There must exist an algorithm for network segmentation in small domains that contribute to remediate security breaches, allowing optimal allocation of k security elements in critical nodes that commute traffic between small zones; to protect and optimize cloud services. It increases network security by reducing the time necessary for threat characterization and at the same time warranty the throughput in links required for optimal connection to cloud services because a deep packet inspection allows to classify traffic in good, malicious and discardable.

5.2 Simulation of Transmission

Before starting the simulation of traffic in selected graph network, it is important to interpreter that exist two scenarios:

1. When traffic is out of control and consequently many different forms of attacks and search of network vulnerabilities are important components of network flows.
2. When traffic is under control and the maximum priority is assigned to business critical and socio business applications, reducing to minimum the traffic of undesired applications and blocking the traffic of malware and other known threats.

The reduction in terms of traffic and consequently the reduction in terms of paying by amount of transmitted data is proportional to the increase of network security, as result of discarded packets, (see Table 4).

Table 4. Simulation of reduction of breaches as consequence of control policies.

Node number	Free average traffic [GBps]	Controlled average traffic [GBps]	Reduction of breaches in terms of attack and exploits
a_1	366.00	299.80	18%
a_2	329.00	279.22	15%
a_3	573.00	406.18	29%
a_4	818.00	594.40	27%
a_5	233.00	218.18	6%
a_6	450.00	396.80	12%
a_7	407.00	311.62	23%
a_8	180.12	160.14	11%

As results of the simulation, it was obtained that most critical nodes are nodes a_3, a_4 and a_7. Therefore, the reduction of traffic in those nodes as consequence of elimination of undesired traffic will contribute significantly to the average hole network reduction of vulnerabilities and network attacks. At the same time, its nodes are optimal for a network segmentation and guarantee the best network performance.

6 Conclusions

Today, the Cloud Service Framework models demand very advanced techniques for network analysis and visibility of all packets, and at the same time it is important to consider the importance of correct classification of applications, services and users.

Nowadays, the cloud service security goes further than the corporate network perimeter, therefore it is necessary to develop a new generation of technologies, that allows to protect the data that flows between users and cloud services (encrypted media).

The respect to user information data, must be seriously warranty in SLA (service legal agreement), because the actual Model of Cloud Computing allows companies to storage information, sensible data and intellectual property in different geographical locations, that may not respect the main Civil Liberties.

References

1. Amman, P., Wijesekera, D., Kaushik, S.: Scalable, graph-based network vulnerability analysis (2002)
2. Cisco: Annual Security Report, pp. 10–28 (2016)
3. IDC: The Role of Virtual WAN Optimization in the Next Generation Datacenter, December 2012
4. NIST: Developing a Framework to Improve Critical Infrastructure Cybersecurity, Submitted on 04 August 2013
5. SecureWorks: The next generation of cybercrime: how it's evolved, where it's going, executive brief (2010)
6. Trend Micro: Security Threats to Business, the Digital Lifestyle, and the Cloud (2013)
7. Weinhardt, C., Anandasivam, A., Blau, B.: Cloud Computing – A Classification, Business Models, and Research Directions. Springer, BISE (2009)
8. De la Torre, A., De La Torre, M.: Network forensic analysis in the age of cloud computing. In: CEPOL Budapest, Conference 2016. https://www.cepol.europa.eu/sites/default/files/18-marco-de-la-torre.pdf

Human-Computer Interface, Multimedia Systems

Using Open Source Libraries for Obtaining 3D Scans of Building Interiors

Adam L. Kaczmarek[1(✉)], Mariusz Szwoch[1], and Dariusz Bartoszewski[2]

[1] Gdansk University of Technology, ul. G. Narutowicza 11/12, 80-233 Gdansk, Poland
{adam.l.kaczmarek,szwoch}@eti.pg.edu.pl
[2] Forever Entertainment S.A., Gdynia, Poland
dariusz.bartoszewski@forever-entertainment.com

Abstract. This paper describes methods for making 3D scans of building interiors. The main application of these methods is the development of first-person view (FPV) perspective games or creating virtual museums. 3D scans can be made with the use of different equipment such as Light Detection and Ranging (LIDAR), time of flight (TOF) cameras and structural light 3D scanners. However, the paper focuses on using stereo cameras for obtaining 3D scans, because of its low cost. It is a significant factor for small and medium size game development studios. The paper considers both the method based on photogrammetry and stereophotogrammetry. In photogrammetry the Structure from Motion technology is used for making 3D scan on the basis of images of an object taken from different locations. Photogrammetry uses stereo vision algorithms for acquiring depth maps and point clouds representing distances between a stereo camera and objects. The paper analyses implementations of these technologies available in programing libraries OpenCV, openMVG and openMVS. The paper shows that the algorithm for Structure from Motion provided in openMVG can be successfully applied for obtaining point clouds from pair of images despite this algorithm is intended for use with a greater number of input images.

Keywords: 3D scanning · Structure from Motion · Stereo vision

1 Introduction

The paper is concerned with acquiring 3D models of buildings interiors, mainly for the purpose of their use in video games. In general, 3D scanning is usually used for making 3D models of individual objects of any size, which usually means reconstruction of their outer surface. However, scanning the inner view of building has also wide range of applications. Obtained 3D scans of interiors can be used for generating a virtual 3D map of rooms or even the entire building. Such models can be used for archiving purposes as the input to CAD (computer aided design) or BIM (building information modeling) systems [1]. Next possible application is to prepare a virtual walk in a real location, e.g. to present the inaccessible or hard to reach interiors to the interested people. This approach can be used to create virtual museums presenting historical interiors or chambers with original equipment [2]. Another possible application is the interior design.

L. Borzemski et al. (Eds.): ISAT 2018, AISC 852, pp. 201–210, 2019.
https://doi.org/10.1007/978-3-319-99981-4_19

Using the techniques of virtual reality (VR) and mixed reality (MR) it is possible to present new architectural ideas, models or solutions that partially or totally replace the existing ones. These new arrangements can be realistically presented to the audience using VR goggles or in a cave automatic virtual environment (CAVE) [3]. Finally, in the last few years, another field of application is gaining popularity and attracting the attention of many researchers. This new application is the use of automatically or semi-automatically scanned interiors in video games. Although, these solutions are usually limited to 3D games with the first-person view (FPV) perspective, which action takes place in real locations, such games are still a large part of market worth billions of dollars [4]. This paper focuses on this application area.

There is a variety of equipment that enables obtaining 3D scans, which can be divided generally into active and passive devices. The most important active devices include Light Detection and Ranging (LIDAR), time of flight (TOF) cameras and structural light 3D scanners. Although, active scanners provide high quality 3D geo-metric models (meshes) in most situations, they usually do not provide information about the objects' texture (material) that has to be manually added later. Moreover, TOF cameras and structural light scanners usually have very limited resolution that is insufficient in many applications. Finally, very important disadvantage of the active scanners is their cost. The equipment dedicated for making 3D scans is usually very expensive, which is particularly significant flaw in case of small game development studios which would like to obtain 3D scans of building interiors. However, 3D scans can be also made with the use of cameras, which are relatively cheaper and widely accessible. A studio making a video game can already be in possession of cameras that can be used for making scans. Therefore, this paper analysis the subject of using cameras for making 3D scans of building interiors.

There are many algorithms that make it possible to obtain a 3D scan from images [5, 6]. Open source implementations of these algorithms are also available. Such an implementation is included in one of the most important programming library in the field of computer vision and image processing, which is OpenCV [7].

Among many algorithms included in OpenCV there are algorithms for processing images in order to obtain 3D images and 3D scans. However, experiments presented in this paper shows that the results of these algorithms are not suitable for scanning interiors of building. The paper also presents experiments with other programming libraries called openMVS and openMVG. These libraries are much less popular than OpenCV however results of using them are much more suitable for the purpose of scanning interiors.

Original contributions of this paper include: (1) The comparison of the results of acquiring 3D scans of building interiors with the use of libraries OpenCV, openMVG and openMVS. (2) The description of the results of applying the technology of Structure from Motion to a pair images from a stereo camera. (3) The proposition of the 3D scanning method for building interiors.

2 Related Work

There are two main methods of acquiring 3D scans and 3D images with the use of cameras. The first one is stereophotogrammetry that uses the technology of stereo vision [6]. The other one is photogrammetry based usually on the Structure from Motion (SfM) algorithm [8].

2.1 Stereo Vision

In the technology of stereo vision 3D images are obtained on the basis of a pair of images taken with a stereo camera that consists of two cameras fixed to each other on a special rig.

Any object, that is visible by both cameras, is located at different coordinates in images taken by these cameras. Let us suppose that a certain object is visible at coordinates x_1, y_1 on the left image and at coordinates x_2, y_1 in right image. The difference between values x_1 and x_2 is called the *disparity* (d). If the object is closer to the camera system, then the disparity is greater. Disparities determined for the entire image form a *disparity map*. A disparity map can be easily transformed to either a *depth map* or a set of points in 3D space (*point cloud*). A depth map contains values of distances between a stereo camera and each point of the objects (so called "pixel with depth"), whereas a point cloud contains 3D coordinates of each point in a certain local 3D coordinate system.

In order to transform a disparity map to a depth map it is necessary to have data related to physical features of a stereo camera set such as distances between cameras b (baseline) and focal length f of lens. These data can be recovered from technical specification of a stereo camera or on the basis of its calibration.

The calibration is a process that is required when a stereo camera is used. The problem with stereo cameras is such that their optical axes are not parallel to each other. In real application there will always be imperfections. Moreover, cameras are also slightly rotated relatively to each other. However, there are also distortions which occur in every single camera. Lens of cameras cause that straight lines visible in real objects became bent in images. This effect occurs in different extent for different kinds of lens. It is the most evident in case of fish-eye lens that cause a barrel kind of a distortion [7]. In images made with this kind of lens, straight parallel lines would look like they were drawn on a sphere.

The problem with distortions is such that they cause deterioration in the quality of disparity maps. Because of distortions, errors occur in these maps. Distortions which occur in images can be neutralized by transforming these images. Different kind of transformations can be used including shifts and rotations. In order to appropriately transform images in order to neutralized distortions it is necessary to resolve which distortions occur in images and to what extent.

These data are obtained in the process of image calibration. The calibration of a stereo camera is based on making images of a specific, regular image pattern. These images are analyzed by an algorithm for calibrating images. The algorithm recognizes characteristic points and on their basis it calculates transformation parameters. A typical kind of an image patters used for such calibration is a chessboard.

The calibration leads to the improvement in the quality of images, but it does not resolve all the problems with obtaining disparity maps from stereo pairs. The main problem is that the pair of images needs to be processed with the use of a stereo matching algorithm in order to determine the location of the same objects in the pair of images. Different kinds of algorithms were developed, which differ in their speed, quality of results and the level of required computer resources to operate.

There are many rankings of stereo matching algorithms. One of the most popular is the ranking provided in Middlebury stereo vision page [9]. Another popular ranking of stereo matching algorithms is KITTY (http://www.cvlibs.net/datasets/kitti/) [10]. These ranking include both well-known algorithms such as those provided in the OpenCV library and many novel algorithms tested only by their authors.

2.2 Structure from Motion

Another approach to obtaining 3D data on the basis of images is the technology of Structure from Motion. In this technology, images of an object are made from many points of view located around the object. The set of images is then analyzed by the SfM algorithm that searches for characteristic (*salient*) points in all images. There are different methods of searching for these landmark points, however one of the most important algorithms designed for this purpose are scale-invariant feature transform (SIFT) and Speeded up robust features (SURF) [11].

On the basis of location of these characteristic points in images the algorithm estimates positions of a camera in the moment, when images were taken. This information is used to determine the location of characteristic points in 3D space. The result of a SfM algorithm is a points cloud reflecting the shape of real objects. This is a significant difference in comparison to algorithms for stereo vision, which produce a depth map acquired from a single point of view. In order to obtain a whole shape of a real object it is necessary to merge disparity maps acquired from different points of view. In case of SfM technology the whole point cloud representing the object is the direct result of the algorithm. The result can also have a form of a triangle mesh used in computer graphics.

There are different rankings of SfM algorithms similarly as there are rankings of algorithms for stereo vision. The Middlebury Stereo Vision Page, apart from the ranking of stereo matching algorithms, contains also a ranking of algorithms for obtaining Structure from Motion (http://vision.middlebury.edu/mview/eval/) [8].

2.3 OpenCV Library

The OpenCV library is open source and it contains implementations of a large number of algorithms used in computer vision and image processing including algorithms for image filtering, recognition and machine learning. The library contains also algorithms for stereo vision and Structure from Motion techniques.

The OpenCV library contains four stereo matching algorithms for obtaining disparity maps from images taken by a stereo camera, which are: Stereo Block Matching (Ste-reoBM), Stereo Semi-Global Block Matching (StereoSGBM), Heiko Hirschmüller algorithm (StereoHH) and Stereo Variational methods (StereoVar)

[12–14]. OpenCV contains also algorithms for stereo camera calibration. The calibration can be performed on the basis of image patterns containing black and white chess-board or circles. The OpenCV calibration makes it possible not only to improve the quality of disparity maps but also to acquire point clouds from these maps. OpenCV library also contains SfM algorithms. They are available in this library starting from version 3.0.

2.4 openMVG and openMVS Libraries

An alternative method to using OpenCV for obtaining 3D scans is taking advantage of Open Multiple View Geometry (openMVG) and Open Multiple View Stereovision (openMVS) libraries [15, 16]. Our experiments showed that these libraries make it possible to acquire a 3D scan that has much higher quality than OpenCV.

OpenMVG library contains computer vision algorithms including those for estimating positions of camera using the SfM technology. The input data to the algorithm are the images taken from different points of view. OpenMVG not only determines camera positions but also produces a sparse point cloud corresponding to real objects visible in images.

Unfortunately, point clouds obtained from openMVG are sparse and not sufficient for using as models in 3D video games. However, this sparse point cloud can be expanded to a dense one with the use of the openMVS library. The output data from openMVG which are the location of cameras and initial point cloud can be provided as input data to openMVS. Taking into account the set of images processed by openMVG the openMVS library can generate a dense point cloud of the scene.

3 Technological Requirements

The research presented in this paper is dedicated for making 3D scans of building interiors. This assumption imposes requirements on the technology, which needs to be used. Foremost, as described in the introduction, only passive scanning technology, based on making images, is concerned. There are also some additional requirements that have to be fulfilled.

In case of building interiors it is not possible to take images from points of view located around objects as is expected by SfM technology. Inside buildings it is only possible to take photos of walls from one, visible side.

Moreover, it would be advisable to make it possible to exclude some images from the set of input images. That is why, a set of input images needs to be prepared manually. It is necessary to select images that will be used. In case of SfM technology, the entire image set is analyzed by the algorithm for obtaining 3D scans, what is time consuming. When a scan of a building interior is acquired, some fragments of the space can be unnecessary. The possibility of adding and deleting some parts of a 3D scan without the necessity to recalculate the entire model would enhance the efficiency of preparing scans.

Another significant technological limitation is such that inside the building space there may appear cavities with very limited access. Such cavities occur in narrow spaces blocked from many sides by building structure or furniture. In these cases it is only possible to make photos from only one point of view.

Because of the requirements, restrictions and constraints listed above, the research presented in this paper is based on using stereo cameras that enables to obtain a point cloud representing objects in 3D space visible from one point of view. These point clouds can be included and excluded from the resulting, complete 3D scan of an interior consisting of many point clouds obtained from different points of view.

Using stereo cameras do not cause that only the technology of stereo vision can be used. The application designed for making 3D scans of interiors can also take advantage of the SfM. In case of scanning building interiors this technology is particularly advantageous when it is applied to pairs of images from a stereo camera.

4 Experiments

The purpose of the experiments was to select a mature technology that would make it possible to obtain 3D scans of building interiors on the basis of images from a stereo camera. We did not concern novel algorithms which were not verified in different circumstances and environment. These algorithms have a potential to become commonly used methods. We have focused on optimized and well-tested implementations of algorithms for processing images. Therefore, for our tests, we have selected algorithms available in open source libraries OpenCV, openMVG and openMVS.

The input data set, that we used in the experiments, consisted of images obtained by ourselves and images provided on Middlebury Stereo Vision Page. We used the Sony RX100 camera for making images. The camera has a 1.0” sensor size.

4.1 Stereo Vision in OpenCV

We have performed experiments with stereo vision using the OpenCV library. We have encountered such a problem that algorithms for calibrating stereo cameras used in OpenCV are intended for use for a specific range of distances between a stereo camera and real object. Image transformations performed during calibration include shifting images, so that they can be matched by stereo matching algorithm for a specific disparity range that corresponds to a range of distances.

The problem with this method is such that it is inconvenient to use this kind of calibration for scanning building interiors. It would be necessary to calibrate cameras with the use of an image pattern which has a size of a building wall. It is problematic to prepare such samples. Cameras calibrated with OpenCV for a certain distance can be used for other distances however it causes significant deterioration in the quality of the results. Moreover, there are also problems with determining point clouds from such obtained disparity maps. Sample calibrated images showing furniture and a part of a wall are presented in Fig. 1. Cameras were calibrated with the use of a chessboard which had a size 78 cm $\times$ 62.5 cm.

Fig. 1. A sample calibrated images from a stereo camera.

Calibrated images are input data to a stereo matching algorithm which obtains a disparity map. Figure 2 present disparity map obtained for the input images presented in Fig. 1. The Stereo Semi-Global Block Matching algorithm was used. Results show that there are many points for which disparities were not obtained appropriately. Although, we did not have ground truth with real values of distances between cameras and objects, we could determine, that the point cloud generated on the basis of the disparity map also did not reflect the shape of objects visible in images.

Fig. 2. A disparity map obtained for images presented in Fig. 1.

Stereo matching algorithms that are available in the OpenCV library, require providing a disparity range which is analyzed by the algorithm. If the range is wide, then the algorithm considers identifying the location of objects that are within a long range of distances from the stereo camera. However, the increase of this range causes the increase in the number of errors occurring in the results. In case of making 3D scans of building interiors, it would be advantageous to make scans of all objects located both close and far from cameras. Using the technology described in this subsection it is not possible, if a high quality of the results need to be achieved.

4.2 Structure from Motion in OpenCV

The authors of this paper have also verified results of using algorithms for SfM implemented in the OpenCV library. The algorithm did not manage to produce readable results for the image set of building interiors prepared by the authors. Therefore, the authors conducted an experiment that involved processing a sample series of images for testing the technology of SfM provided in Middlebury Stereo Vision Page. It was the Temple data set. In the experiment the authors obtained a 3D scan on the basis of 5 images acquired from this set. These images are presented in Fig. 3.

Fig. 3. Images from the temple data set provided in Middlebury Stereo Vision.

The results are presented in Fig. 4. The image shows location of cameras retrieved by the algorithms included in the library (marked with yellow color) and the 3D scan obtained from these images.

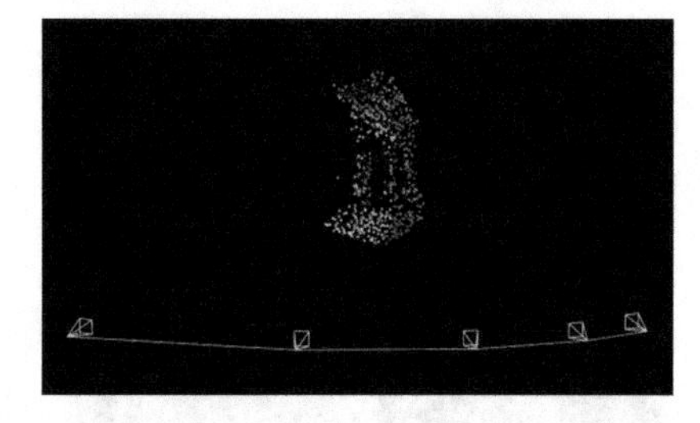

Fig. 4. The location of cameras retrieved by the SfM algorithms and the obtained 3D scan.

The scan contains a small amount of points representing the object. Ground truth for this data set is available in Middlebury Stereo Vision Page and results of using OpenCV are included in the ranking presented in this page. However, the algorithm did not manage to obtain realistic results when it was executed with only two input images. Even without comparing them with ground truth the authors could determine that they did not met our expectations. As described in Sect. 3 obtaining results for two images was a desired technology for scanning building interiors.

4.3 Structure from Motion in openMVG and openMVS

We have conducted experiments in order to verify the performance of these libraries for the purpose of obtaining 3D scans of building interiors. Experiments were focused on obtaining data from a pair of images taken with the use of a stereo camera. Experiments included making images of a parking lot, a conference room and a room resembling medieval chambers.

Figure 5(a) shows a sample input image from a pair of images. The image shows a conference room with a blackboard, modern chairs and a hanger. Figure 5(b) presents the result of using algorithms available in openMVG and openMVS.

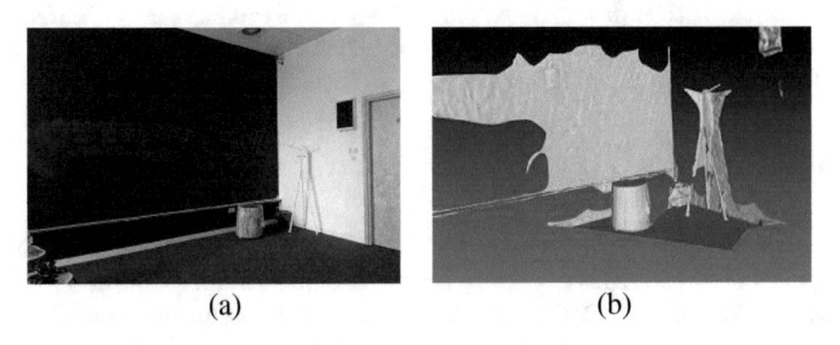

(a) (b)

Fig. 5. An image from the input set (a) and a 3D scan obtained on the basis of openMVG and openMVS libraries.

Our experiments showed that openMVG leads to correct results, even if there are only two input images. In case of this data set the authors did not have ground truth, however in contrast to the results obtained from OpenCV, the results acquired from openMVG were not blurred and it was possible to recognize shapes of real objects in the generated point cloud. The general shape was reconstructed, though, there are some areas for which a 3D model was not obtained. This observation applies to all kinds of areas considered in experiments. However, in case of reconstructing building interiors there are other stereo images taken from other locations and different angles. A complete scan will consists of many point clouds obtained with the use of a stereo camera from different points of view. Thus, blank areas can be filled in, when stereo images from other camera locations are taken into account. The greatest advantage of this technology is such that it provides reliable results for the area of a scene for which a 3D scan is obtained even if there are areas which are not included in the scan.

This kind of areas occurs in particular in case of monochrome flat surfaces. These elements of the scene can be relatively easily added to the 3D scan manually in order to obtain a complete scan. There are also other kinds of surface which are problematic for 3D scanning like reflective or transparent materials. The scan of a building interior containing such materials will have to be edited manually in order to be correct, however the technology of 3D scanning will provide the base of the reconstruction of buildings interiors.

5 Summary

One of the most significant conclusions from the research presented in this paper is such that the technology of Structure from Motion can be used for pairs of images obtained with the use of stereo camera. In general, SfM is not dedicated for stereo cameras, but for a large set of input images from different points of view. However, algorithms implemented in openMVG and openMVS libraries make it possible to acquire 3D data from image pairs. This feature is crucial for developing an application for making 3D scans of building interiors.

Acknowledgment. This work was supported by the Sectoral Programme GAMEINN within the Operational Programme Smart Growth 2014–2020 under the contract no POIR. 01.02.00-00-0140/16.

References

1. Jung, J., Hong, S., Jeong, S., Kim, S., Cho, H., Hong, S., Heo, J.: Productive modeling for development of as-built BIM of existing indoor structures. Autom. Constr. **42**, 68–77 (2014)
2. Gomes, L., Regina, O., Bellon, P., Silva, L.: 3D reconstruction methods for digital preservation of cultural heritage: a survey. Pattern Recogn. Lett. **50**, 3–14 (2014)
3. Lebiedz, J., Szwoch, M.: Virtual sightseeing in immersive 3D visualization lab. ACSIS-Ann. Comput. Sci. Inf. Syst. **8**, 1641–1645 (2016)
4. Szwoch, M., Kaczmarek, A., Bartoszewski, D.: STERIO - reconstruction of 3D scenery for video games using stereo-photogrammetry. In: Proceedings of the Conference on Game Innovations (CGI), Lodz (2017)
5. Kaczmarek, A.L.: Improving depth maps of plants by using a set of five cameras. J. Electr. Imaging **24**(2), 023018 (2015)
6. Kaczmarek, A.L.: Stereo vision with equal baseline multiple camera set (EBMCS) for obtaining depth maps of plants. Comput. Electr. Agric. **135**, 23–37 (2017)
7. Bradski, D.G.R., Kaehler, A.: Learning OpenCV, 1st edn. O'Reilly Media Inc., Sebastopol (2008)
8. Seitz, S.M., Curless, B., Diebel, J., Scharstein D., Szeliski, R.: A comparison and evaluation of multi-view stereo reconstruction algorithms. In: 2006 IEEE Computer Society Conference on Computer Vision and Pattern Recognition (CVPR 2006), vol. 1, pp. 519–528. IEEE (2006)
9. Scharstein, D., Szeliski, R.: A taxonomy and evaluation of dense two-frame stereo correspondence algorithms. Int. J. Comput. Vis. **47**(1/2/3), 7–42 (2002)
10. Geiger, A., Lenz, P., Urtasun, R.: Are we ready for autonomous driving? The KITTI vision benchmark suite. In: Conference on Computer Vision and Pattern Recognition (CVPR), pp. 3354–3361. IEEE (2012)
11. Bay, H., Ess, A., Tuytelaars, T., Van Gool, L.: Speeded-Up Robust Features (SURF). Comput. Vis. Image Underst. **110**(3), 346–359 (2008)
12. Konolige, K.: Small vision systems: hardware and implementation. In: Shirai, Y., Hirose, S. (eds.) Robotics Research, pp. 203–212. Springer, London (1998)
13. Kosov, S., Thormhlen, T., Seidel, H.P.: Accurate real-time disparity estimation with variational methods. In: Advances in Visual Computing. LNCS, vol. 5875, pp. 796–807, Springer, Heidelberg (2009)
14. Hirschmuller, H.: Stereo processing by semi-global matching and mutual information. IEEE Trans. Pattern Anal. Mach. Intell. **30**, 328–341 (2008)
15. openMVG Homapage. http://openmvg.readthedocs.io/en/latest/. Accessed 16 May 2018
16. openMVS Homapage. http://openmvg.readthedocs.io/en/latest/software/MVS/OpenMVS/. Accessed 16 May 2018

Performance Comparison of Two Head-Controlled Computer Interaction Systems with a Multi-directional Tapping Task

Marcin Kuliński(✉) and Katarzyna Jach

Faculty of Computer Science and Management, Wrocław University of Science and Technology, Wybrzeże Wyspiańskiego 27, 50-370 Wrocław, Poland
marcin.kulinski@pwr.edu.pl

Abstract. The performance comparison of two head-controlled interaction systems was made using a multi-directional tapping test on the basis of Fitts' paradigm, as recommended by ISO 9241-9 standard. Both investigated input systems were based on head position and movements visual recognition by a camera and a special software. The study was made on the sample of 14 healthy subjects without motoric deficiencies, with counterbalanced within-subject experiment design. Both systems were marked by similar error rates (average 23.45%), while their throughput values were significantly different (0.67 versus 1.92 bit/s). Additionally, a comparison of performance characteristics with several other head activated systems was made by means of a literature review.

Keywords: Human-computer interaction · Usability · Fitts' paradigm Head tracking · Face Controller · Camera Mouse

1 Introduction

The most commonly used method accepted for several decades as a standard in human-computer interaction (HCI) performance research is Fitts' paradigm [1]. Starting from a general basis of the information theory, the paradigm assumes that performing a task of a motoric nature requires processing of a certain amount of information in the receptor-nerve-effector system. Due to the limited capacity (throughput) of this biological channel a more difficult tasks, i.e. these requiring a larger amount of information to be transmitted and processed, are performed slower than the easier ones. At the same time, the level of performance, which can be determined experimentally by multiple measurements of a motor task execution times in various variants of difficulty, remains relatively constant, thus constituting an universal metric characterizing any given man-machine interface, regardless of its distinctiveness and constraints. Therefore, Fitts' paradigm gives a solid ground for an objective, quantitative comparison of different pointing devices [2], which found its expression in the ISO standard on the evaluation of non-keyboard input devices [3–5].

In the case of a graphical user interface controlled by direct manipulation a user performs targeted movements in the screen space. Consequently, the difficulty level of such a motor task is influenced by the distance which a pointer has to be moved by, as

L. Borzemski et al. (Eds.): ISAT 2018, AISC 852, pp. 211–221, 2019.
https://doi.org/10.1007/978-3-319-99981-4_20

well as the size of the target area it must stop within. The performance level is determined on the one hand by the psychomotor skills of the user, and on the other by the characteristics of the particular pointing device. In other words, the type of pointing device, and thus the way in which direct manipulation is performed, has a direct impact on the speed of the same person performing the same tasks.

In the study, the performance (throughput, TP) was measured as the ratio of the task difficulty (index of difficulty, ID) to its execution duration (movement time, MT). The difficulty index is determined by the target area size (W) and its distance from the starting point (D), measured in screen units (pixels). The higher W value, the easier a task is; analogously, the higher D value, the more difficult a task becomes. The difficulty of the motor task ID is expressed in bits, while the performance TP values are in bits per second [2, 4].

The main goal of the presented study was a performance comparison of two different systems that allow a human-computer interaction with the use of head tracking, namely Face Controller and Camera Mouse. Details about both investigated systems, the task and its parameters, used software and its modifications, as well as about participants involved are reported below. Next, obtained accuracy and performance measures are presented and discussed, including a wider comparison of various head-driven systems, based on data retrieved from the literature review. Finally, conclusions and some further exploration possibilities are outlined.

2 Research Details

One of the performance tests proposed in the ISO standard is multi-directional tapping task [4]. We used it in a form enriched with guidelines formed by Soukoreff and MacKenzie [2]. The task measures speed and accuracy of reciprocal movements performed between alternating circular-shaped areas, which are distributed in equal angular distances on a circular plan. Each single movement is conducted roughly along a diameter of this circle of circular targets. A trial consists of a series of sequential movements that involve all visible target areas, thus ensuring that various directions and angles of approach are being accounted. Every trial begins from the uppermost target, which is also the last one targeted

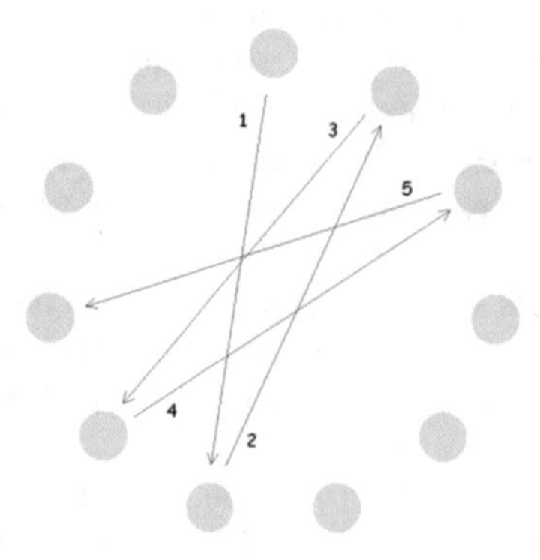

Fig. 1. Scheme of movements sequence and directions during a multi-directional tapping task used in the study

in any given sequence. The general diagram of the tapping order and the direction of consecutive movements is presented in Fig. 1.

2.1 Experiment Description

Two different target sizes W (30 and 150 pixels in diameter) and two different distances D between subsequently targeted circles (250 and 550 pixels) were chosen, which in turn formed 4 different ID values, i.e. 1.42, 2.22, 3.22 and 4.27 bits. The highest W and D values (ID = 2.22) were chosen in a way that all the task area occupied the entire height of a computer screen with the resolution of 1366 × 768, which was used in the study. All four variants are presented in Fig. 2.

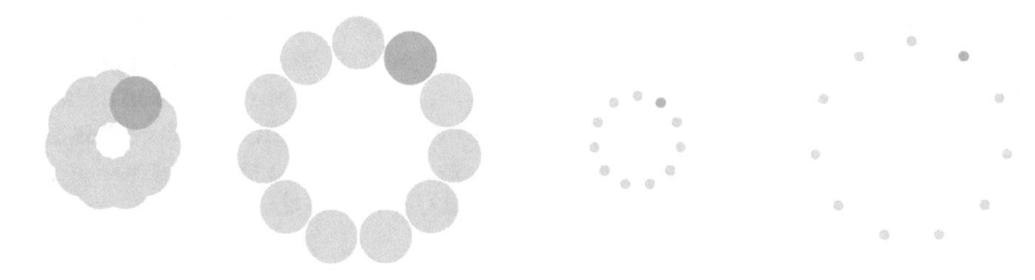

Fig. 2. Four ID variants used in the study. From the left side: W = 150, D = 250, ID = 1.42; W = 150, D = 550, ID = 2.22; W = 30, D = 250, ID = 3.22; W = 30, D = 550, ID = 4.27

All the used variants contained 11 target areas, thus required 11 movements and 12 pointing actions. Each of them was repeated twice, while maintaining a fully random order of administering. In a consequence, quantitative data from a total of 88 targeted movements for every examined subject was collected.

It was decided to use a universal method of target selection confirmation, that is appropriate for any hands-free computer interaction system, regardless of its additional capabilities, e.g. an ability to detect facial gestures like eye blink or mouth opening. The screen pointer should stay immobile for a specified period of time, called dwell time (DT), at the moment when a subject is going to confirm the end of the movement and the target selection. Taking into account the act of pointing with different body parts, including eyes, Müller-Tomfelde [6] indicates a suitable range of DTs from 300 to 2000 ms. For interaction with the use of eyetracking devices values of ~500 ms can be regarded as suitable for advanced users [7], while ~1000 ms is suggested for beginners [8]. DT values in the range of 1000 ms were applied both in studies of systems designed for fast text typing with eye movements [9–11], as well as interfaces controlled by head movements [12, 13]. Based on the findings and guidelines from our literature review, a DT value of 1000 ms was chosen and positively validated in a pilot study.

2.2 Interaction Systems Compared

Camera Mouse [14] is one of the most popular solutions dedicated for disabled people. The system has been developed at Boston College since 2001 [15] and allows individuals with severe motor impairments, who still are able to move their head, to control a computer with Windows operating system in XP version or above. It is software based, uses any RGB camera built into a computer or connected via USB, and is available for free. The software is able to track the position of individually chosen part of the face, like the top of a nose or a left eye, in order to control the movement of a mouse pointer, while clicking events are emulated using dwell time. Numerous settings suitable for fitting the system to a particular user, as a face characteristic to be tracked, movement sensitivity and dwell time duration, can be applied. For the study described here, Camera Mouse version 2016 was used.

Face Controller is the software created by Polish enterprise Risenbit [16]. Its operation relies entirely on Microsoft Kinect for Windows v2 sensor connected to a Windows 10 computer. Similarly to Camera Mouse, Face Controller allows for a mouse pointer control by head movements. Clicking actions can be provided with different facial gestures (a smile, one or both eyes blinking, mouth opening). The assignment of gestures to different actions (e.g. left and right mouse click), sensitivity of recognition of head movements as well as response for gestures can be adjusted individually. Moreover, the software is capable of voice commands recognition, which can be used to navigate the screen in a special grid mode or assigned to standard UI events, including mouse buttons activity.

In combination with Microsoft Windows 10 system visual keyboard or a third party replacement, both systems allow for a fully-fledged computer interaction, including text typing, with the use of head movements only.

2.3 Software Used

All experiments were conducted using FittsStudy software [17], version 4.2.4.0. The software was developed by J. O. Wobbrock in cooperation with S. Harada, E. Cutrell and I. S. MacKenzie, and is distributed under the New BSD License in the form of binary executable and source code. It is suitable for conducting both one- and two-dimensional reciprocal pointing tasks in accordance with ISO 9241-9 standard [4]. Any combination of movement amplitudes and target sizes can be administered in ordered or fully randomized manner, as well as any number of trials per every condition (i.e., an ID value) chosen can be performed. FittsStudy saves output data in xml files which can be opened after the experiment for further investigation. For example, trajectories for each movement can be analyzed and graphically recreated, including some metrics about movement speed, acceleration/deceleration and jerkiness. Though, the main advantage is automatic calculation of the most crucial values, namely error rates, mean MT and mean TP, and computing of linear regression fit models in accordance with Fitts' paradigm.

The software was modified in certain ways to facilitate our experimental needs. Specifically, all text messages seen by a subject during the test were translated into Polish. Next, a textbox that is being displayed after every finished condition was altered to present a feedback information about the total number of conditions passed and remained, as well as about current and cumulative pointing success rates obtained so far. The latter stays in contrast to the original application, which displayed somewhat less motivating messages about error rates. Moreover, this textbox was converted into a clickable widget, which in turn allowed for a fully hands-free operation during the test. In the absence of a possibility to change, save and reuse experiment defaults, all needed conditions, number of trials and repeats, and other parameters were hard-coded into a new compiled executable, thus providing a way to deploying the test without the need of any manual configuration.

2.4 Data Preparation

When calculating mean TP values, all measured MT values were, according to the recommendations of Soukoreff and MacKenzie [2], corrected for a time constant DT equal 1000 ms. Thanks to this, the MT values are independent of the accepted method of confirming the end of targeted movement, which consequently gives the opportunity to directly compare the TP values obtained in different research and using various pointing devices.

The obtained data was presented both in one-dimensional (1D) and two-dimensional (2D) approach. The difference lies in the method of calculating the standard deviation of the end points cloud coordinates for a given combination of amplitude A and target size W, thus it affects TP values obtained. In 1D approach, the standard deviation is calculated along the approach axis (determined by a straight line drawn between start point and end point of a movement), with omission of the Y coordinate value [18]. Therefore the information about the accuracy of particular pointing takes the form of a signed scalar value: negative one for a movement that ended before the geometric center of the target area, and a positive one for a movement that ended beyond it.

In 2D approach, a centroid (center of mass) of the end points cloud is identified first by means of its X-Y coordinates calculation, which in turn gives the basis for calculating the scalar value of a two-dimensional standard deviation [19]. It should be noted that in the case of the vast majority of studies on pointing devices the calculation of TP value

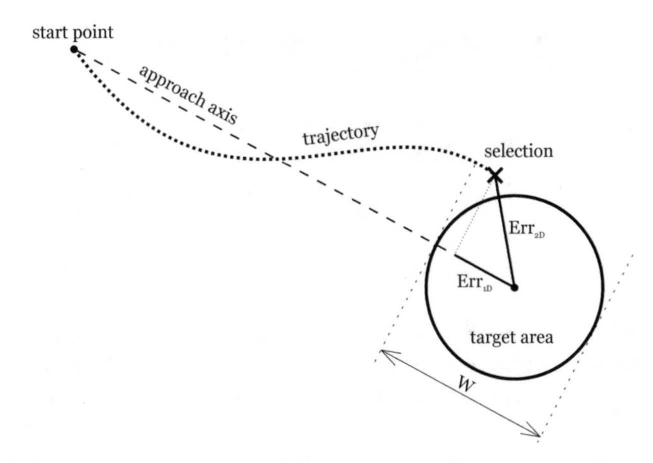

Fig. 3. One-dimensional (1D) and two-dimensional (2D) approach for the accuracy calculation

is made with aforementioned 1D approach. The implications of both approaches on the accuracy calculation are illustrated in the Fig. 3.

2.5 Participants

The full results were obtained for 14 able-bodied subjects aged 17 to 68 years ($M_{age} = 30.0$, $SD_{age} = 14.46$), mostly women (10 vs 4 men), with good computer skill level ($M_{comp} = 4.07$, $SD_{comp} = 0.47$, measured on a five–point scale). A counterbalanced within-subject experiment design was used. Half of the subjects started the performance test with Face Controller, while the other half began with Camera Mouse system. In any case the performance test was preceded by other tasks as well as free system using in order to familiarize participants with given system and its interaction specifics. Overall, the familiarization with both systems took over one hour for each participant.

3 Results

As two measured outcomes demonstrate different features of investigated interaction systems, data for error rates and throughput values are presented and discussed separately.

3.1 Error Rates

The pointings were counted as errors when the end point was outside of a given target area, as indicated by the movements sequence. The metric gives information about the interaction system accuracy, which is especially important in the context of its effectiveness [20]. The error rates for both systems are shown in Fig. 4. Even if Face Controller system generated less errors during the test, the difference is not statistically significant ($F_{1,26} = 2.59$, $p = .12$).

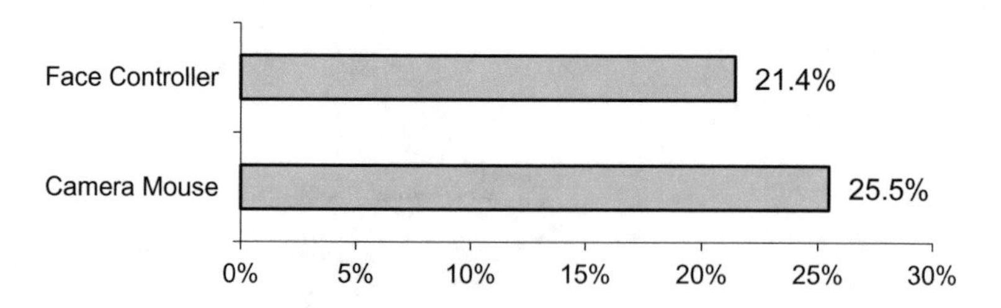

Fig. 4. Error rates obtained for Face Controller and Camera Mouse systems

The error rates results are high, also in comparison with other studies. The errors occurred mainly in more difficult test variants, with ID values of 3.22 and 4.27 bits. Error rates observed for Camera Mouse in similar investigation was 8.1% [12], however it can be explained by lower values of index of difficulty used in the experiment (the highest ID = 3.7). On the other hand, the review of research provided on the basis of ISO 9241-9 standard made by Soukoreff and MacKenzie [2] showed some considerable discrepancies in error rates measured even for the same type of input device. Similarly, high differences for various input devices were noticed by Thomas [21].

3.2 Throughput Values

As the basic system performance measure [20], throughput (TP) values for both systems are shown on Fig. 5. These were separately calculated with 1D and 2D accuracy approaches described earlier.

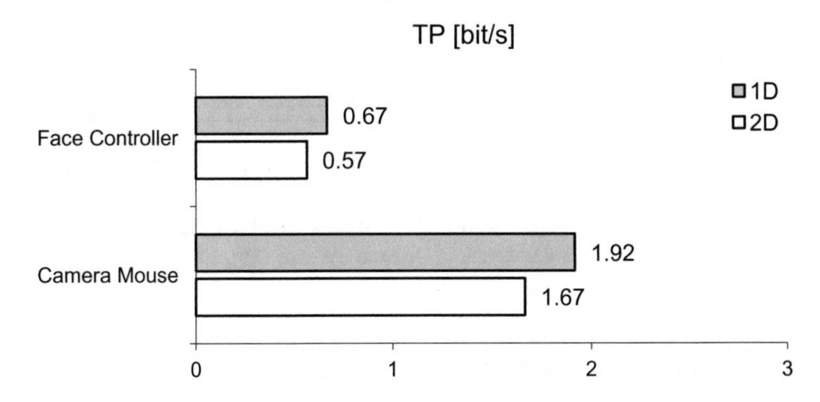

Fig. 5. Throughput values obtained for Face Controller and Camera Mouse (1D and 2D accuracy calculation approach)

For both methods of accuracy calculating, Camera Mouse generated significantly higher TP values ($F_{1,26} = 30.09$, $p < .0001$ for 1D; $F_{1,26} = 28.76$, $p < .0001$ for 2D). Camera Mouse results are comparable with Bian and others research [23]. In their experiment made on 16 participants these authors obtained an average TP value equal to 1.52 bits/s for the system.

Despite the standardization efforts, inconsistency in the calculation of throughput is still common, therefore between-study comparisons should be done with caution [5, 22]. However, according to their descriptions, the researches presented on Fig. 6 were conducted on the basis of the ISO standard.

The comparison of results obtained with other head controlled interaction systems and with healthy subjects presented on the Fig. 6 shows that the throughput values generated by Face Controller are relatively low. As it was discussed in the context of error rates, these effects can be partially caused by relatively high range of difficulty variants used for the experiment, with the highest ID = 4.27 bits, while i.e. ENLAZA system was evaluated for a maximum of ID = 2.58 [26]. However, in Bian et al. research a comparable value of ID = 4.23 was used [23].

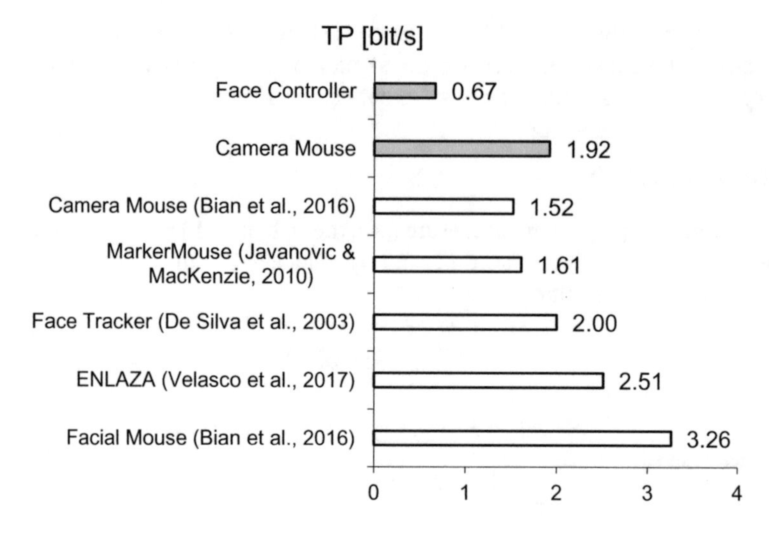

Fig. 6. Comparison of throughput values for different head-controlled pointing systems, including available data for MarkerMouse [24], Face Tracker [25], ENLAZA [26], and Facial Mouse [23]

4 Discussion and Closing Remarks

Comparing both investigated systems, Camera Mouse showed its advantage, with throughput value (TP) 2.8 times higher than Face Controller result (see Fig. 5) and a comparable error rate (see Fig. 4). Before the experiment we assumed that Kinect sensor utilized by Face Controller system gives a big advantage in comparison to a regular RGB camera used by Camera Mouse in terms of the ability to precisely determine the position of the head in 3D space, and face plane and its normal vector calculations used to resolve the desired position of the screen pointer. Contrary to a RGB camera, depth camera used in Kinect sensor works with its own infrared light source, therefore is not sensitive to changes in the amount and direction of visible light from windows or any artificial sources. This kind of problems were, in fact, observed during longer work with Camera Mouse by some participants in our study, so it should be taken into account in the process of designing and assessment of face/head controlled input systems [27].

Face Controller as a system oriented to enable the use of a computer for people with disabilities uses various modes of interaction (voice recognition, facial gestures, head movements). Perhaps due to this versatility it does not implement the most optimal algorithms for controlling the screen pointer by head movements (for example, the motion mapping function used, a dead zone size, absence of an automatic recalibration during operation, although it is possible to manually activate it), as opposed to specialized systems that allow interaction only by means of head tracking. Moreover, Face Controller, using a sensor developed for motion games which was designed for tracking the entire human silhouette, not just the head, can be potentially limited by the quality of standard libraries provided by the manufacturer (Microsoft).

For standardization reasons dwell time was chosen in both investigated systems as the activation method. However, whereas for Camera Mouse dwell time is default and the only possible way to emulate crucial UI events, Face Controller allows for individual choice of activation method, including voice commands and several facial gestures. Experiments with different ways of interacting with mobile phones on the basis of facial tracking with front-facing camera use shown that dwell time generated the least error rates (1.8%), while facial gestures were much more error productive (smile – 19.9%; blink selection – 28.7%) [28]. Similar observations were made by Tuisku et al., who found dwell time technique produced the least mean error percentages (15.7%) in comparison to frowning (21.7%) or smiling technique (16.1%) [29]. Nevertheless, the efficiency measured by throughput values could be underestimated for Face Controller. In the research of facial tracking made by Cuaresma and MacKenzie, throughput was 26% higher for smile selection than for dwell selection [28].

Camera Mouse is a mature system developed for nearly 20 years (the first publication appeared in 2000), created and improved by scientists [15]. The new version released in 2018 can be even more accurate and efficient than the previous one, used in the experiment described here. Furthermore, the system is available free of charge and is very tolerant for various software and hardware configurations. Contrary to that, Face Controller was created as a hobby project and it is a very young system, with a large field for improvements both in the configuration area of its working parameters and in the mode of operation itself. Furthermore, it relies on the hardware that is already discontinued – Microsoft has ended the Kinect product line on October 2017.

The presented research has its limitations. The participants' interaction time with both systems was relatively short (2 h maximum). Probably this does not impact the comparison data, however after longer training lower error rates and higher throughput values should be expected. Only one method of activation (dwell time) was applied, even if Face Controller gives many options of choice in this particular area. One of the things that are planned to incorporate into our future studies is some sort of user satisfaction assessment conducted along with accuracy and performance measurements. As for users who are the main recipients of such alternative interaction systems, that is, individuals with a broad range of motion impairments, we already conducted a study aimed at measuring interaction performance with subjects from this particular group, but the results are yet to be published.

Considering that our approach was positively validated with both physically capable and impaired subjects, results and conclusions reported in this paper may present a certain value to usability researchers interested in non-standard methods and techniques of human-computer interaction, which are beneficial for a wide variety of users in different possible contexts. Additionally, creators of head-controlled interaction systems can use the methodology described here to evaluate quality of their products.

Acknowledgement. We thank the Polish society "Association for Your New Opportunities" (Stowarzyszenie "Twoje nowe możliwości") for making Face Controller system available for our usability testing. Special thanks for students Karolina Bochenek and Sandra Trzeciak for their cooperation.

References

1. Fitts, P.M.: The information capacity of the human motor system in controlling the amplitude of movement. J. Exp. Psychol. **47**(6), 381–391 (1954)
2. Soukoreff, R.W., MacKenzie, I.S.: Towards a standard for pointing device evaluation, perspectives on 27 years of Fitts' law research in HCI. J. Hum.-Comput. Stud. **61**(6), 751–789 (2004)
3. Douglas, S.A., Kirkpatrick, A.E., MacKenzie, I.S.: Testing pointing device performance and user assessment with the ISO9241, Part 9 standard. In: Proceedings of the ACM Conference on Human Factors in Computing Systems CHI 1999, pp. 215–222. ACM, New York (1999)
4. International Organization for Standardization (ISO): ISO 9241–9. 2000. Ergonomics Requirements for Office Work with Visual Display Terminals (VDTs) – Part 9: Requirements for Non-Keyboard Input Devices. International Organization for Standardization (2002)
5. International Organization for Standardization (ISO): ISO 9241–411. 2012. Ergonomics of Human-System Interaction – Part 411: Evaluation Methods for the Design of Physical Input Devices. International Organization for Standardization (2012)
6. Müller-Tomfelde, C.: Dwell-based pointing in applications of human computer interaction. In: Baranauskas, C., Palanque, P., Abascal, J., Barbosa, S.D.J. (eds.) Human-Computer Interaction – INTERACT 2007, vol. 4662. Springer, Berlin (2007)
7. Majaranta, P., Räihä, K.J.: Text entry by gaze: Utilizing eye-tracking. In: MacKenzie, I.S., Tanaka-Ishii, K. (eds.) Text Entry Systems: Mobility, Accessibility, Universality, pp. 175–187. Morgan Kaufmann, San Francisco (2007)
8. Majaranta, P., Ahola, U.K., Špakov, O.: Fast gaze typing with an adjustable dwell time. In: Proceedings of the SIGCHI Conference on Human Factors in Computing Systems (CHI 2009), pp. 357–360. ACM, New York (2009)
9. Špakov, O., Miniotas, D.: On-line adjustment of dwell time for target selection by gaze. In: Proceedings of the Third Nordic Conference on Human-Computer Interaction (NordiCHI 2004), pp. 203–206. ACM, New York (2004)
10. Majaranta, P., MacKenzie, I.S., Aula, A., Räihä, K.J.: Effects of feedback and dwell time on eye typing speed and accuracy. Univ. Access Inf. Soc. **5**(2), 199–208 (2006)
11. Räihä, K.J., Ovaska, S.: An exploratory study of eye typing fundamentals: dwell time, text entry rate, errors, and workload. In: Proceedings of the SIGCHI Conference on Human Factors in Computing Systems (CHI 2012), pp. 3001–3010. ACM, New York (2012)
12. Magee, J., Felzer, T., MacKenzie, I.S.: Camera Mouse + ClickerAID: Dwell vs. single-muscle click actuation in mouse-replacement interfaces. In: Antona, M., Stephanidis, C. (eds,) Universal Access in Human-Computer Interaction. Access to Today's Technologies. UAHCI 2015, vol. 9175. Springer, Cham (2015)
13. Zuniga, R., Magee, J.: Camera Mouse: dwell vs. computer vision-based intentional click activation. In: Antona, M., Stephanidis, C. (eds.) Universal Access in Human–Computer Interaction. Designing Novel Interactions. UAHCI 2017, vol. 10278, pp. 455–464. Springer, Cham (2017)
14. Camera Mouse website. http://www.cameramouse.org/. Accessed 20 May 2018
15. Betke, M., Gips, J., Fleming, P.: The camera mouse: visual tracking of body features to provide computer access for people with severe disabilities. IEEE Trans. Neural Syst. Rehabil. Eng. **10**(1), 1–10 (2002)
16. Face Controller website. http://www.face-controller.com/. Accessed 10 May 2018
17. FittsStudy https://depts.washington.edu/madlab/proj/fittsstudy/. Accessed 19 May 2018

18. Wobbrock, J.O., Jansen, A., Shinohara, K.: Modeling and predicting pointing errors in two dimensions. In: Proceedings of the SIGCHI Conference on Human Factors in Computing Systems, pp. 1653–1656. ACM (2011)
19. Wobbrock, J.O., Shinohara, K., Jansen, A.: The effects of task dimensionality, endpoint deviation, throughput calculation, and experiment design on pointing measures and models. In: Proceedings of the SIGCHI Conference on Human Factors in Computing Systems, pp. 1639–1648. ACM (2011)
20. International Organization for Standardization (ISO). ISO 9241-11:2002, Ergonomic requirements for office work with visual display terminals (VDTs) - Part 11: Guidance on usability (2002)
21. Thomas, P.R.: Performance, characteristics, and error rates of cursor control devices for aircraft cockpit interaction. Int. J. Hum.-Comput. Stud. **109**, 41–53 (2018)
22. Mackenzie, I.S.: Fitts' Throughput and the remarkable case of touch-based target selection. In: Artificial Intelligence and Lecture Notes in Bioinformatics, vol. 9170, pp. 238–249 (2015)
23. Bian, Z.P., Hou, J., Chau, L.P., Magnenat-Thalmann, N.: Facial position and expression-based human-computer interface for persons with tetraplegia. IEEE J. Biomed. Health Inform. **20**(3), 915–924 (2016)
24. Javanovic, R., MacKenzie, I.S.: MarkerMouse: mouse cursor control using a head-mounted marker. In: Miesenberger, K., Klaus, J., Zagler, W., Karshmer, A. (eds.) Computers Helping People with Special Needs. ICCHP 2010, vol. 6180. Springer, Berlin (2010)
25. De Silva, G.C., Lyons, M.J., Kawato, S., Tetsutani, N.: Human factors evaluation of a vision-based facial gesture interface. In: Computer Vision and Pattern Recognition Workshop, CVPRW 2003, vol. 5. IEEE, Madison (2003)
26. Velasco, M.A, Clemotte, A., Raya, R., Ceres, C., Rocon E.: Human-computer interaction for users with cerebral palsy based on head orientation. Can cursor's movement be modeled by Fitts's law? Int. J. Hum.-Comput. Stud. **106**, 1–9 (2017)
27. Manresa-Yee, C., Varona, J., Perales, F.J., Salinas, I.: Design recommendations for camera-based head-controlled interfaces that replace the mouse for motion-impaired users. Univ. Access Inf. Soc. **13**(4), 471–482 (2014)
28. Cuaresma, J., Mackenzie, I.S.: FittsFace: exploring navigation and selection methods for facial tracking. In: Artificial Intelligence and Lecture Notes in Bioinformatics, vol. 10278, pp. 403–416 (2017)
29. Tuisku, O., Rantanen, V., Špakov, O., Surakka, V., Lekkala, J.: Pointing and selecting with facial activity. Interact. Comput. **28**(1), 1–12 (2016)

Analysis of Brain Activity Changes Evoked by Virtual Reality Stimuli Based on EEG Spectral Analysis. A Preliminary Study.

Martyna Wawrzyk(✉), Kinga Wesołowska, Małgorzata Plechawska-Wójcik, and Tomasz Szymczyk

Lublin University of Technology, Lublin, Poland
{martyna.wawrzyk,kinga.wesolowska}@pollub.edu.pl,
{m.plechawska,t.szymczyk}@pollub.pl

Abstract. The purpose of the paper is to apply EEG spectral analysis to compare brain activity during Virtual Reality (VR) stimulation. This preliminary study covers analysis of EEG data from three participant at the age of 20–25. In the examination 10–20 EEG system was applied. The examination consists of resting state recording as well as five intervals with different VR simulation intended to evoke different emotions such as anger, fear or excitement. These scenarios were presented for participant with special goggles designed for displaying VR. Collected data has been subjected to preprocessing covering filtering and artifacts elimination and spectral analysis. As a result maps of power spectral from each intervals and wave band distribution were obtained. Changes were observed in alpha/theta ratio for various emotional sates. What is more, the amplitude value of alpha wave indicated strong changes among particular intervals.

Keywords: EEG · Spectral analysis · Emotion recognition · Virtual reality

1 Introduction

Virtual Reality (VR) is a picture of artificial reality created based on informatics technologies. VR enables to generate a faithfully rendered, three-dimensional scenery [1]. This technology might be easily applied in research scenarios, in order to, for example, evoke particular emotions or fears [2]. There are many well-known phobias which might be simulated and analysed with VR. VR technique enables to generate and present objects evoking fears such as spiders, dogs, wolves, swarms [3, 4]. It is also possible to generate environment such as infinite, open spaces or small, cramped rooms[t2] and examine the impact of their projection triggering natural fears such as fear of heights, acrophobia or anxiety of space [5, 6].

For example in [7] VR helmet was used to analyse the fear of spiders. This aim of the study was to show whether the therapy of fear of spiders using virtual reality was effective. Participants in the group treated with VR underwent four one-hour therapy sessions. According the survey, Eighty-three percent of patients from the group of person treat using VR showed significant clinical improvement what confirms the effectiveness

L. Borzemski et al. (Eds.): ISAT 2018, AISC 852, pp. 222–231, 2019.
https://doi.org/10.1007/978-3-319-99981-4_21

of VR in therapy for fears. Analysis of Virtual Reality Exposure Therapy (VRET) shows that it is more and more popular in research and treatment of phobias. Searches from electronic databases gave 52 studies, including 21 studies (300 people) applying VRET. Although the meta-analysis showed significant decreases in anxiety symptoms after VRET, moderate analyzes were limited due to inconsistent reporting in VRET literature. This indicates the need for further scientific research that provides uniform and detailed information on the presence, intensity, anxiety and duration of phobias and demographics [8].

Regardless such strong emotions like phobias, every day people experience dozens of emotions caused by various stimuli. Despite this, it is very difficult to define them. They affect the behaviour and decisions that people make, they affect the mood, social life and vulnerability, affect the physiological and mental state. Awareness of the emotional state and the current level of attention is of great importance not only in medicine or psychology, but also for example in marketing, where knowing the user mood would allow for better customization of displayed content, i.e. advertising or proposed films [9].

Emotional state might be analysed based on surveys and tests, but this solution is subjective. As objective measure, biomedical signals might be applied, such as EEG.

Electroencephalogram (EEG) is a recording of brain waves, which is the electrical impulses generated by the brain. Traditionally it is used to diagnose diseases of the central nervous system, however its modern applications cover brain-computer interfaces, cognitive load analysis or emotion recognition. The EEG signal is the average of electrical activity from a certain area of the brain. EEG data is characterized by very low amplitude recorded write about 100 μV. The highest levels of voltage EEG are recorded at the time of the epileptic seizure, and usually have a few hundred microvolts.

Traditionally, EEG is used to diagnose diseases of the central nervous system, however its modern applications cover brain-computer interfaces, cognitive workload analysis or emotion recognition. Including classification of emotions.

Electroencephalography is one of the most widespread and most commonly used methods of studying brain work. The present paper focuses on checking the impact of visual stimuli on the level of brain wave activity by analyzing EEG signal. The spectral power maps and a table of brain wave activity of three participants are presented. This preliminary study was performed to adjust analysing procedure before the extended research.

2 Related Work

Emotions evoked by visual and sound stimuli were analyzed based on EEG data for many years. First item confirm that change of frontal midline(Fm) theta wave activity is related to type of music (pleasant or unpleasant] and it proves Fm theta is strongly associates with different emotional states [8]. Also familiarity of song has a meaning influence on power spectra of EEG and the brain functional in certain levels [10]. In case of average alpha rhythm frequency compared with the baseline signal increases for emotions of joy and anger and decrease for sorrow and fear [11]. Also other research

[12] confirms that alpha is related with variable emotions (increase alpha-coherence for aggression and joy, decrease for anxiety and sorrow).

Additionally analysis of EEG signals is used to recognize emotional stress states evoked by visual stimuli [13] and mechanisms of initialization disgust and phobia [14, 15]. The level of stress and effects of treatment were also widely analyzed [16] as well as tracking the trajectory of emotional changes [17]. Many research covers also attempts of emotion classification based on pictures or short video clips [18, 19]. Research show the relation between activity of the frontal brain part in the case of experiencing negative and positive emotions and changes in alpha and beta rhythm power related to changes in grade hemisphere's inactivity [20].

3 Research Scenario and Data Acquisition

The study was conducted with the participation of three students of the Lublin University of Technology of the same sex, between the ages of 20–25. The interview did not revealed any phobias of subjects. The participants' task was to view the several VR presentations displayed on the dedicated goggles (Samsung Gear). The single study consisted of six scenarios preceded and ended with a resting EEG signal recording. Such one minute signal was gathered in order to obtained baseline signal to compare. Applied scenarios were as follows:

- RELAX('1') - 60s of resting, calming music was played, no graphical stimuli were used;
- DINOSAUR('2') - 90s of VR simulation of prehistoric world with a herbivorous dinosaur appearing suddenly very close to the user;
- AVENGERS ('3') - 45s fragment of popular movie presented in 3D covering fast changing scenery, combat effects, shots, explosions as well as huge open spaces and high heights;
- ROLLERCOASTER('4') - 120s of VR simulation of extreme rollercoaster ride covering fast movement at high altitudes as well as sharp corners and steep inclines causing rapid acceleration;
- SPIDERS ('5') - 60s display of VR application enabling moving around a room with different size spiders walking on the floor and walls;
- EYES OPEN ('6') - 60s of resting, no graphical stimuli were used.

Scenarios were adjusted to evoke different feelings and emotions of subjects, varying from pleasure to fear. During the examination, after each scenario 30s minute of break was introduced.

In the experiment, dedicated virtual reality goggles (Samsung Gear VR) and synchronized Samsung phone were applied. The Oculus Rex application was used in the study is. WinEEG was used to record and monitor EEG data. What is more, ECG signal was measured to check whether the visualizations presented can contribute to the increase of blood pressure.

4 Data Analysis

4.1 Data Obtained in the Study

Mitsar EEG 201amplifier was used to record the EEG data. It is a 21 channel amplifier enabling data recording and real-time signal monitoring. The data was transmitted to the computer with the EEG Studio software.

The study used 19 electrodes arranged in accordance with the 10–20 EEG standard. Two reference electrodes (A1, A2) were placed on the ears. The ground electrode was placed in the middle part of the frontal lobe. The recording frequency was 500 Hz.

4.2 Pre-processing

Pre-processing of the EEG signal was performed in the WinEEG software. The first step covered noise reduction using notch filter and basic band pass filter (1 Hz to 50 Hz). The next step covered signal filtration with standard high-pass filter to remove noised signal below 3 Hz. Filters might be applied to eliminate unnecessary components or to extract a given characteristic wave.

EEG signal is highly exposed into artifacts. For example, the disturbance of the received signal might come from muscle activity, such as breathing, speaking, blinking. Therefore, the patient should remain at rest and refrain from speaking while taking measurements. The problem might be also caused by the lack of proper adhesion of the electrode to the skin of the subject, or sweating.

Removal of artifacts from the EEG signal was performed by independent component analysis (ICA). Three ICA components were removed to obtain the target signal. The analysis of EEG data was made on the Monopolar 1- Average [A1/A2] montage.

4.3 Feature Extraction

The features extracted to the needs of the analysis cover spectral power values reflecting the activity of brain waves under the influence of various visual stimuli derived from VR glasses. Alpha, beta-1 (low beta) and beta-2 (regular beta) and theta waves were extracted as well as alpha/theta ratio. Theta component contains waves with a frequency in the range of 4–7 Hz. The alpha band contains waves with a frequency of 8–13 Hz. The beta-1 component corresponds of frequencies between 13–22 Hz and beta-2 23–30 Hz [21]. Alpha/theta ratio determines the relationship of alpha waves responsible for the state of relaxation to theta waves associated with intense emotions. What is more, maps of spectral power were also generated.

The spectral analysis was performed in WinEEG. The recording of the EEG signal consisted of scenarios in which various visual stimuli originating from the VR glasses were displayed. The analysis was performed separately for each of the examined persons. A 10-second fragments of the EEG signal were selected for each scenario. Selected fragments were subjected to spectral analysis in order to isolate different levels of brain wave activity to analyze changes in the brain activity in response to presented visual stimuli. The results of the analysis were presented in form of spectral power maps and power table.

5 Results

Analysis of EEG data covered boxplots for each scenario and for each subject. The boxplots were prepared for values of alpha, theta, beta1 and beta2 waves, for amplitude and power. Data were received from four groups of electrodes (I-F7,F8,F3,Fz,F4,C3,Cz,C4, II-T5,P3,Pz,Cz,P4,T6,O1,O2, III-F3,C3,Cz,T5,P3, IV-F4,C4,Cz,P4,T6). The most significant differences were noticed for amplitudes of alpha wave from second group of electrodes. Figure 1 illustrates the change of alpha amplitude depending on presented scenarios. For all participant similar values, without statistical differences of amplitude observed the for first scenario (RELAX STATE) representing the base of the signal (obtained for resting state). The boxplots of scenario no 5 (SPIDERS) for each participant presents a wider spread compared to others scenarios. It may indicates about occurrence of phobia to spiders. Changes of alpha wave are individual, each participant show different tendency. This proves the varies reception of emotions by each person, but the most significant differences between boxplots were observed at the P1.

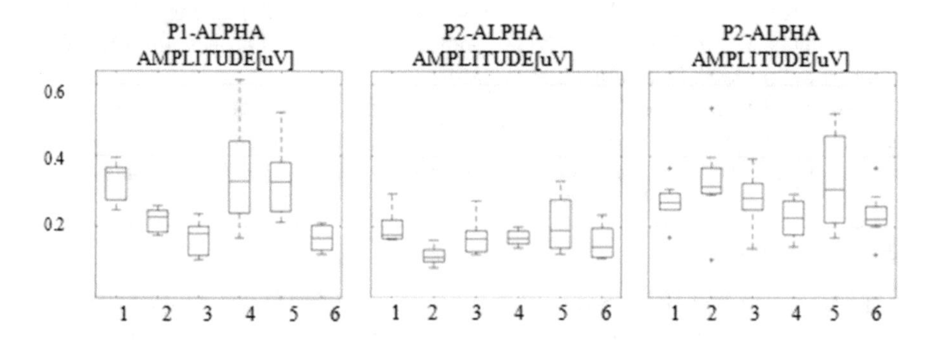

Fig. 1. Boxplots for scenarios for participants P1, P2 and P3

FTT spectral analysis method was used to generate maps of spectral power for alpha/theta ratio and for amplitude value of alpha wave. Figures 2 and 3 present maps of a various activity of brain's waves in each scenario. Alpha/theta ratio is widely known as a focus factor. Increase of its value is determined with state of relaxation and calmness, decrease with state of focus and attention. Figure 2 presents changes of alpha/theta ratio for all scenarios. For all examined persons maps of spectral power are discriminable what proves the significant differences between them. Also maps clearly proves results presented by boxplots (Fig. 1). For P1 changes are the most visible. These results are also confirmed in maps of alpha amplitude (Fig. 3). Figures 2 and 3 show distinctly noticeable changes in maps for scenarios 2 and 3 maps(P1 and P2).

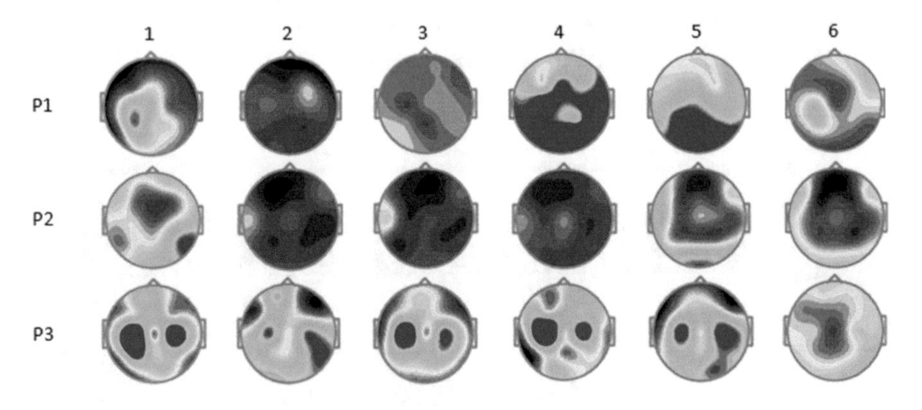

Fig. 2. Alpha/theta ratio maps for all participant

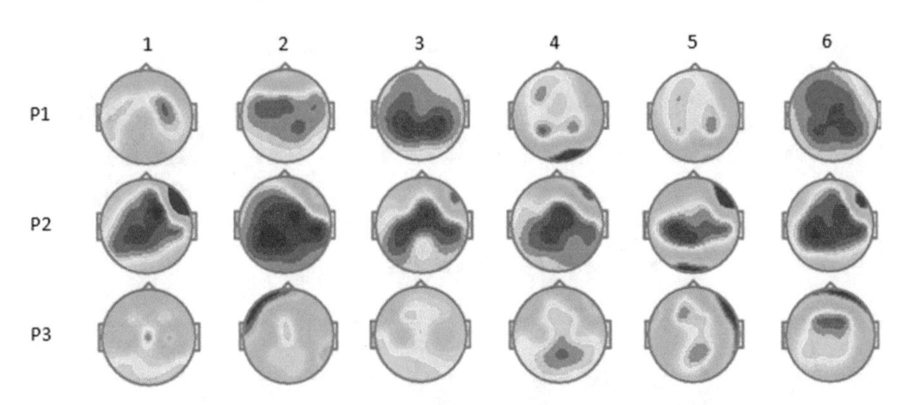

Fig. 3. Alpha amplitude maps for all participant

Additionally alpha and theta band distribution from selected electrodes (F3, F4, C3, C4, Cz, P3, P4, O1, O2) are presented (Figs. 4 and 5) illustrating differences between scenarios and between participants. These differences are more noticeable for alpha wave graphs.

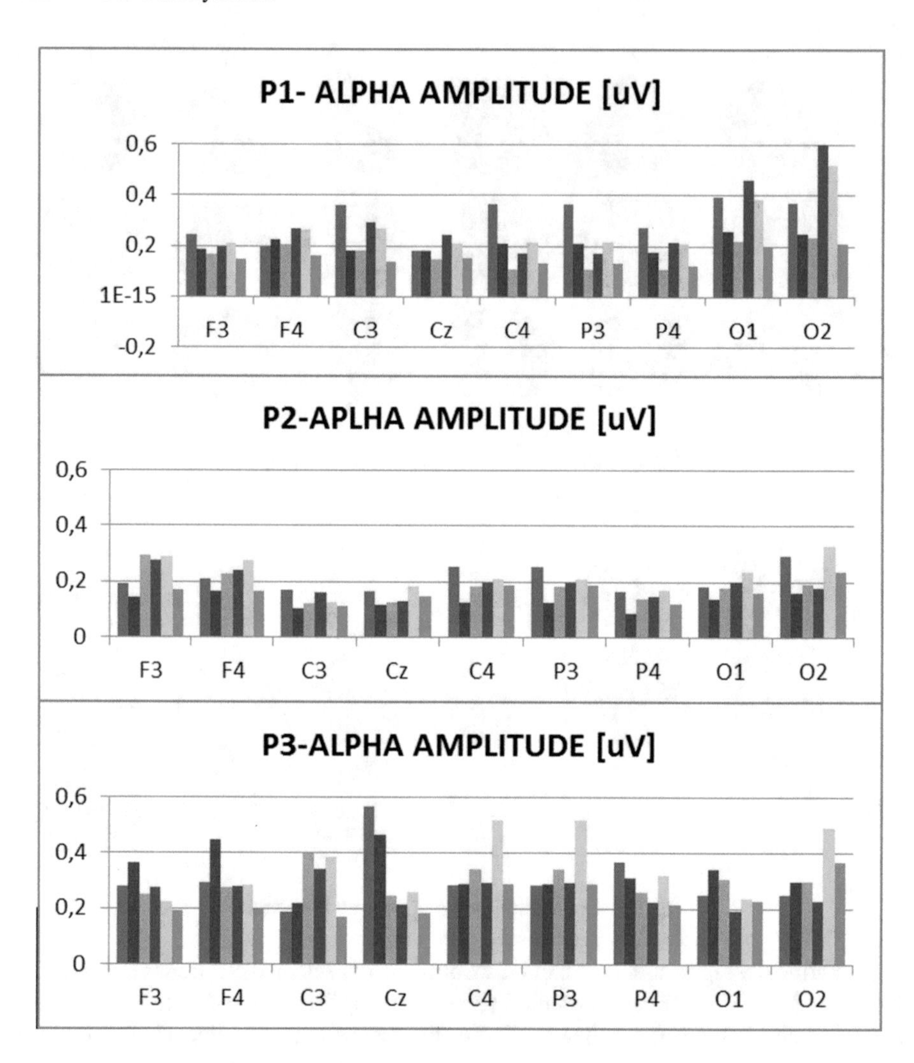

Fig. 4. Alpha band distribution for P1, P2 and P3

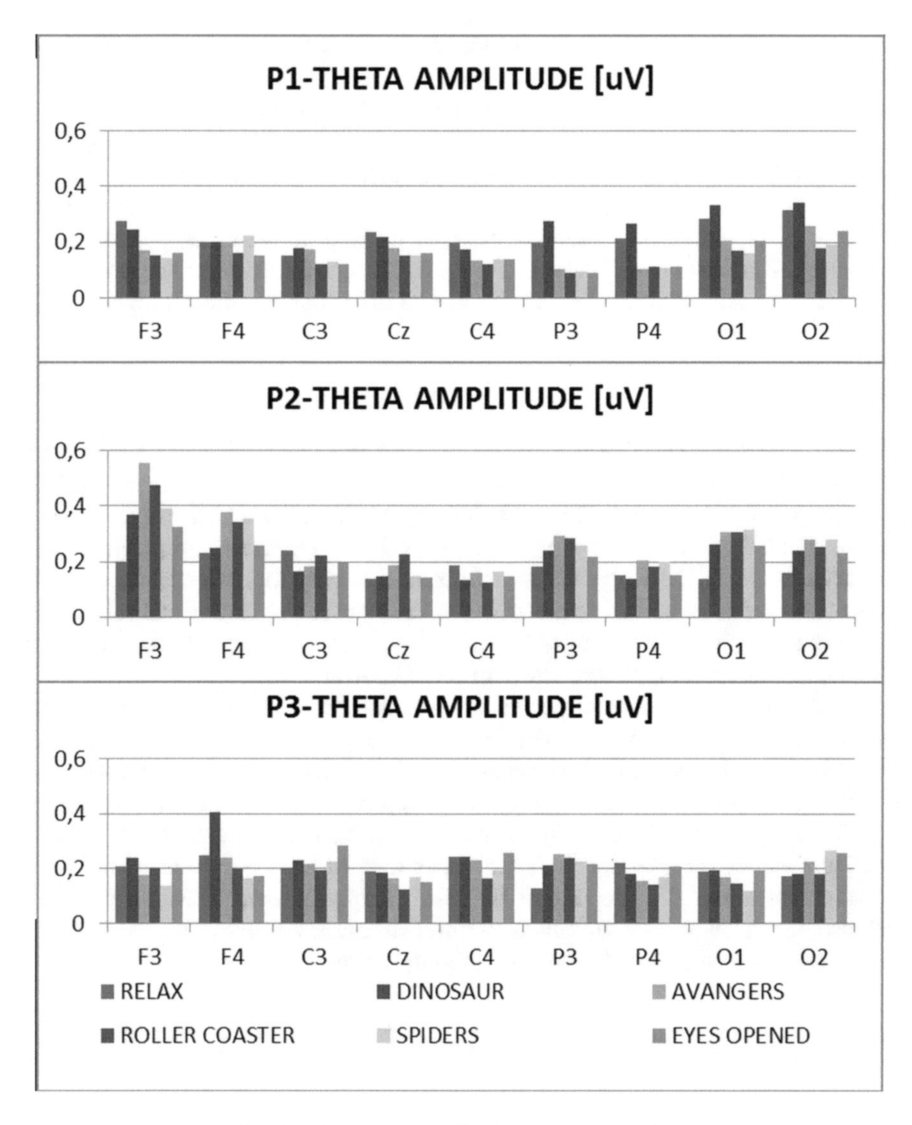

Fig. 5. Theta band distribution for P1, P2 and P3

6 Discussion

The paper discussed the possibility of using EEG spectral analysis as method to compare various emotional sates. Analysis of subsequent parts showed the impact of visual impressions provided by the VR system.

Obtained research results introduce the changes of amplitude and power alpha and theta depending on particular scenarios. The most noticeable changes are in alpha waves.

Based on results, alpha wave can be considered as a strong indicator to compare different emotional states.

Obtained results are preliminary. Only three participants were analysed, that is why no statistical analysis was presented. This study was performed to assess the possibility to combine VR and EEG analysis as well as to indicate the most promising features. Extended research with more subjects should be done in order to confirm the presented hypothesis. As individual variations between participants exist, it is possible that extended analysis will reveal changes for beta1 and beta 2 waves. In the present study differences for beta1 and beta2 for P1, P2 and P3 exists but they are not regular.

7 Conclusion

The aim of the work was to apply EEG spectral analysis to compare brain activity during Virtual Reality (VR) stimulation. The study was conducted on three subjects aged 20–25. The examination consisted in recording the resting state and five intervals with various simulations of virtual reality that are to evoke different emotions, such as anger, fear or excitement. The analysis was performed separately for each examined person. For each scenario, 10-second fragment of the EEG signal was chosen and subjected to spectral analysis to isolate levels of brain wave activity. The results of the analysis are presented in the form of spectral power maps and power table.

The resting state for all three people is at a similar level. Visible differences noticeable in scenarios no 4 and 5. The results for person P1 in scenario 4 (ROLLERCOASTER) shows the highest amplitude value in comparison to other scenarios. Results for person P2 shows the approximate amplitude value in each scenario, while the P3 the highest amplitude value in the scenario number 5 (SPIDERS). Different amplitude values for all subjects in particular scenarios are the result of individual variation and different perception of the visual stimuli presented. Future work will include a larger number of subjects to perform statistical analysis. The results for the person P1 gives perspective for obtaining significant results.

Acknowledgement. We thank Małgorzata Bernat, Jakub Kopczyk, Monika Mańko for assistance in conducting research.

References

1. Szymczyk, T., Montusiewicz, J., Skulimowski, S.: An educational historical game using virtual reality. In: INTED 2018 Proceedings 2018, pp. 5964–5971 (2018)
2. Plechawska-Wojcik, M., Semeniuk, A.: Usage analysis of VR headset in simulated stress situations. In: INTED 2018 Proceedings (2018)
3. Sarlo, M., Palomba, D., Angrilli, A., Stegagno, L.: Blood phobia and spider phobia: two specific phobias with different autonomic cardiac modulations. Biol. Psychol. **60**, 91–108 (2002)
4. Chamove, A.S.: Therapy toy for spider phobics. Int. J. Clin. Health Psychol. (2006)

5. Stanica, I.C., Dascalu, M.I., Moldoveanu, A., Bodea, C.N., Hostiuc, S.: A survey of virtual reality applications as psychotherapeutic tools to treat phobias. In: Conference Proceedings of: eLearning and Software for Education (eLSE), vol 01, pp. 392–399 (2016)
6. Ibrahim, N., Muhamad Balbed, M.A., Yusof, A.M., Mohammed Salleh, F.H., Singh, J., Shahrul, M.S.: Virtual Reality approach in acrophobia treatment. In: 7th WSEAS International Conference on Applied Computer & Applied Computational Science (ACACOS 2008), Hangzhou, China, pp. 194–197 (2008)
7. Parsons, T.D., Rizzo, A.A.: Affective outcomes of virtual reality exposure therapy for anxiety and specific phobias: a meta-analysis. J. Behav. Ther. Exp. Psychiatry **39**, 250–261 (2008)
8. Sammler, D., Grigutsch, M., Fritz, T., Kolesch, S.: Music and emotion: electrophysiological correlates of the processing of pleasant and unpleasant music. Psychophysiology **44**(2), 293–304 (2007)
9. Picard, R.: Affective Computing, Brain Informatics. MITpress, Cambridge (2000)
10. Thammasan, N., Moriyama, K., Fukui, K., Numao, M.: Familiarity effects in EEG-based emotion recognition. Brain Inform. **4**(1), 39–50 (2017)
11. Kostyniuna, M.B., Kulikov, M.A.: Frequency characteristics of EEG spectra in the emotions. Neurosci. Behav. Physiol. **26**(4), 340–343 (1996)
12. Hinrichs, H., Machleidt, W.: Basic emotions reflected in EEG coherence. Int. J. Psychophysiol. **13**(3), 225–232 (1992)
13. Hosseini, S.A., Naghibi-Sistani, M.B., Khalilzadeh, M.A., Niazmand, V.: Higher order spectra analysis of EEG signals i emotional stress states. In: 2010 Second International Conference on Information Technology and Computer Science, pp. 60–63 (2010)
14. Lee, M., Kim, H., Kang, H.: EEG-based analysis of auditory stimulations generated-from watching disgust-eliciting videos. J. Korea Multimedia Soc. **19**(4), 756–764 (2016)
15. de Jong, P.J., Merckelbach, H., Muris,P., Pool, K.: Resting EEG asymmetry and spider phobia, Anxiety Stress & Coping (1998)
16. Haak, M., Bos, S.,Panic, S., Rothkrantz, L.J.M.: Detecting stress using eye blink and brain activity from EEG signals. In: Proceeding of the 1st driver car interaction and interface (DCII 2008), pp. 35–60 (2009)
17. Lu, B.L., Nie, D., Wang, X.: Emotional state classification from EEG data using machine learning approach. Neurocomputing **129**, 94–106 (2014)
18. Israsena, P., Jirayucharoensak, S., Pan-Ngum, S.: EEG-Based emotion recognition using deep learning network with principal component based covariate shift adaptation, Sci. World J. (2014)
19. Kwon, M., Giyoung, L., Minho, L., Sri Swathi, K.: Emotion recognition based on 3D fuzzy visual and EEG features in movie clips. Neurocomputing **144**, 560–568 (2014)
20. Niemic, C.P.: Studies of emotion: a theoretical and empirical review of psychophysiological studies of emotion. J. Undergraduate Res. **1**, 15–18 (2002)
21. Zarjam, P., Epps, J., Chen, F.: Spectral EEG features for evaluating cognitive load. In: Annual International Conference of IEEE Engineering Medicine Biology Society, pp. 3841–3844 (2011)

Big Data, Knowledge Discovery and Data Mining, Software Engineering

Application of Quantum k-NN and Grover's Algorithms for Recommendation Big-Data System

Marek Sawerwain(✉) and Marek Wróblewski

Institute of Control & Computation Engineering, University of Zielona Góra,
Licealna 9, 65-417 Zielona Góra, Poland
{M.Sawerwain,M.Wroblewski}@issi.uz.zgora.pl

Abstract. The growing size of modern databases and recommendation systems make it necessary to use a more efficient hardware and also software solutions that will meet the requirements of users of such systems. These requirements apply to both the size of databases and the speed of response, quality of recommendation. The evolving techniques of the quantum computational model offer a new computing possibilities. This chapter presents an approach based on the quantum algorithm of k-nearest neighbours, and Grover's algorithm for building a recommendation system. The algorithmic correctness of the proposed system is analysed. The advantages of the presented solution are also indicated such as exponential capacity system and response speed which are independent of the amount of classic data stored in the quantum system. The final computational complexity does not depend on the amount of features but only on the length of the feature.

Keywords: Quantum k-NN · Grover's algorithm
Recommendation systems · Big-data · Quantum algorithms

1 Introduction

The continuous inflow large amounts of new data requires the optimization the way of gathering and organizing into structures. The recommendation system is a set of features a mechanisms, and algorithms whose task is to give a result in the form of predictions about a particular set of elements based on the provided input data of the feature vector [14,16]. The construction of systems is based on traditional structural as well as unstructured databases. Their task is to issue prediction recommendations to the end-user of the system based on the data collected and analysed a set of input features with a specific user in the context of the data set [9].

Currently known techniques and algorithms of the quantum computational model allow as shown in this chapter to construct a correctly operating recommendation system for classical data sources. The chapter proposes and discusses

L. Borzemski et al. (Eds.): ISAT 2018, AISC 852, pp. 235–244, 2019.
https://doi.org/10.1007/978-3-319-99981-4_22

the construction of such a system based on the currently known quantum algorithm (a book [10] contains wide introduction to quantum computation models) of k-nearest neighbors [17,18] and in order to improve the quality of recommendation, the Grover algorithm to searching for the key in an unstructured database [3].

The document was organized in the following way. Section 2 discusses an example of a classic database and the construction of a quantum register that can be successfully used to carry out the recommendation process based on the characteristics specified by the user [1,2,15].

Algebraic correctness, i.e. a description of individual probability distributions obtained through the implementation of the recommendation system is described in the Sect. 3. In addition, this part presents schematic of the recommendation algorithm and general diagram of quantum circuit implementing the proposed solution.

The Sect. 4 presents numerical example illustrating how the probability distribution in quantum register behaves after the implementation of the main calculation steps. Naturally, it has an illustrative character. However, it shows that the system works correctly and the recommended database element will be proposed for recommendation with the highest probability of measurement.

The completion of work is the point Sect. 5, where briefly summarized the chapter and placed a brief discussion on the computational complexity of the proposed system. The last element of the chapter are acknowledgements and bibliography.

2 Hybrid Quantum-Classical Approach to Recommendation Systems

The solution presented in this chapter is a hybrid system. Because the full database from which items are recommended based on the user's suggestions and needs is naturally stored in the classic system. For the purposes of practical verification of the proposed system a database from the OMDB was used [11]. Wherein, of course it is not necessary that all of the classic database stored within the quantum system. Only two data columns, the identifier of the recommended elements, and the element's feature expressed as a binary string are relevant. Therefore, the two columns id and feature are used and encoded with a suitable quantum register.

Diagram of the most important database elements presented in Fig. 1. In addition, its schematic within the quantum system is shown in Fig. 2.

The exemplary database includes over 12,000 items. However, in the case of quantum calculations, the quantum register stores an exponential amount of data [13]. Therefore for the referenced movie database, to save indexes of individual films, you need to apply only 15 qubits: $2^{15} > 12,000$.

Remark 1. Currently, experimental installations of quantum computing systems [8] offer access to 20 qubits. This is not a sufficient number, however, the technological progress seems to soon allow the construction of the system by more

Id	Year	Title	Time	Feature	Language	Rating	Votes	Type	Feature in bits
266	1993	The Legend	100 min	Action, Comedy, History	Cantonese	7.3	5,475	movie	101100011100100000
334	1994	Princess Caraboo	97 min	Comedy, Drama, History	English	5.9	2,073	movie	111001100001000000
351	1998	Frank Lloyd Wright	146 min	Documentary, Biography, History	English	7.8	470	movie	100010011110100000
371	2001	The Inner Tour	97 min	Documentary, History	Arabic	7.5	105	movie	100010100000000000
426	1988	Young Einstein	91 min	Comedy, History	English	5.0	7,034	movie	111001000000000000
500	1987	Matewan	135 min	Drama, History	English, Italian	7.9	5,739	movie	110000100000000000
545	2004	Japan: Memoirs of a Secret Empire	123 min	Documentary, History	English	7.6	110	series	100010100000000000
551	1998	America's Journey Through Slavery	360 min	History	English	7.4	55	series	100000000000000000
850	1981	Quest for Fire	100 min	Adventure, Drama, History	French	7.4	15,079	movie	100110110000100000
857	1935	Mutiny on the Bounty	132 min	Adventure, Drama, History	English, Polynesian	7.8	14,972	movie	100110110000100000
898	1978	Gray Lady Down	111 min	Adventure, Drama, History	English, Norwegian	6.2	1,87	movie	100110110000100000
994	1961	King of Kings	168 min	Biography, Drama, History	English	7.1	4,3	movie	111101100001000000
1048	2004	Ben Franklin	210 min	Documentary, History	English	7.4	30	movie	100010100000000000
1061	1921	Orphans of the Storm	150 min	Drama, History	English	7.9	3,565	movie	110000100000000000
1070	2000	Uncle Saddam	63 min	Documentary, Comedy, History	English	6.4	221	movie	100010011100100000
1086	1997	Waco: The Rules of Engagement	165 min	Documentary, History	English	8.0	2,514	movie	100010100000000000
1147	1996	The Crucible	124 min	Drama, History	English	6.8	26,826	movie	110000100000000000

Fig. 1. The structure of classical database. Only two columns, "id" and "feature" are used in quantum part of recommendation system. The mentioned columns represent also primary key in the classical database

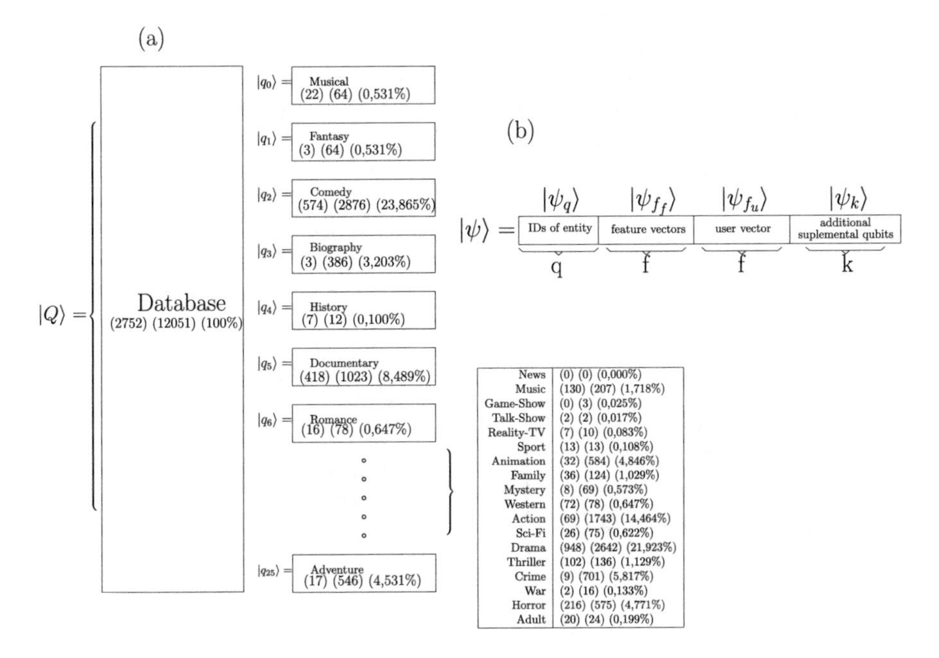

Fig. 2. Division diagram of quantum sub-register (a) for exemplary database from OMDB and detailed structure (b) of quantum register used in proposed recommendation system

than 50 qubits and this number will allow for the full implementation of the proposed solution.

In general, the use of this algorithm recommendation system, can be described by the following steps [6,12]:

(I) the user determines which features should possesses recommended elements,

(II) based on a specific feature, the appropriate sub-register of the quantum database is selected,
(III) a recommendation process is performed using quantum algorithm of k - nearest neighbours,
(IV) the obtained probability distribution of the recommended elements from step (III) is amplified with the Grover's algorithm in order to improve the probabilistic properties of the best recommended elements,
(V) performing a measurement on quantum register representing database, to finally indicate the recommended element.

3 Algebraic Correctness of Quantum Recommendation System

The presented analysis follows the general control flow in the proposed approach to the recommendation system. It was presented in the previous Sect. 2, and the corresponding diagram was presented on Fig. 3.

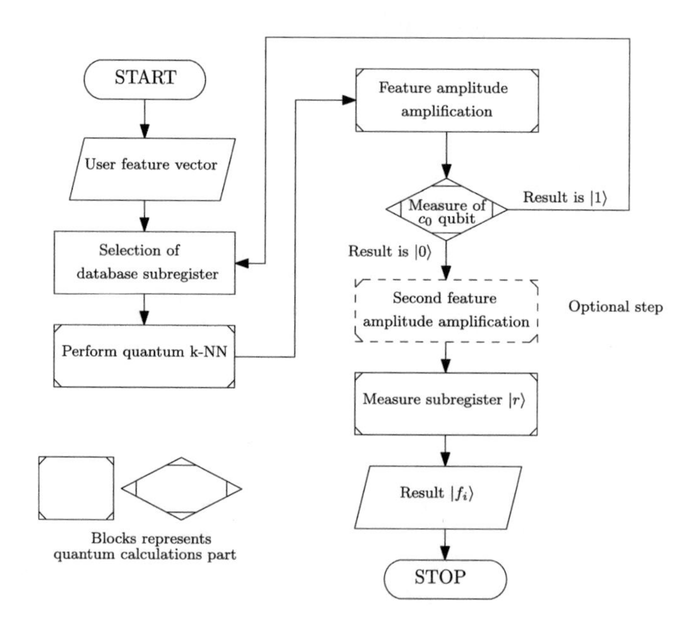

Fig. 3. The control flow of proposed the quantum recommendation system

We will begin the analysis of algebraic correctness by determining the initial state of the quantum register during the first stage, i.e. preparing the database. For simplicity, qubits describing the id field of given database entry will be omitted [7]:

$$|\psi_0\rangle = |0\rangle^{\otimes 2l+1}. \tag{1}$$

The first l qubit represents the features from database file, the second l qubits denotes the user's feature vector. Initializing the sub-register with the features database gives us the following status:

$$|\Psi_1\rangle = \frac{1}{\sqrt{L}} \sum_{p=1}^{L} |r_1^p, \ldots, r_l^p\rangle, \tag{2}$$

where $L = 2^l$, and r_p^k is a bit description of the features of the selected entry in the database. Naturally, for clarity, the part $|\psi_q\rangle$ is omitted from Fig. 2. The introduction of the user's feature vector gives the following state:

$$|\Psi_2\rangle = \frac{1}{\sqrt{L}} \left(\sum_{p=1}^{L} |r_1^p, \ldots, r_l^p\rangle \right) \otimes |t_1, \ldots, t_l\rangle \otimes \frac{1}{\sqrt{2}}(|0\rangle + |1\rangle). \tag{3}$$

At this point, it is now possible to calculate the Hamming distance between the user's feature and the features stored in the database. This operation is described by unitary operation U:

$$U = e^{-\mathbf{i}\frac{\pi}{2l}\hat{H}}, \quad \hat{H} = \begin{bmatrix} 1 & 0 \\ 0 & 0 \end{bmatrix}^{\otimes l} \otimes I_{l \times l} \otimes \begin{bmatrix} 1 & 0 \\ 0 & -1 \end{bmatrix}. \tag{4}$$

Remark 2. It should be emphasized that the computational complexity of this step depends only on the width of the pattern. In contrast to the classical approach, where the number of elements should be taken into account in the analysis of computational complexity.

After applying the operation U described by Eq. (4) we get the state:

$$\begin{aligned} |\psi_4\rangle = \tfrac{1}{\sqrt{L}} \textstyle\sum_{p=1}^{L} \big(& e^{\mathbf{i}\frac{\pi}{2l} d(t, r^p)} |d_1^p, \ldots, d_l^p\rangle \otimes |t_1, \ldots, t_l\rangle \otimes |0\rangle \\ & + e^{-\mathbf{i}\frac{\pi}{2l} d(t, r^p)} |d_1^p, \ldots, d_l 1^p\rangle \otimes |t_1, \ldots, t_l\rangle \otimes |1\rangle \big) . \end{aligned} \tag{5}$$

After the Hadamard operation on the additional qubits c_0, we get a state after realisation of the quantum k-NN algorithm:

$$\begin{aligned} |\psi_5\rangle = \tfrac{1}{\sqrt{L}} \textstyle\sum_{p=1}^{L} \big(& \cos\left(\tfrac{\pi}{2l} d(t, r^p)\right) |d_1^p, \ldots, d_l^p\rangle \otimes |t_1, \ldots, t_l\rangle \otimes |0\rangle \\ & + \sin\left(\tfrac{\pi}{2l} d(t, r^p)\right) |d_1^p, \ldots, d_l^p\rangle \otimes |t_1, \ldots, t_l\rangle \otimes |1\rangle \big) . \end{aligned} \tag{6}$$

Probability of measuring zero on the state of c_0, i.e. success, that we will go through the desired probability distribution with the correct indication of the recommended element:

$$P(c_0) = \frac{1}{L} \sum_{p} \cos^2 \left(\frac{\pi}{2l} d(t, r^p) \right). \tag{7}$$

If simplifying by P we denote the probability of receiving a particular recommendation, then obtaining probability distribution with accuracy of ε generally requires $O(P \cdot (1-P) \cdot \frac{1}{\varepsilon^2})$ repetitions of the implementation of the quantum algorithm of the k-nearest neighbours. It should be emphasized, however, that we can use many quantum machines to solve the same task, for example to obtain a linear time calculation of the probability distribution for the exponential amount of data L, described by Eq. (2).

Remark 3. Unfortunately, if the features indicated by the user result in choosing one compatible element and a few relatively close ones, then the probability distribution for the type of recommendation proceeding is not satisfactory. But, it is possible to use the Grover algorithm to improve the final probability distribution [4].

Therefore, assuming that zero was obtained by measuring the state c_0, we get the state $|\psi_6\rangle$ described in the following way:

$$|\psi_6\rangle = \frac{1}{\sqrt{L}} \sum_{p=1}^{L} m_p |\psi_{rcmd}\rangle, \tag{8}$$

where $|\psi_{rcmd}\rangle = |d_1^p, \ldots, d_l^p\rangle \otimes |t_1, \ldots, t_l\rangle \otimes |0\rangle$, while m_p represent the amplitudes of probability obtained after the measurement. In general, the register will contain the g amplitudes for recommended movies with the highest compatibility of the feature m_p^r and $L-g$ and with the lower compatibility of the feature m_p^{nr}:

$$|\psi_7\rangle = \frac{1}{\sqrt{L}} \left(\sum_{p=1}^{g} m_p^r |\psi_{rcmd}\rangle + \sum_{p=g+1}^{L} m_p^{nr} |\psi_{rcmd}\rangle \right). \tag{9}$$

Based on the article [3] it is possible to introduce the mean for amplitudes and the variance for the probability amplitudes for the state $|\psi_7\rangle$:

$$\begin{gathered} \overline{m^r}(t) = \tfrac{1}{g} \textstyle\sum_{p=1}^{g} m_p^r(t), \\ \overline{m^{nr}}(t) = \tfrac{1}{g} \textstyle\sum_{p=g+1}^{L} m_p^{nr}(t), \\ \sigma_{nr}^2(t) = \tfrac{1}{L-g} \textstyle\sum_{p=g+1}^{L} |m_p^{nr}(t) - \overline{m^{nr}}(t)|^2, \end{gathered} \tag{10}$$

where t is the iteration number (time) of the Grover algorithm. For $t = 0$ we have the initial distribution of probability amplitudes. The maximum probability of measuring recommended elements from the database is defined as:

$$\begin{aligned} P_{max} = 1 - (L-g)\sigma_{nr}^2 - \tfrac{1}{2}\left((L-g)|\overline{m^{nr}}(0)|^2 + g|\overline{m^r}(0)|^2\right) \\ + \left(\tfrac{1}{2}|(L-g)\overline{m^{nr}}(0)^2 + g\overline{m^r}(0)^2|\right), \end{aligned} \tag{11}$$

for performing $O(\sqrt{\frac{L}{g}})$ iterations.

Figure 4 shows the schematic of the quantum circuit implementing the recommendation system. It contains three main elements, a quantum part responsible

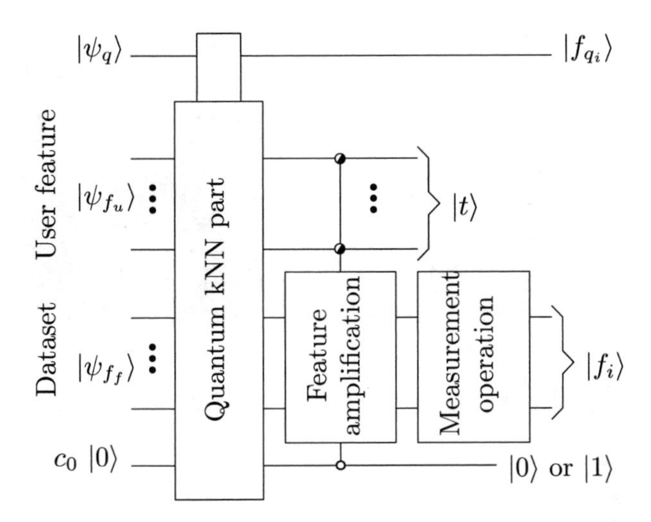

Fig. 4. General diagram of the quantum circuit recommendation system. In the diagram omitted data preparation process, this process involves the use of controlled gates negation, to build layout features for each element of the database.

for k-NN, i.e. indication of the nearest elements to the features defined by the user's requirements. The second element is amplification of the preferred features using the Grover algorithm and the third part represented by the final measurement, which allows to read the recommended element.

Remark 4. In presented description of recommendation system only a pure vector state formalism has been used. However, is not a problem to use a density matrix language to describe state of quantum system in more realistic case. Such approach is necessary in the case when someone try to use the presented approach in environment with noise which directly affect on the system which conducts recommendation algorithms.

4 Numerical Experiment

To present of a proposed solution a numerical experiment was carried out, presenting the main stages of the discussed recommendation system. These stages are the initial probability distribution, the distribution after the execution of the k-NN algorithm and the possibility of observing depending on the t variable as recommended elements are amplified. Figure 5 presents charts showing the state of the registry for a sample database with eight elements. Indicated by the user feature is the f_1.

After the execution of the k-NN part, the obtained probability distribution (Fig. 5 graph (b)) is not the best. Indeed, for searched feature f_1 we have the highest probability of measuring, but due to the fact that the difference in Hamming distance for the remaining features was not significantly large, the

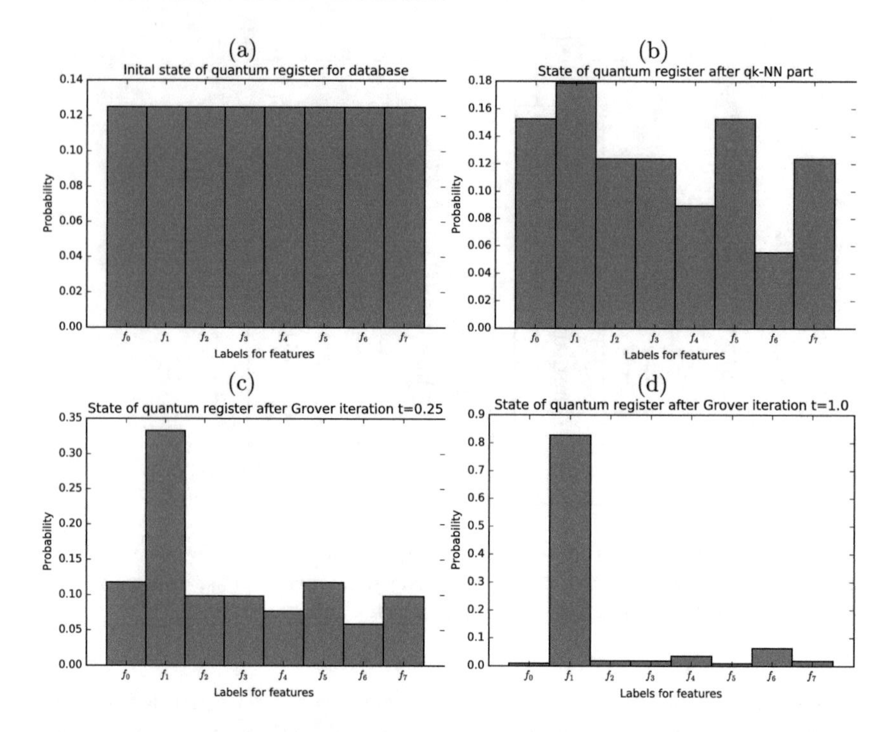

Fig. 5. An exemplary database (only eight elements for good readability) in four main stages in the proposed recommendation system. Case (a) describes the initial probability distribution for the initial state of the database. Case (b) shows the state of the register after the k-NN algorithm. Case (c) and (d) represent the process of probability amplifying while using the Grover's algorithm

remaining elements are also characterized by high probability. If we stop the process of recommendation at this stage, we should repeat the implementation of steps (I), (II) and (III), in order to collect the measurement results to approximate the probability distribution described by the $|\psi_5\rangle$ quantum register.

Therefore, it is reasonable to use the Grover algorithm to amplify the probability for the element indicated by k-NN. This was done in cases (c) and (d), where it can be observed that the use of the Grover algorithm for two exemplary time values of t amplifies the preferred amplitude. For $t = 1$ we already have a very high probability, close to the maximum theoretical probability value of measuring the recommended element [5]. However, even for smaller values of t, the gain gives the desired effect, while maintaining the probability distribution obtained after the quantum part k-NN. It also means reducing the circuit's operating time, because you do not have to do many iterations for the Grover algorithm.

The presented experiment describes the case of recommending only an one element. However, the quality of the gain will be equal to the good for two or more elements that will be indicated by user-specified characteristic. The value of the probability of separation is divided into recommended elements, so that together it will be as high as for one element.

Remark 5. At this point, ask yourself whether the Grover algorithm alone is enough to amplify the recommended element. Yes, if there is an element with the same vector of features as the user's requirements. If, however, there is no such element, then in this case the Grover algorithm will not be able to amplify elements similar to those defined by the user. Such a possibility is given to us by the quantum algorithm of the k-nearest neighbors. Therefore, it seems that the combination of two quantum k-NN algorithms and amplification of amplitudes with the Grover algorithm allows for the construction of a recommendation system.

5 Summary

The chapter presents a recommendation system based on the quantum algorithm of k-nearest neighbours and the Grover algorithm. The algebraic correctness of the proposed solution is discussed and explicit formulas for probability distributions are given, which are obtained during the operation of the proposed system. Despite the probabilistic nature of the recommendation system, it is possible to correctly indicate recommended elements from the database system with a high probability. Numerical tests were performed to illustrate the proposed solution for quantum recommendation system.

The most important advantage of the proposed solution is the ability to work with an exponential amount of data, but the complexity finally depends on the Grover part of proposed solutions. It should be emphasized that we store an exponential amount of data, using a linear number of quantum units of information. However, it should be noted that no additional operations are required to prepare the database structure to get the assumed complexity. In the case of quantum, it is enough to directly place the features and identifiers in the quantum register. In the case of classic approach, high performance for large data sets naturally requires initial processing and data optimization to allow, for example, logarithmic or more real poly-logarithmic computational complexity. The quantum approach provides a sub-linear class of complexity without any additional data processing.

Acknowledgments. We would like to thank for useful discussions with the *Q-INFO* group at the Institute of Control and Computation Engineering (ISSI) of the University of Zielona Góra, Poland. We would like also to thank to anonymous referees for useful comments on the preliminary version of this chapter. The numerical results were done using the hardware and software available at the "GPU μ-Lab" located at the Institute of Control and Computation Engineering of the University of Zielona Góra, Poland.

References

1. Alpaydin, E.: Introduction to Machine Learning. MIT press, Cambridge (2004)
2. Armbrust, M., Fox, A., Griffith, R., Joseph, D.A., Katz, R., Konwinski, A., Lee, G., Patterson, D., Rabkin, A., Stoica, I., Zaharia, M.: A view of cloud computing. Commun. ACM. **4**, 50–58 (2010)
3. Biham, E.O., Biron, D., Grassl, M., Lidar, D.: Grover's quantum search algorithm for an arbitrary initial amplitude distribution. Phys. Rev. A **60**, 2742 (1999)
4. Brassard, G., Hoyer, P.: An exact quantum polynomial-time algorithm for Simon's problem, pp. 12–23. IEEE Computer Society Press (1997)
5. Busemeyer, J.R., Bruza, P.D.: Quantum Models of Cognition and Decision. Cambridge University Press, New York (2012)
6. Erdal, A.: An information-theoretic analysis of grover's algorithm. In: Quantum Communication and Information Technologies, pp. 339–347. Springer, Netherlands (2003)
7. Hechenbichler, K., Schliep, K.: Weighted k-nearest-neighbour techniques and ordinal classification, p. 399. Sonderforschungsbereich (2004)
8. IBM Q Homepage. https://quantumexperience.ng.bluemix.net/. Accessed 28 Apr 2018
9. Nielsen, P.: Big data analytics – a brief research synthesis. In: Information Systems Architecture and Technology: Proceedings of 36th International Conference on Information Systems Architecture and Technology - ISAT 2015, Part I, pp. 3–9 (2015)
10. Nielsen, M.A., Chuang, I.L.: Quantum Computation and Quantum Information. 10th Anniversary Edition. Cambridge University Press, New York (2010)
11. OMDb Homepage. http://www.omdbapi.com/. Accessed 21 Apr 2018
12. Pinkse, P.W.H., Goorden S.A., Horstmann M., Skoric B., Mosk A.P.: Quantum pattern recognition. In: Conference on Lasers and Electro-Optics Europe (CLEO EUROPE/IQEC), and International Quantum Electronics Conference, Munich, p. 1 (2013)
13. Schuld, S., Sinayskiy, I., Petruccione, F.: Quantum computing for pattern classification. In: PRICAI 2014: Trends in Artificial Intelligence, pp. 208–220. Springer (2014)
14. Stefanowski, J., Krawiec, K., Wrembel, R.: Exploring complex and big data. Int. J. Appl. Math. Comput. Sci. **27**, 669-679 (2017)
15. Trugenberger, C.A.: Quantum pattern recognition. Quantum Inf. Process. **1**(6), 471–493 (2002)
16. Veloso, B., Malheiro, B., Carlos Burguillo, J.: A multi-agent brokerage platform for media content recommendation. Int. J. Appl. Math. Comput. Sci. **25**, 513–527 (2015)
17. Wiebe, N., Kapoor, A., Svore, K.M.: Quantum algorithms for nearest-neighbour methods for supervised and unsupervised learning. Quantum Inf. Comput. **15**(3–4), 316–356 (2015)
18. Wiśniewska, J., Sawerwain, M.: Recognizing the pattern of binary hermitian matrices by a quantum circuit. In: Intelligent Information and Database Systems, pp. 466–475. Springer, Cham (2017)

Survey Analyser: Effective Processing of Academic Questionnaire Data

Damian Dudek(✉)

Department of Information Technology,
The President Stanislaw Wojciechowski State University of Applied Sciences
in Kalisz, ul. Nowy Swiat 4, 62-800 Kalisz, Poland
d.dudek@pwsz.kalisz.pl

Abstract. Many universities conduct survey research in order to evaluate and improve quality of their education processes. Surprisingly, replacing traditional paper-based questionnaires with electronic ones usually decreases response rates and thus negatively affects overall reliability of results. In the paper, this issue is addressed and an effective approach to academic surveying is described, including a recommended strategy, matrix-based questionnaire representation, methods and technical framework for data handling. A crucial part of this method is the *Survey Analyser* – an innovative application, which automates the processing of questionnaire results using information extraction and transformation. The proposed surveying method and the corresponding software were successfully implemented into production environments at two universities, showing good performance based on real-life results.

Keywords: Survey research · Student's evaluation of teaching
Information extraction · Automated processing of questionnaire data
Survey response rate

1 Introduction

Universities commonly use survey research to evaluate and enhance teaching quality based on feedback gathered from participating students [5,8,15,18,20,26]. Nowadays it seems obvious to employ computer systems for this purpose – instead of traditional paper-based questionnaires – in order to raise the availability of surveys and automate processing of resulting data [3,7,10,12,13,17]. It is however surprising that in many cases response rates[1] of online surveys are low [2,6,11,22] – often much worse than in classical research with printed sheets [1,13,17,23]. Another problem is incompleteness of some assessment methods, which means that survey participants do not have to evaluate all their classes, but they may select only some of them, leaving the rest with no grades [27]. Poor

[1] Response rate is the number of responders, who actually answered a survey, divided by the overall number of authorized participants.

L. Borzemski et al. (Eds.): ISAT 2018, AISC 852, pp. 245–257, 2019.
https://doi.org/10.1007/978-3-319-99981-4_23

feedback, which negatively affects reliability of collected data, may be caused by a number of reasons, such as too complex and extensive questionnaires, *survey fatigue* (alternatively called *over-surveying* or *survey saturation*), lack of participant's anonymity or just their insufficient motivation for responding to surveys [1,6,8,26].

In this paper, an effective surveying method is described, which incorporates: general strategy, concise matrix-based questionnaire representation, algorithms for survey data extraction, and the title *Survey Analyser* – an innovative application for automated survey response processing.

The remainder of the paper is organized as follows. In Sect. 2 the proposed approach to surveying is introduced. The title *Survey Analyser* program is described in Sect. 3. In the next section, we present real-life results of the practical implementation of the method and the software into production environments in two universities.

2 The Effective Approach to Surveying

A number of strategies have been proposed in order to improve response rates of survey research [1,8,17,18,26]. A common observation is that using a computer system and electronic-based questionnaires alone definitely does not ensure good results, but it needs to be a part of a broader, effective surveying procedure, which should be implemented by a university. Thus, in consecutive subsections, we introduce the proposed strategy, the matrix-based survey structure, and methods for resulting response data processing.

2.1 Survey Strategy – How to Make It Work?

Inefficient surveying strategies, which result in poor response rates, commonly share the same mistake, which is relying on fully unsupervised activity of participants. University workers responsible for survey research prepare and publish online questionnaires in web applications, send individual e-mail invitations to entitled users (e.g. students, who completed a given set of classes) and wait for their voluntary response, possibly sending reminder messages from time to time. This approach usually fails, because most participants do not want to spend extra time filling out questionnaires and they may not feel anonymous if they are assigned individual access codes – whether or not a university would really use them to track responses and their authors. The situation becomes even worse if surveys are too extensive, complicated and spanned over many pages.

The alternative, effective surveying process may be summarized as follows.

1. Coordinators prepare and publish secured online surveys for appropriate groups of students. Each questionnaire should contain all the classes, which are valid for a given group of participants – with selective options if necessary.
2. The coordinators plan filling out questionnaires during students' classes at university. Access to surveys is restricted to the appointed periods.

3. Students fill out questionnaires within due time, during their group classes. One survey access key is shared by the whole group of students (providing participant's anonymity) and delivered by an academic teacher.
4. No responder's data (e.g. account name, IP address, response date and time) is logged during answering a survey.
5. Survey responses are collected and analysed – preferably using software, which provides data processing automation.

The above sequence is valid for a single surveying round and it is intended to be repeated periodically after every academic term.

In order to reduce *survey fatigue* [1], electronic questionnaires should be placed on single pages and contain concise sets of salient questions so that participants could answer them within a short period (e.g. 10–20 min). One of the practical solutions is using a one-page matrix survey, which is described in the next subsection.

2.2 Structure of Matrix-Based Surveys

The detailed formal model of matrix-based surveys and their processing methods were introduced in the previous paper [8]. A sample questionnaire is shown in Fig. 1. The names of assessed classes and their teachers are placed in rows, the evaluation criteria (survey questions) are put in columns and the number grades within a given scale are selected in intersections (in the example four marks are available: 5 – *very good*, 4 – *good*, 3 – *satisfactory*, 2 – *unsatisfactory*). In order to obtain reasonable results, grading scales have to be uniform in all questionnaires within a given survey research.

Evaluation criteria – survey questions

Classes and their teachers

	Teacher's command of the course content.	Effectiveness of teaching methods.	Interestingness of classes.	Alignment of course scope and final assessment.	Objectivity of students' assessment.	Respect and good manners in interaction with students.	Teacher's punctuality and conscientiousness.	Teacher's availability to students during office hours.
Computer graphics - lecture (dr Thomas Fisher)	4	5	5	4	5	4	5	4
Computer graphics - exercises (dr Thomas Fisher)	5	4	4	3	4	5	3	5
Databases - lecture (dr John Smith)	4	5	4	4	5	4	4	3
Databases - project (dr John Smith)	5	4	5	4	3	4	3	5
Project management - lecture (prof. Mary Newman)	5	4	3	5	4	5	3	5
Web development - lecture (prof. Mary Newman)	5	5	4	3	5	5	4	5
Web development - workshop (dr Andrew Woodhill)	4	3	5	4	4	5	5	4

Fig. 1. Structure of a matrix-based survey implemented using the *LimeSurvey* tool.

In practice, more than one such matrix may be used within one online questionnaire – especially when there are too many evaluated classes for a single one or there are some elective courses, which need to be selected by a participant. All survey parts can be easily combined into a single result set, anyway. It is also strongly recommended to provide an optional text field with sufficient length (e.g. 2000 characters) as some responders might want to leave valuable feedback as free-form comments [21].

2.3 Questionnaire Data Processing and Reporting

The introduced survey model enables processing of questionnaire data and generating various statistical reports – as needed by an end user. Examples of useful summaries are described below and their formal definitions can be found in the previous paper [8].

Report 1. The *summary of average grades* for particular evaluation criteria and a given set of classes. The total average for each class is equivalent to the *course average* introduced in [16]. An example, referring to all the classes attended by a group of students during one academic term, is presented in Fig. 2. This kind of report is also useful for examining grades awarded to all classes taught by one faculty member.

Classes (format)	Teacher	Responses	Evaluation criteria (questions)								Average
			q_1	q_2	q_3	q_4	q_5	q_6	q_7	q_8	
Computer graphics (lecture)	Prof. Mary Newman	28	4.13	2.94	3.00	3.88	3.69	3.38	4.19	4.00	**3.65**
Computer graphics (lab)	Dr Thomas Fisher	12	4.13	2.69	3.19	3.63	3.69	3.25	4.19	4.13	**3.61**
Computer graphics (lab)	Dr Andrew Woodhill	16	3.94	3.81	3.56	4.19	4.25	4.25	4.31	3.88	**4.02**
Databases (lecture)	Dr John Smith	28	4.57	4.36	4.07	4.39	4.57	4.71	4.79	4.79	**4.53**
Databases (project)	Dr Thomas Fisher	28	4.06	3.38	3.25	4.25	4.19	4.69	4.75	4.81	**4.17**

Fig. 2. An example summary of average grades for the classes attended by a group of students.

Report 2. The *weighted average* $\bar{a}_w(C_t, R)$ of all the grades awarded within the survey response set R to all the classes C_t taught by a given faculty member t, is defined by the following formula [8]:

$$\bar{a}_w(C_t, R) = \sum_{i=1}^{k} \bar{a}_i \cdot |r_{ci}| \Big/ \sum_{i=1}^{k} |r_{ci}|, \tag{1}$$

where: k – is the number of classes taught by a given academic staff member t, graded in r_c responses; $\bar{a}_i$ – is the mean grade of a single class $c_i \in C_t$; $|r_{ci}|$ is the number of responses for the class c_i. Considering the previous example shown in Fig. 2, we could calculate the weighted average for Dr Thomas Fisher as follows: $(3.61 \cdot 12 + 4.17 \cdot 28)/(12 + 28) \approx 4.00$. This summary accurately reflects grades awarded by individual students and is useful for creating teachers' rankings.

Report 3. The *average score* of all the classes graded within one surveying round is an average value of all the grades awarded by participants in their responses. It may be considered an equivalent to the *departmental average* presented in [16]. This summary may be helpful for evaluating the general satisfaction of survey participants. It can be also used for tracking changes of overall grades in subsequent surveying rounds or comparing parallel ones (e.g. for different subjects of studies).

3 Survey Analyser – Questionnaire Data Processing

3.1 Overview

The general architecture of the solution is pictured in Fig. 3. Matrix-based surveys (Fig. 1) are implemented and managed using *LimeSurvey* – a versatile, open-source surveying platform [19]. Access to each online survey is time-restricted and requires an access key (token). Tokens are delivered by teachers during students' classes at a university and shared by the whole groups of participants, providing sufficient anonymity. An SSL certificate is used for securing data transmission between responders' browser sessions and the surveying system. After participants fill out questionnaires, their responses are exported using CSV (*Comma Separated Values*) text files and they are ready for further processing. The *Survey Analyser* application imports the CSV files, extracts and transforms questionnaire data, loads it into the database and generates reports, which can be used for evaluating and improving teaching quality.

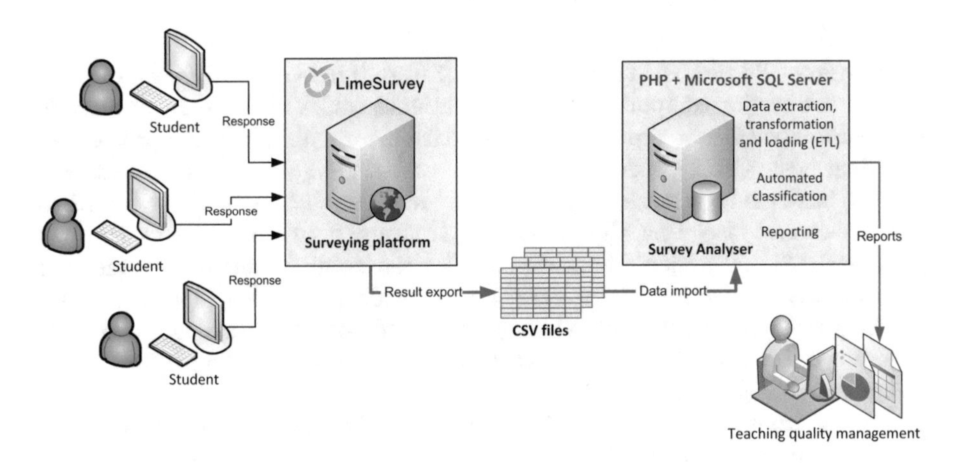

Fig. 3. The high-level architecture of the surveying solution [8].

3.2 Technical Architecture

The latest version 3.0 of the *Survey Analyser* is based on the n-tier client-server architecture (Fig. 4). The data layer is deployed on the *Microsoft SQL Server*

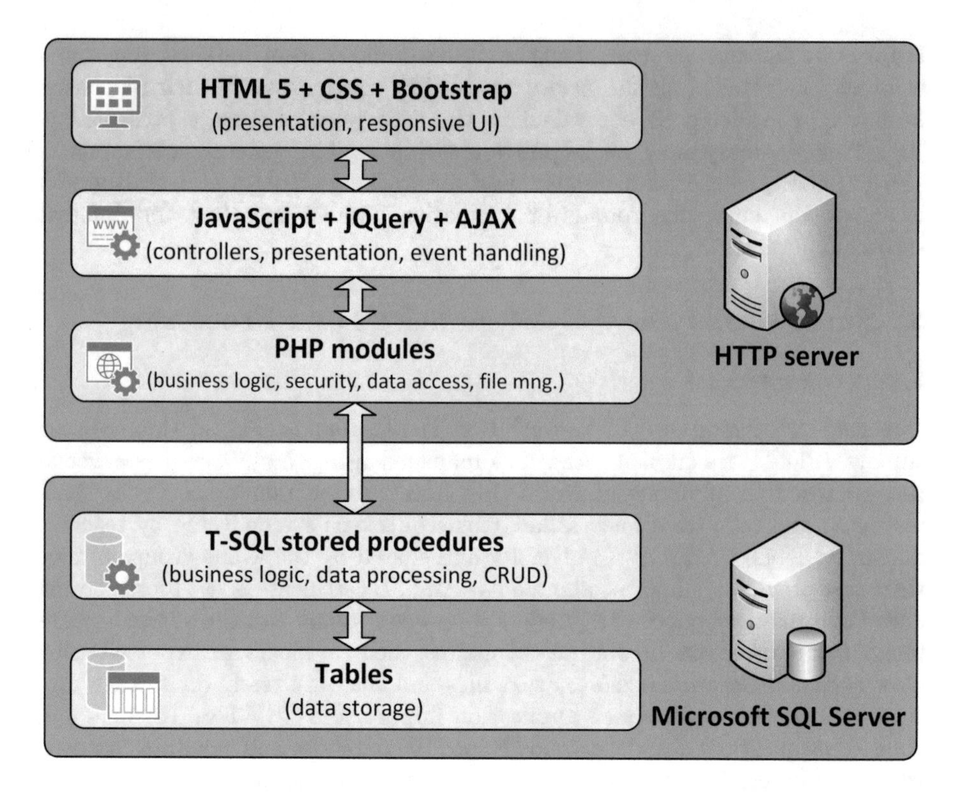

Fig. 4. The technical architecture of the *Survey Analyser* 3.0 application.

2014 Express Edition x64 (version 12.0.5207.0) database system [14] and it consists of two major component groups: relational tables where data are saved, and *Transact-SQL* stored procedures, responsible for data access and processing, including basic CRUD operations (*create*, *read*, *update*, *delete*), complex transactions, business logic rules, and aggregated reports (such as the ones described in Subsect. 2.3). Isolating the relational structure from upper layers through parametric *T-SQL* objects improves application security and management of data-intensive operations. The PHP 7 server-side technology [25] is used for handling data access, disk file management, and core business logic processes including automated data extraction, transformation and loading (ETL) based on the A-QDX algorithm [8].

On the front-end side, a user is offered a comfortable, fully responsive web interface based on the HTML 5 language, CSS styles and the *Bootstrap 3.3.7* framework [4] (Fig. 6). User's HTTP requests are handled by JavaScript controllers, which communicate with the PHP modules using the AJAX (*Asynchronous JavaScript And XML*) techniques and the JSON (*JavaScript Object Notation*) data format. The *jQuery* library [24] is intensively used within the controllers for dynamic page content delivery and interaction with the DOM (*Document Object Model*) elements, including handling modal windows and dialogue boxes.

3.3 Data Structures

The high-level schema of the table structure – the lowest application layer stored and managed by the *Microsoft SQL Server* relational engine – is shown in Fig. 5 in the form of a UML class diagram. The presented database design enables adjusting the software to many universities with different sets of survey questions (criteria). Data for some objects, including *University*, *UnivFaculty*, *QuestionSet*, *SurveyQuestion*, *StudiesSubject*, *StudentsGroup*, *SurveyResearch*, and *Survey*, need to be provided by a user during a preparation phase, before the actual analysis. In contrast, the remaining tables – *SurveyClasses*, *Classes*, *ClassesFormat*, *Course*, *Teacher*, *SurveyResponse*, *SurveyClassesResponseIndex*, *SurveyQuestion*, *Response* – are automatically populated with data, which are extracted and transformed from CSV text files with questionnaire responses.

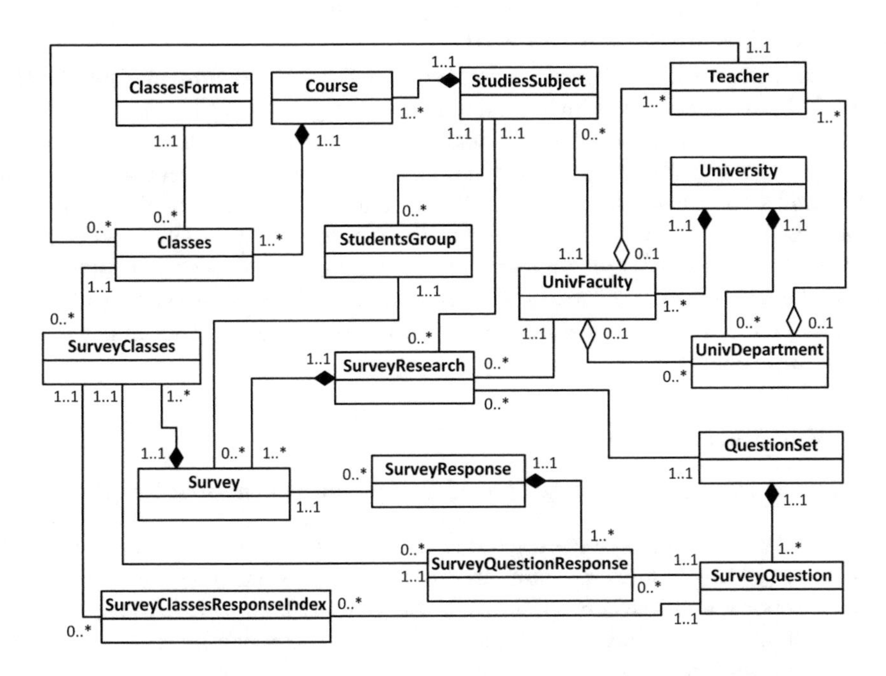

Fig. 5. The relational database layer of the *Survey Analyser* 3.0 (UML class diagram).

3.4 Automated Survey Data Extraction

An essential part of the *Survey Analyser* is the complex ETL *(extraction, transformation, loading)* process, which turns semi-structural survey responses stored in CSV files into fully structural form, conforming to the model of matrix-based surveys (introduced in Subsect. 2.2). The core procedure is based on the A-QDX (*Automated Questionnaire Data eXtraction*) algorithm, which was introduced in the previous work [8]. There are two consecutive phases of this process: (1) in the first stage header data are extracted and loaded, (2) in the second phase main response data are imported into relational tables. Headers in CSV files imported from the *LimeSurvey* program contain concatenated textual data, which need to be accurately decomposed into separate entities. An example header cell is shown below.

Classes – common teachers for all students. [Web development – lecture(Dr John Smith, Assoc Prof)][Respect and good manners in interaction with students.]

The sample text string consists of the following elements: (a) a question group name (*Classes – common teachers for all students.*); (b) a course name (*Web development*); (c) a classes format name (*lecture*); (d) teacher's title (*Dr*); (e) teacher's first and last name (*John Smith*); (f) teacher's position (*Assoc Prof*), which is optional; (g) survey question (criterion) text (*Respect and good manners in interaction with students*).

Header texts are decomposed using the PCRE regular expressions [9] and the extracted parts are loaded into database tables (e.g. *Course*, *Classes*, *Teacher*). During the first phase, the A-QDX algorithm also determines and saves indexes of columns with grades referring to particular teachers and their classes. Then, in the second stage consecutive rows of data are read from the CSV file and separate values are inserted into the *SurveyQuestionResponse* table with respect to the indexes, which were found and stored before. Important features introduced in the 3.0 version of the *Survey Analyser* are: improved tolerance to different variants of input data and auto-correction of some user's errors made during survey implementation.

3.5 End User Operation

The *Survey Analyser* version 3.0 provides the following end-user functionality:

- survey research management: general data, registering surveys, sending CSV files, importing responses, summarizing results, generating aggregated statistical reports and teachers' rankings[2], exporting results to CSV files;
- participant data management: fields of study, groups of students;
- survey question sets management;
- user session management: user accounts, authentication, access control;
- tools and auxiliary functions: generating survey questions for *LimeSurvey* questionnaires, creating backups of the database and all data files (source and result), recovering the database and files; program settings.

The user interface of the *Survey Analyser* version 3.0 is responsive and it can be easily operated using different devices (Fig. 6). The syntax of the web pages was successfully tested with the *W3C Markup Validation Service* (https://validator.w3.org). These features have a positive impact on the general availability of the application for clients using various work environments.

[2] Including the reports described in Sect. 2.3.

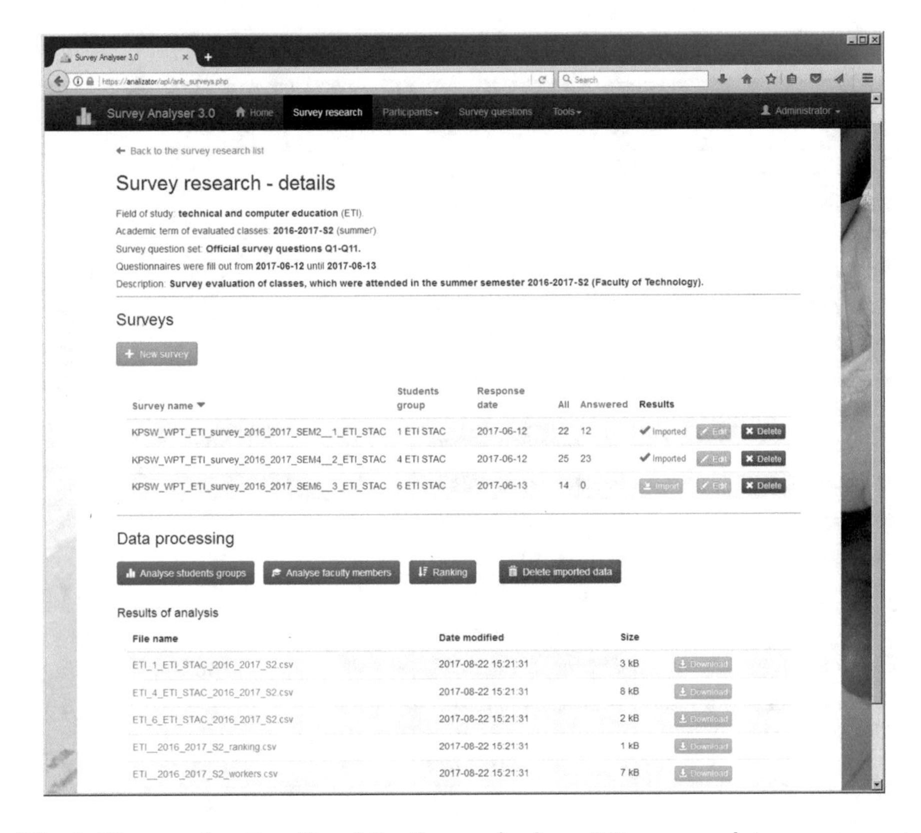

Fig. 6. The core functionality of the *Survey Analyser* 3.0 – survey data management and response processing.

4 Practical Implementation at Universities

Reliability and accuracy of data import and processing done by the *Survey Analyser* – with respect to the introduced matrix-based survey model (Sects. 2.2 and 2.3) – were positively verified in the previous work [8]. The proposed surveying strategy and the related data analysis software were practically implemented into production environment at two Polish universities: the President Stanislaw Wojciechowski State University of Applied Sciences in Kalisz (SUAS) and the Karkonosze College in Jelenia Góra (KCJG). The summarized results for the KCJG and the SUAS are presented in Table 1 and Table 2, respectively. The average response rates of 71% (KCJG) and 78% (SUAS) significantly outperform corresponding results obtained at some other universities [2,11,18,22,27] – sometimes even by an order of magnitude [2].

Table 1. Practical implementation of the surveying system at the KCJG in Jelenia G.

Participants' field of study	Academic term of evaluated classes	Responses acquired	Total number of students	Response rate
Physiotherapy	Summer 2014–2015	81	110	74%
Technical and computer education	Summer 2014–2015	47	58	81%
Nursing	Winter 2015–2016	46	105	44%
Pedagogy	Winter 2015–2016	139	186	75%
Philology	Winter 2015–2016	69	128	54%
Technical and computer education	Winter 2015–2016	73	86	85%
Dietetics	Winter 2016–2017	88	107	82%
Journalism	Winter 2016–2017	24	42	57%
Nursing	Winter 2016–2017	72	96	75%
Pedagogy	Winter 2016–2017	64	102	63%
Philology	Winter 2016–2017	72	98	73%
Physical education	Winter 2016–2017	48	70	69%
Physiotherapy	Winter 2016–2017	92	116	79%
Technical and computer education	Winter 2016–2017	50	67	75%
Dietetics	Summer 2016–2017	67	107	63%
Journalism	Summer 2016–2017	28	39	72%
Nursing	Summer 2016–2017	83	95	87%
Pedagogy	Summer 2016–2017	72	102	71%
Philology	Summer 2016–2017	66	95	69%
Physical education	Summer 2016–2017	58	67	87%
Physiotherapy	Summer 2016–2017	86	117	74%
Technical and computer education	Summer 2016–2017	43	61	70%
Total		**1468**	**2054**	**71%**

Table 2. Practical implementation of the surveying system at the SUAS in Kalisz.

Participants' field of study	Academic term of evaluated classes	Responses acquired	Total number of students	Response rate
Computer science	Summer 2013–2014	116	154	75%
Computer science	Winter 2014–2015	130	182	71%
Computer science	Summer 2014–2015	95	116	82%
Computer science	Winter 2015–2016	121	155	78%
Computer science	Summer 2015–2016	74	97	76%
Computer science	Winter 2016–2017	109	145	75%
Computer science	Summer 2016–2017	76	91	84%
Computer science	Winter 2017–2018	111	132	84%
Total		**832**	**1072**	**78%**

5 Conclusions

In the paper, the *Survey Analyser* – an original application for automated processing of academic questionnaire data was presented. The program transforms semi-structural survey responses into fully structured relational tables using advanced ETL procedures and generates aggregated statistical reports, which may be helpful in evaluating and improving teaching quality. The application was successfully implemented into production environments at two universities and validated on real questionnaire data entered by participants representing different fields of study.

However, the contribution of this work goes far beyond developing a software solution for automated data analysis, as the essential part of the picture is the effective surveying strategy, which a university needs to implement before any survey research starts. This may often mean re-designing inner rules, procedures, and workflows of quality management in order to achieve satisfying response rates of surveys. Otherwise, data analysis – regardless of its sophistication – may appear to be futile due to lack of representativeness. In order to address this issue, in the paper the partly supervised surveying approach was described, which proved to be very effective in real life, resulting in average response rates over 70%.

An interesting continuation of this research would be extending the *Survey Analyser* by functions of electronic questionnaire management, which would be directly integrated with the online surveying platform using its API.

Acknowledgments. The author wishes to thank the Karkonosze College in Jelenia Góra for sharing the results of survey research, which were presented in this paper.

References

1. Adams, M.J.D., Umbach, P.D.: Nonresponse and online student evaluations of teaching: understanding the influence of salience, fatigue, and academic environments. Res. High. Educ. **53**(5), 576–591 (2012). https://doi.org/10.1007/s11162-011-9240-5
2. Bienkowska, A.: Teaching Quality Assurance in the Faculty of Computer Science and Management - A Report for the Academic Year 2014–2015. Wroclaw University of Technology, Wroclaw (2016). (in Polish)
3. Blair, E., Inniss, K.: Student evaluation questionnaires and the developing world: an examination of the move from a hard copy to online modality. Stud. Educ. Eval. **40**, 36–42 (2014). https://doi.org/10.1016/j.stueduc.2014.01.001
4. Bootstrap Core Team: Bootstrap. https://getbootstrap.com
5. Cano-Hurtado, J.J., Carot-Sierra, J.M., Fernandez-Prada, M.A., Fargueta, F.: An Evaluation Model of the Teaching Activity of Academic Staff. Valencia University of Technology, Valencia (2009)
6. Ciesielski, P.: Remarks for Students' Questionnaire Evaluation of Teaching. Jagiellonian University, Krakow (2012). (in Polish)
7. Cork, D.L., Cohen, M.L., Groves, R., Kalsbeek, W. (eds.): Survey Automation: Report and Workshop Proceedings. The National Academies Press, Washington, DC (2003). https://doi.org/10.17226/10695
8. Dudek, D.: Automated information extraction and classification of matrix-based questionnaire data. In: Swiatek, J., Tomczak, J.M. (eds.) Advances in Systems Science, Proceedings of the International Conference on Systems Science 2016 (ICSS 2016), Advances in Intelligent Systems and Computing, vol. 539, pp. 109–120. Springer, Heidelberg (2017). https://doi.org/10.1007/978-3-319-48944-5_11
9. Hazel, P.: PCRE – Perl Compatib. Regular Expressions (2018). http://www.pcre.org
10. Inter-University Computerization Center: USOS – University Study-Oriented System (2018). http://muci.edu.pl. (in Polish)
11. Izdebski, A., et al.: Report of the all-university teaching quality assessment survey. University of Warsaw, Poland (2015). (in Polish)
12. Kirchner, A., Norman, A.D.: Evaluation of electronic assessment systems within the USA and their ability to meet the National Council for Accreditation of Teacher Education (NCATE) Standard 2. Educ. Assess. Eval. Acc. **26**(4), 393–407 (2014). https://doi.org/10.1007/s11092-014-9204-3
13. Leung, D.Y.P., Kember, D.: Comparability of data gathered from evaluation questionnaires on paper and through the internet. Res. High. Educ. **46**(5), 571–591 (2005)
14. Mistry, R., Misner, S.: Introducing Microsoft SQL Server 2014: Technical Overview. Microsoft Press, Redmond (2014)
15. Moreno-Murcia, J.A., Torregrosa, Y.S., Pedreno, N.B.: Questionnaire evaluating teaching competencies in the university environment. Evaluation of teaching competencies in the university. New Approaches Educ. Res. **4**(1), 54–61 (2015). https://doi.org/10.7821/naer.2015.1.106
16. Nikolaidis, Y., Dimitriadis, S.G.: On the student evaluation of university courses and faculty members' teaching performance. Eur. J. Oper. Res. **238**, 199–207 (2014). https://doi.org/10.1016/j.ejor.2014.03.018
17. Nulty, D.D.: The adequacy of response rates to online and paper surveys: what can be done? Assess. Eval. High. Educ. **33**(3), 301–314 (2008)

18. Przybylski, W., Rudnicki, S., Szwed, A. (eds.): Evaluating Teaching Quality in Higher Education - Methods, Tools, Best Practice. Tischner European University, Krakow (2010). (in Polish)
19. Schmitz, C.: LimeSurvey: An Open Source Survey Tool. LimeSurvey Project Team, Hamburg, Germany (2018). https://www.limesurvey.org
20. Stein, S.J., Spiller, D., Terry, S., Harris, T., Deaker, L., Kennedy, J.: Tertiary teachers and student evaluations: never the twain shall meet? Assess. Eval. High. Educ. (2013). https://doi.org/10.1080/02602938.2013.767876
21. Stupans, I., McGuren, T., Babey, A.M.: Student evaluation of teaching: a study exploring student rating instrument free-form text comments. Innovative High. Educ. **41**(1), 33–42 (2016). https://doi.org/10.1007/s10755-015-9328-5
22. Szaban, D., Kolodziej, T.: Report of academic teachers assessment 2016/2017. University of Zielona Gora, Poland (2017). (in Polish)
23. Tarasiuk, J.: Survey research in AGH - the system and sample results. The Fourth All-Poland Meeting of the Rectors' Proxies for Education Quality, Wroclaw, Poland (2017). (in Polish)
24. The jQuery Foundation: jQuery. https://jquery.com
25. The PHP Group: PHP page. http://php.net
26. Thielsch, M.T., Brinkmoeller, B., Forthmann, B.: Reasons for responding in student evaluation of teaching. Stud. Educ. Eval. **56**, 189–196 (2018). https://doi.org/10.1016/j.stueduc.2017.11.008
27. UMCS: Report of the evaluation survey of classes in the summer semester 2015–2016. Maria Curie-Sklodowska University, Lublin (2016). (in Polish)

Applying Basket Analysis and RFM Tool to Analyze of Customer Logs

Jolanta Wrzuszczak-Noga(✉)

University of Science and Technology, Wyb. Wyspiańskiego 27, 50-370 Wrocław, Poland
Jolanta.Wrzuszczak-Noga@pwr.edu.pl

Abstract. In this paper a customer business analysis using data mining tools and marketing performance index will be presented. Mechanisms based on machine learning principles will be described and discussed. Customer logs will be analyzed using two different tools and compared for decision making process.

Keywords: Data mining · RFM index · Business analysis

1 Introduction

Nowadays the decision making process based on customer behaviour as recorded in server logs is becoming more popular, which helps users and data analysts. Big and small companies discover that keeping regular customers is a very important process towards increasing the profit. That is why the introduction of special offers should be based on current customer demand.

Collecting and analyzing client shopping logs is the main process in customer relationship management. The aim is to identify regular customers and to predict customer behaviour i.e. by suggesting dedicated and personalized offers.

A lot of investigations show that more that 80% of shop revenues are generated by regular customers [1, 2]. This paper presents two mechanism for analyzing clients activity logs as a decision making process. The first policy focuses on a basket in an e-shop, the second analyses products from the e-shop history including the RFM (Recency Frequency Monetary) – the loyalty customer index.

The aim of this paper is to present and compare two different tools for customer behaviour analysis in the context of business aspects.

The contribution of this paper is to exam support tools for business to business decisions. Scientific research of decision making process based on stored customer logs are very important for assuring customer satisfaction and carry information about the purchased products. One tool reports information about purchased products, the second one distinguishes groups of customers based on purchases they made. The customer valuation information can be implemented for preparing dedicated offers using i.e. One-to-One Marketing or Direct Marketing strategies.

L. Borzemski et al. (Eds.): ISAT 2018, AISC 852, pp. 258–266, 2019.
https://doi.org/10.1007/978-3-319-99981-4_24

2 Background

The data mining process is used for sorting large data sets to identify some patterns, behaviours, relations or trends [3, 4]. There could be distinguish some types of mining: usage mining, profile mining or web mining. The last type - web mining concentrates on finding patterns observed in customer relationship management.

Figure 1 presents the five major steps of the data mining process. The first based on preparing data in the right format, the second - analyzing data for finding some failures, some corrupted data could be deleted or replaced. The third most important data mining step uses some algorithms (i.e. path analysis, classification, clustering, predictive analysis). The postprocessing stage formats the output information, charts and rules. The last presents the information picked out from the data mining algorithms.

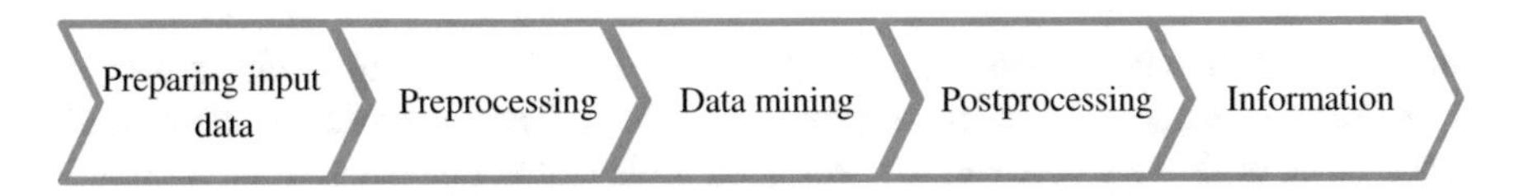

Fig. 1. The major steps of data mining process

Many works in the literature [5, 6] are concerned with regular customers, their behaviour and predicting their shopping patterns. Some indexes are introduced to measure customers' behaviour and loyalty. It is not possible to analyze and compare current trends in connection chain of bought and visited products without automatization. The cost of acquiring new customers is relatively high comparing to supporting the existing customers. The data mining techniques are introduced in the area of customer relations for analyzing and discovering some benefits based on proposed special offers [2].

The relationship of satisfaction to profits was examined in the 1980s and described in [7]. Since 1990s the loyalty has been measured and analyzed, some investigation has been made into how the customer satisfaction is related to financial impact [8]. In [8] a framework was proposed, how customer satisfaction can impact the market share. Due to the market concept a spending level of predicted shopping list was presented [8].

In papers [9, 10] authors describe the benefits of CRM (Customer Relationship Management), namely: increased customer retention and loyalty, higher customer profitability, creation value for the customer, customization of products and services, higher quality products and services. In [9] a relation between customer satisfaction, customer value, customer interaction and customer knowledge was presented.

Paper [11] describes predictive modelling. Recent advances have led to the newest trends in data mining – text mining and web mining (based on textual comments from survey research or e-mails and log files from web servers). Authors in [11] present tools for data mining process.

Work [12] presents impact of new media (like shopping using bots or mobile devices) in the context of CRM and a conceptual pinball framework. In the paper [13] data mining process based on personalisation mechanism was depicted and the paper [14] describes product ranking, features of e-shop.

Paper [15] presents a new approach called New Product Development (NPD) and a conceptual framework based on customer participation and NPD performance.

In the data mining tools numerous machine learning algorithms are implemented as solutions for analyzing of growing data [2] in the domain of:

- shopping analysis,
- insurance analysis,
- bank analysis (credit decision support) [16, 17].

The next section describes machine learning tools, the fourth presents details of log analyzing process, the next presents results and summary.

3 Machine Learning Tools

Machine learning tools are programs used as quick decision making mechanisms based on artificial intelligence and correlation techniques. They are used in data mining process. The difference between the presented tools is that one is based on transaction without customer information and the second one involves the analysis of customer information about the bought products in detail. Two approaches will be presented and discussed. The first one is based on basket analysis and is available without fee (plugin for Sql Server - academic licence). The second tool - IBM SPSS Statistics (available trial version) - uses the common customer scoring index (RFM mechanism).

3.1 Data Mining Ad Ins for Microsoft SQL Server

The Data Mining Ad-ins 2012 are designed for excel sheet. It consists of prediction, classification and market basket analysis (MBA) algorithms. The functionality of the mechanism is based on detection of some rules regarding the bought products stored in logs.

This market basket analysis plugin is an a priori mechanism and is used to identify correlation between products that were bought. The transactions are analyzed and as output some sets of groups of products are given.

Some terminology will be introduced to present the process [18]:

- Items (products) will be used for identifying associations. *I* is a set of items

$$I = \{ \mathrm{i}_1, \mathrm{i}_2, \mathrm{i}_3, \ldots, \mathrm{i}_\mathrm{n} \}$$

- Transactions are instances of groups of items. For each transaction an item set is given

$$t_n = \{\mathrm{i}_\mathrm{i}, \mathrm{i}_\mathrm{j}, \ldots, \mathrm{i}_\mathrm{k}\}$$

- Association rules are statements presented as

$$\{\mathrm{i}_1, \mathrm{i}_2, \ldots\} => \{\mathrm{i}_\mathrm{k}\}$$

The rule will be formulated as:

If in the basket product i_1 and i_2 are given, than the customer will be interested in product i_k.

3.2 IBM SPSS Statistics

This tool contributes an analytical process and consists of clustering, prediction or customer management policies. This software contributes the market mechanisms based on RFM index (loyalty index). This technique is used to classify an entire customer list.

RFM Analysis helps to reach and focus on the real customer that drives the business profit [19]. In the RFM algorithm a user is assigned a score based on the last date of purchase, frequency of purchase and shopping value. These mechanisms use a simple ranking of calculated index. Finally customers are ranked within the RFM index in a decreasing manner.

The vector elements are calculated as follows [20]:

- **Recency** - refers to how long ago a customer placed last order. This metric is based on the date of last shopping. The ranking is usually defined by an interval from 1 to 5.
- **Frequency** - refers to how many times a customer bought a product in the e-shop over customer's lifetime, calculated based on an interval from 1 to 5.
- **Monetary value** - refers how much money the customer spent in the e-shop, transformed into a set from 1 to 5.

Having the three defined elements R, F, and M, the model will be presented. Each customer has a three-number score, with each category (R, F, or M) as an ordered list of customers based on the value of the metric [20]. For example, customers who order most frequently will all receive a 5, customers who ordered only one time will get a 1. The other customer group will get 2, 3, or 4. The set will be created proportionately to the R, F, M elements.

The other elements will be calculated in the same manner (the ranking system). Every e-shop user has a three-number score, such as 115, 235, or 535.

4 Analysis of Logs

A set of server logs containing 32 000 transactions from the e-shop was given. The stream contains information about products, prices and transactions. To prevent identification of actual products, both - the products and categories were made anonymous as Book_A, Book_B,…

The first step in the web mining was to detected some failures and inconsistent data. The second step based on importing the text file into a calculating tool with defined separators.

The data set was imported simultaneously into Excel (Ad-ins 2012 for data mining for MS SQL Server) and into the IBM SPSS Statistics (19. 0 19.0) tool.

The data set was characterized by the following attributes:

- Transaction ID,
- Item,
- Item category,
- Item value (price),
- Transaction Date.

Two algorithms were tested:

- Market Basket Analysis,
- RFM analysis.

The aim of the performed test was to detect some business information about selling, trends and benefits, and also to present the customer scoring.

5 Results

Results of performed log analysis will be presented for Market Basket Analysis algorithm and for IBM SPSS Statistics tool.

5.1 Results of the Market Basket Analysis Algorithm

The Fig. 2 presents the output of the MBA algorithm, which built some bundle of items. It is to notice, that some bundle of 2 or 3 defined books were observed. The highest number of sales was 805.

Shopping Basket Bundled Items

Bundle of items	Bundle size	Number of sales	Average Value Per Sale	Overall value of Bundle
Book_L, Book_A	2	805	548	441140
Book_S, Book_A	2	752	580	436160
Book_G, Book_A	2	644	580	373520
Book_V, Book_A	2	536	546	292656
Book_D, Book_A	2	510	545,49	278199,9
Book_N, Book_A	2	462	547	252714
Book_F, Book_A	2	351	548,49	192519,99
Book_T, Book_A	2	340	533,99	181556,6
Book_W, Book_A	2	299	580	173420
Book_C, Book_A	2	314	545,98	171437,72
Book_B, Book_A	2	292	573,99	167605,08
Book_T, Book_G, Book_A	3	277	589,99	163427,23
Book_H, Book_A	2	288	545,49	157101,12
Book_Z, Book_A	2	262	559	146458
Book_Z, Book_S, Book_A	3	231	615	142065
Book_M, Book_G, Book_A	3	224	601,49	134733,76

Fig. 2. Bundled Items for Market Basket Analysis algorithm

The algorithm Market Basket Analysis introduced new rules (Fig. 3). One of them means that the Book_A should be recommended to customers who bought the Book_F, because in the analysed logs there were 842 transactions consisting of Book_F and Book_A.

Shopping Basket Recommendations

Selected Item	Recommendation	Sales of Selected Items	Linked Sales	% of linked sales	Average value of recommendation	Overall value of linked sales
Book_F	Book_A	842	351	042%	218,4370546	183924
Book_T	Book_G	1190	988	083%	46,49411765	55328
Book_M	Book_G	1005	897	089%	49,98208955	50232
Book_Z	Book_S	810	645	080%	44,59259259	36120
Book_M	Book_L	1005	552	055%	13,18208955	13248
Book_U	Book_S	515	216	042%	23,48737864	12096
Book_O	Book_D	529	359	068%	14,58395085	7714,91
Book_R	Book_D	458	322	070%	15,10868996	6919,78
Book_P	Book_S	129	73	057%	31,68992248	4088

Fig. 3. Shopping Basket Recommendations for Market Basket Analysis algorithm

From the business point of view it is clear how to prepare a personal special offer.

5.2 Results of the IBM SPSS Statistics

The score of the IBM SPSS Statistics program is presented in Fig. 4. Some groups of clients were categorized based on the last transaction date, frequency of product purchase and monetary value. A customer map was created. It was noticed that every group was characterized by a similar value of performed purchases.

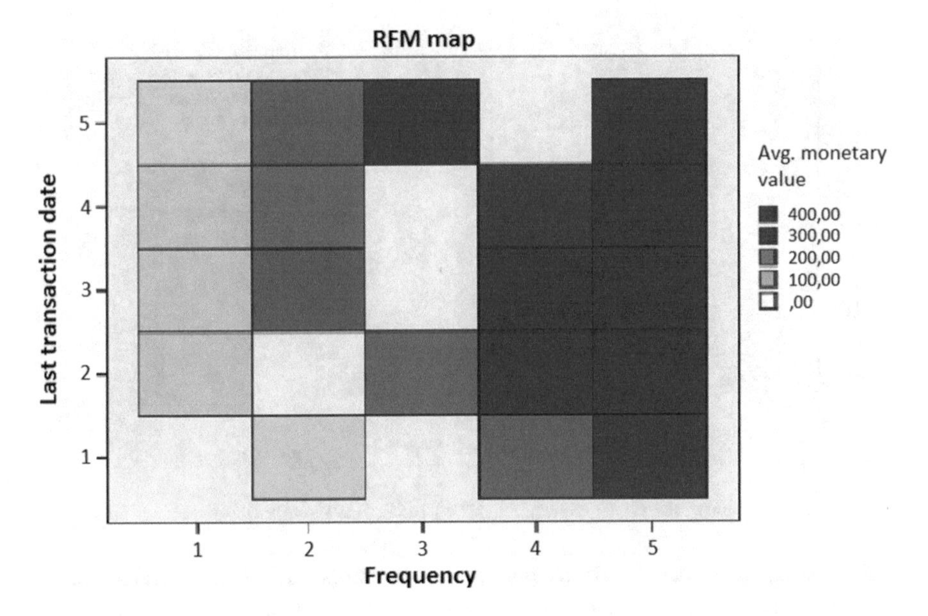

Fig. 4. RFM map in SPSS Statistics

The scoring process is depicted in (Fig. 5), where the ranking model can be found.

1: TrID SO61269

	TrID	Last_transac_date	Transaction_frequency	Transaction_value	Score_based_on_last_transact_date	Score_based_on_frequency	Gr
1	SO61269	16-Jul-2016	2	573,99	3	2	
2	SO61270	18-Nov-2015	1	21,98	1	2	
3	SO61271	05-Aug-2014	2	77,49	1	4	
4	SO61272	13-Aug-2016	2	112,00	4	2	
5	SO61273	02-Jul-2016	1	56,00	3	1	
6	SO61274	07-Jul-2016	2	91,48	3	2	
7	SO61275	05-Jan-2016	1	524,00	2	1	
8	SO61276	10-May-2016	2	74,49	2	3	
9	SO61277	20-Apr-2016	3	127,49	2	4	
10	SO61278	09-May-2016	4	154,49	2	5	
11	SO61279	08-Sep-2016	2	44,98	4	2	
12	SO61280	04-Aug-2016	1	524,00	3	1	
13	SO61281	02-Oct-2016	1	156,00	4	1	
14	SO61282	16-Apr-2016	1	56,00	2	1	
15	SO61283	18-Nov-2016	3	566,98	5	3	
16	SO61284	08-Feb-2016	2	77,49	2	3	
17	SO61285	23-Oct-2016	3	578,09	5	3	
18	SO61286	28-Jun-2016	2	112,00	3	2	
19	SO61287	28-Oct-2016	3	118,99	5	3	
20	SO61288	11-Jun-2016	3	636,00	3	4	
21	SO61289	12-Jan-2016	3	636,00	2	4	
22	SO61290	22-Nov-2016	2	65,99	5	2	
23	SO61291	08-Nov-2015	1	69,99	1	2	
24	SO61292	28-Sep-2015	2	65,99	1	4	

Fig. 5. Example of scoring process in SPSS Statistics

In the Fig. 6 a histogram of three measurements elements is presented. It was observed that in the last observation period the number of transactions increases and that the average value of transaction was 240 $.

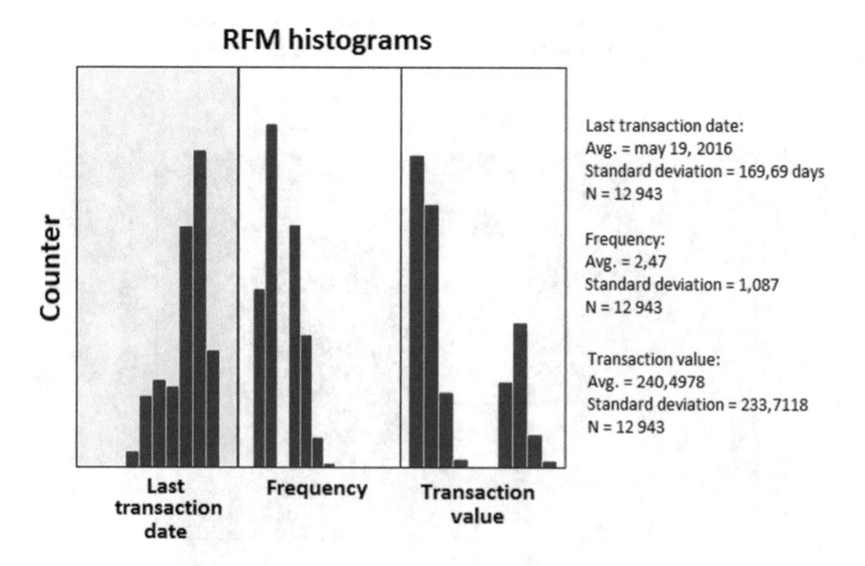

Fig. 6. Histogram of the RFM index in SPSS Statistics

Analyzing dependencies from purchase logs supports business decisions. The presented mechanisms can be used to evaluate new campaigns (e.g. by comparison of the store revenue before and after campaign). The research result shows that the process of preparing special offers can base on items or be targeted to dedicated clients (special benefits or discounts).

6 Summary

The performed analyses of logs shown that web mining process is a very important mechanism from the point of view of the seller as well as for the user. Based on historical knowledge and detecting some rules special offers can be prepared. Some offers could be personalized, so that the regular customer will be satisfied. Furthermore it could be propose to sell some items with dynamic price (among auction rule).

The results of Market Basket Analysis algorithm show that it is a very quick and useful mechanism for behaviour detection, for making recommendations and for sales measurement.

The RFM based algorithm distinguish groups of customers, which bring the highest income to the e-shop. It is easy to recognize them and to suggest some special products to them on special conditions (e.g. lower price).

The presented mechanisms could be consider separately or one after another, i.e. by applying at first the RFM algorithm for distinguish key customers, and then for the "top N" group applying the Market Basket Analysis algorithm to finding product correlations. The knowledge could be used for promotion of correlated products.

Collecting and analysing logs based on customer relationship management is a very important process especially in the business aspects of detecting recommendations and market trends.

The general conclusion is, that both tools help in the decision making process as supported by data mining techniques. The building of correlation models improves the business process. The tools presented are quick and business oriented mechanisms for supporting customer management.

Furthermore a scoring model based on prediction (income, education, gender, marital state, region, age, place residence) can be used to measure the output of the RFM algorithm.

References

1. Anderson, D., Sweeney, D., Williams, T., Camm, J., Cochran, J.: Quantitative Methods for Business, 13th edn. Cengagne Learning, USA (2015)
2. RFM Algorithm. http://searchdatamanagement.techtarget.com/definition/RFM-analysis. Accessed 18 Dec 2017
3. Wamba, S.F., Akter, S., Edwards, A., Chopin, G., Gnanzou, D.: How 'big data' can make big impact: findings from a systematic review and a longitudinal case study. Int. J. Prod. Econ. **165**, 234–246 (2015)
4. Feelders, A., Daniels, H., Holsheimer, M.: Methodological and practical aspects of data mining. Inf. Manag. **37**(5), 271–281 (2000)
5. Watson, G., Beck, J., Henderson, C., Palmatier, W.: Building, measuring, and profiting from customer loyalty. J. Acad. Mark. Sci. **43**(6), 790–825 (2015)

6. Jean, R., Kim, D., Bello, D.: Relationship-based product innovations, evidence from the global supply chain. J. Bus. Res. **80**, 127–140 (2017)
7. Hsiao, Y.H., Chen, L.F., Choy, Y.L., Su, C.T.: A novel framework for customer complaint management. Serv. Ind. J. **36**, 675–698 (2016)
8. Chen, C., Liu, H.: The moderating effect of competitive status on the relationship between customer satisfaction and retention. Total Qual. Manag. Bus. Excellence, 1–24 (2017)
9. Kim, Y., Suh, E., Hwang, H.: A model for evaluating the effectiveness of CRM using the balanced scorecard. J. Interact. Mark. **17**(2), 5–19 (2003)
10. Jutla, D., Craig, J., Bodorik, P.: Enabling and measuring electronic customer relationship management readiness. In: Proceedings of the 34th Annual Hawaii International Conference on System Sciences, Organizational Systems and Technologies Track, pp. 1–10. IEEE Computer Society Press, Big Island (2001)
11. Bose, R.: Advanced analytics: opportunities and challenges. Ind. Manag. Data Syst. **109**(2), 155–172 (2009)
12. Hennig-Thurau, T., Malthouse, T., Friege, E., Gensler, C., Lobschat, S., Rangaswamy, L., Skiera, B.: The impact of new media on customer relationships. J. Serv. Res. **13**(3), 311–330 (2010)
13. Kwon, K., Kim, C.: How to design personalization in a context of customer retention: who personalizes what and to what extent? Electron. Commer. Res. Appl. **11**(2), 101–116 (2012)
14. Filieri, R., Fraser, M.F.: E-WOM and accommodation: an analysis of the factors that influence travelers' adoption of information from online reviews. J. Travel Res. **53**(1), 44–57 (2014)
15. Chang, W., Taylor, S.: The effectiveness of customer participation in new product development: a meta-analysis. J. Mark. **80**(1), 47–64 (2016)
16. Market basket analysis – rules. http://discourse.snowplowanalytics.com/t/market-basket-analysis-identifying-products-and-content-that-go-well-together/1132. Accessed 20 Jan 2018
17. Market basket analysis. http://www.statisticshowto.com/market-basket-analysis/. Accessed 20 Jan 2018
18. SPSS Statistics. https://www.ibm.com/support/knowledgecenter/en/SSLVMB_21.0.0/com.ibm.spss.statistics.help/rfm_intro.xml.html. Accessed 05 Feb 2018
19. SPSS Statistics – algorithms. http://algomine.pl/tag/ibm-spss. Accessed 18 Dec 2017
20. RFM Model. https://www.ibm.com/developerworks/library/ba-direct-marketing-spss/index.html. Accessed 02 Feb 2018

15 Interpretation Problems in the Semantics of UML 2.5.1 Activities

Karolina Rączkowska(✉) and Anita Walkowiak-Gall

Faculty of Computer Science and Management,
Wroclaw University of Science and Technology, Wrocław, Poland
{karolina.raczkowska,anita.walkowiak-gall}@pwr.edu.pl

Abstract. The semantics of UML presented in the OMG standard is defined informally in plain language. The lack of formal semantics brings ambiguity problems, crucial especially in case of automation of system development process and design of tools supporting the process (that implement validation of the systems' specification, model-checking, transformations, or code generation). The aim of the paper is to discuss a list of interpretation problems related to part of UML, i.e. Activity. Authors indicate inconsistencies and problems caused by a lack of information in the UML specification, which were identified by them during an attempt of formalization of the Activity semantics.

Keywords: UML activity · Semantics · Inconsistency · Indefiniteness

1 Introduction

The Unified Modeling Language (UML) [9] is a widely used language for creating models within an object-oriented software development processes, especially in the case of processes based on the Model Driven Architecture (MDA) [7] approach and its transformations. Improving and/or automating of MDA transformations demands well-defined models expressed in a precise, consistent and unambiguous language.

The abstract syntax of the UML is specified using the Meta Object Facility (MOF) [6] - based metamodel and syntax rules expressed in the Object Constraint Language (OCL) [8]. The UML semantics is defined informally in natural language. The lack of formal semantics brings ambiguity problems, crucial especially in case of automation of system development process and design of tools supporting the process by models transformations and code generation.

Since the UML specification was firstly published by the Object Management Group (OMG), it has been extended and enhanced multiple times, among other things, to eliminate unclarities and errors. However, it is still not error-free.

The newest version of the specification (i.e. 2.5.1) is partitioned into packages containing related metamodel elements. Each package is divided into five parts: parts of the MOF-based metamodel shown in figures, the semantics of the metamodel elements written in plain text, description of elements containing e.g. operations and constraints stated using OCL, graphical notation and its description, and examples.

L. Borzemski et al. (Eds.): ISAT 2018, AISC 852, pp. 267–276, 2019.
https://doi.org/10.1007/978-3-319-99981-4_25

Splitting interrelated knowledge among multiple parts of the specification is the next, besides informal definition of the semantics, reason of ambiguity problems.

Using UML Activities for specifying requirements (scenarios of use cases) in the MDA approach (and, as a consequence, defining transformations and generating the source code), which is authors' main research area, requires semantics of Activities to be formally specified. The formalization was the subject of the master thesis one of the co-authors [10]. During an attempt of the formalization, a number of ambiguities were identified.

The paper discusses the catalogue of identified problems with UML 2.5.1 Activities, including inconsistencies and issues related to the lack of information. Few problems found in the specification were omitted in the paper because they were previously reported in [5] (however, they have not been fixed yet). Authors also verified the syntax correctness of all OCL constraint and operations defined in the specification. Unlike previous versions of the specification, version 2.5.1 does not contain OCL syntax errors.

The rest of the paper is structured as followed. Section 2 briefly describes the semantics of the UML 2.5.1 Activities, taking into consideration only elements which problems are concerned with. Identified ambiguities are presented in the Sect. 3. Section 4 summarizes the paper and discusses directions for further works.

2 Semantics of UML 2.5.1 Activities

UML *Activity* may be used to model business and systems processes and algorithms. Fundamental units of an *Activity* are *ActivityNodes*, which are connected by *ActivityEdges*. A progress of the execution of an *Activity* is designated by terms of tokens – a control token indicates that a node is executed whereas an object token is used to transfer some data between nodes. *ActivityNodes* may represent an executable functionality of a process (*ExecutableNode*), manage a flow of tokens (*ControlNode*) or act like a container for objects (*ObjectNode*). *ActivityEdges* specify the order of the execution of nodes – tokens traverse an edge from a source node to a target node. The source node offers tokens to the one or more outgoing edges. A token shall be accepted by the edge before it can traverse. The edge may have a *guard* (which is especially used with outgoing edges of a *DecisionNode*) and a *weight* which are both *ValueSpecifications*. A guard specifies conditions that shall be satisfied for a token to allow it to traverse. If an edge has no guard, it means that there is a guard which evaluates to true for each offered token. Tokens cannot traverse an edge if the number of them is lower than the value of a weight of this edge.

There are two kinds of *ActivityEdges*: *ControlFlow* and *ObjectFlow*. Control tokens traverse *ControlFlows* and object tokens traverse *ObjectFlows*. An *ObjectFlow* may have a *transformation* and a *selection Behaviors* defined. The *transformation* can return, for example, values of an attribute of passed objects whereas the *selection* returns up to one object from the passed ones that satisfy defined condition. When a token is offered to an *ObjectFlow*, a *transformation* is evaluated (if defined) and then, the result of the *transformation* is passed to a *selection* (if defined) and the result of the *selection* is passed to a target node.

As mentioned above, *ControlNodes* manage a flow of tokens, so they act like traffic switches for tokens. An *InitialNode* is a starting point of an *Activity* – when the *Activity* is invoked, a control token is placed on all *InitialNodes* (if any exist) and offered to outgoing edges of this nodes. When a token reaches a *FinalNode*, it is destroyed and, if this node is a kind of *ActivityFinalNode*, execution of the *Activity* is terminated.

A *DecisionNode* uses guards of edges to route tokens. When a token comes to the *DecisionNode*, it is offered to all outgoing edges of this node and can be directed only to the edge whose *guard* evaluates to true for this token. *Guards* can refer to data coming from an additional edge of a *DecisionNode*, i.e. from a *DecisionInputFlow*.

A flow can be split into multiple concurrent flows using a *ForkNode*. A token that comes to the *ForkNode* is offered to the outgoing edges; if any of edges accepted an offer, the token is copied and the node which has accepted the offer receives that copy. A copy of the token which was not accepted by an edge as a result of rejecting an offer by the target *ActivityNode* is held by the *ForkNode* and can be accepted later. If all offered tokens are rejected, none of them is copied and they still wait for an acceptance.

Multiple flows of tokens can be merged into one with synchronization (using a *JoinNode*) or without it (using a *MergeNode*). A *JoinNode* synchronizes flows using a *joinSpec* property, which is a *ValueSpecification* that shall be evaluated to true to emit a token on the outgoing edge of the node. The *joinSpec* may refer to the names of the incoming edges (to mark the ones that a token shall be offered from) or to values of objects (if any of the incoming edges are *ObjectFlows*).

While an *Activity* is a representation of a process, *ExecutableNodes* are the steps that are performed during the execution of this process. They can represent, for example, dealing with events that come from outside of the *Activity*, reading and modifying objects or calling another *Activity*. *Actions* are the only kind of *ExecutableNodes*, and one of them is a *OpaqueAction* whose behaviour is expressed as a text.

Objects produced by *Actions* or passed to *Activity* from the outside and transferred through *ObjectFlows* are held by *ObjectNodes*. An *ObjectNode* may have a type which means that the type of every object in a token held by this node shall conform to the type of the node. Tokens held by an *ObjectNode* are offered to outgoing edges in the order specified for this node; by default, they are offered in the order they have come into the node (FIFO – first in, first out). The *ObjectNode* can accept offered tokens up to the limit specified by its *upperBound* property.

One of the kinds of an *ObjectNode* is *Pin*, which is connected with an *Action*. An *OutputPin* connected with an *Action* holds object tokens produced by this *Action* whereas an *InputPin* holds object tokens that are consumed by its *Action*. An *upper* and a *lower* properties (which are properties of a *Pin*) specify the maximum and minimum number of objects that can be produced (in case of an *OutputPin*) or consumed (in case of an *InputPin*) by an *Action*.

An *ActivityGroup* is a container for nodes and edges of the *Activity*; each of those nodes and edges can be contained in more than one *ActivityGroup*. One of specific kinds of *ActivityGroup* is *InterruptibleActivityRegion*. It has one or more *interruptingEdges*. An execution of the nodes which belongs to this region is aborted when a token is accepted by any *interruptingEdge*.

3 Ambiguities in UML 2.5.1 Activities

Problems described in this section are caused at most by missing information in the UML specification and inconsistency between parts of this specification. During our research, misspellings and incorrect figures were also identified, but due to lack of space in this paper, we skipped them.

1. *Is it really the case that a token can be accepted by multiple targets?*
   As described on page 375 of [9], a token can be accepted by only one of targets which it was offered to. Nonetheless, on page 390 of [9] there is information that a token can be accepted by multiple targets, but can traverse only one edge (which leads to one of the nodes that accepted the token). So, it is unclear how many targets can accept the same token in the same time.
2. *Can a token be accepted by a ControlNode?*
   As described on page 374 of [9], ObjectNodes accept object tokens while ExecutableNodes accept control tokens; ControlNodes only manage a flow of accepted tokens, acting like traffic switches between them.
   Nonetheless, a few parts of the specification mention about accepting control tokens by ControlNodes, i.e. parts which describe the semantics of FinalNodes (page 388 of [9]) or the semantics of DecisionNodes (page 390 of [9]). Moreover, according to the page 374 of [9], accepting an offer by an edge depends, at least, on accepting the offer by a target ActivityNode. That means that in a situation when a target node of an edge is a ControlNode and a token is offered to this edge, there is no possibility of accepting the token by the edge because the target node cannot accept it, as it is a ControlNode.
3. *Where is a token held after rejecting an offer by a ForkNode?*
   A token which was not accepted by any outgoing edge of a ForkNode because of rejecting an offer by a target ActivityNode waits for the next attempt in which it can be accepted. However, it is not obvious where the token is actually stored. As described on page 388 of [9], this token rests on the outgoing edge of the ForkNode and it is an exception from the rule that tokens cannot rest at ActivityEdges if they cannot move downstream. In contrast with that information, it is said on page 375 of [9] that a ControlNode cannot hold tokens and a ForkNode is an exception from that rule. In this part, it is not explicitly given when the ForkNode can hold tokens. However, it must be related to the situation when a token is blocked from moving downstream as there is no other situation described in the specification in which the token can rest (at a node or edge). Thus, as said before, a token rests in a ForkNode, not on the outgoing edge of this node, which is in contradiction to the page 388 of [9].
4. *Can an ActivityFinalNode be executed?*
   On page 375 of [9], there is information that an execution of an ActivityNode is indicated by placing a token on this node. According to page 375 of [9], InitialNodes and ForkNodes are the only kinds of ControlNodes which a token can be placed on; therefore they are the only kinds of ControlNodes which can be executed. Nevertheless, in a description of the semantics of an ActivityFinalNode, it is

mentioned several times that this node is executed. On the one hand, a token shall be placed on an ActivityFinalNode when executing (as an indication of that execution), but on the other hand, an ActivityFinalNode is a ControlNode and does not belong to the set of nodes that a token can be placed on. Thus, it is unclear if an ActivityFinalNode can be executed.

5. *How many times an ExecutableNode is executed, if multiple tokens will come to it?* Consider the situation shown in Fig. 1. According to the page 393 of [9], if only one of the actions Action1 or Action2 completes, the node Action3 is executed once, but if both complete, Action3 is executed twice.

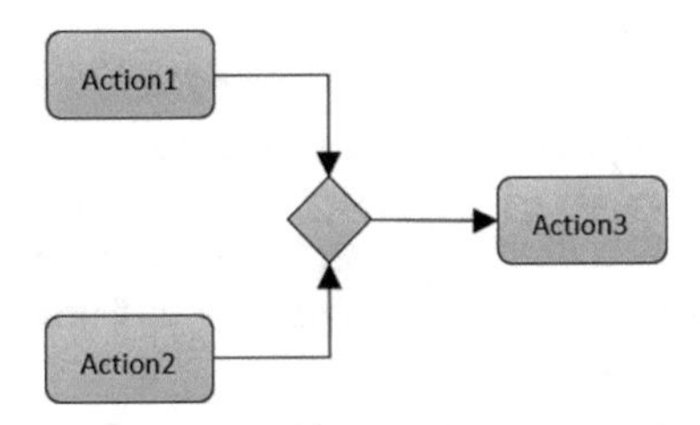

Fig. 1. An example of the usage of a MergeNode.

The description is undoubted when we considering the completion of individual actions at another time. However, if we take into account the possibility of completing the action at the same time, there are doubts as to the number of action executions indicated in the description:
If Action1 and Action2 complete, two tokens are offered to the MergeNode; because all tokens offered to the MergeNode are offered to the outgoing edge of this node (according to page 389 of [9]), two tokens are offered to Action3.
On page 403 of [9], there in information that "before an ExecutableNode begins executing, it accepts all tokens offered on incoming ControlFlows. If multiple tokens are being offered on a ControlFlow, they are all consumed. (…) While the ExecutableNode is executing, it is considered to hold a single control indicating it is execution."
According to above, two tokens are offered to Action3 simultaneously, after being accepted they are consumed and a single token is placed on Action3. That means that the node executes one (not twice) in this situation.

6. *When evaluation of a guard on an outgoing edge of an InitialNode is fired again?* Consider the situation shown in Fig. 2. When an activity starts, a token is placed on an InitialNode and is offered to an outgoing edge of this node. The only condition which the acceptance of the token depends on is guard on the edge. If guard1 evaluates to true, the token traverses edge; otherwise, the token remains on the InitialNode. As mentioned before, it is unclear when an evaluation of a guard is fired - there is only information that it is evaluated for every token that is offered to the edge.

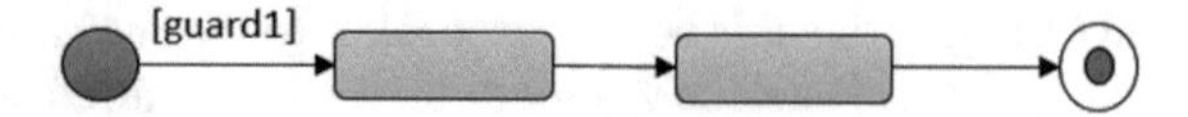

Fig. 2. Illustration of the problem with re-verifying offer from an InitialNode

7. *Can an outgoing edge of a ForkNode accept a token if this token has been rejected by a target?*
   According to the page 374 of [9], an ActivityEdge can accept offered token only if the target ActivityNode accepts it also, which can require acceptance of this token by downstream nodes and edges. The token can traverse an ActivityEdge only if it is accepted by that edge.
   Page 388 of [9], where the semantics of a ForkNode is described, says that copies of tokens are accepted by an ActivityEdge in spite of non-accepting original tokens by the target ActivityNode. After accepting, copies move to this edge and stand there until the target ActivityNode accepts it also. Hence, is it really the case that a token can be accepted by an outgoing edge of a fork node, even if it is not accepted by a target node of this edge?
8. *Is it really the case that ObjectNodes connected by an ObjectFlow (with optionally ControlNodes on this path) shall have compatible types?*
   Types of ObjectNodes connected by an ObjectFlow (with optionally ControlNodes on a path between this nodes) shall be compatible – the downstream ObjectNode type must be the same or a supertype of the upstream ObjectNode type (page 429 of [9]). However, there can be a transformation Behavior defined on an ObjectFlow between two ObjectNodes, which can retrieve e.g. attribute value of an object passed through this edge (as shown in Fig. 3).

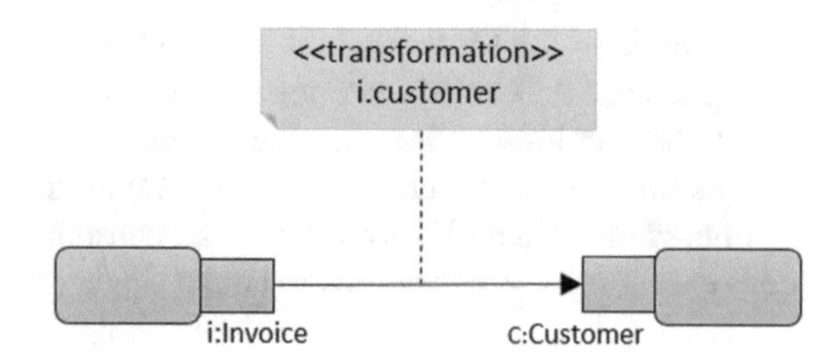

Fig. 3. An example of a transformation behavior.

Moreover, there is a constraint on page 428 of [9] which says that the type of an input parameter of a transformation must be compatible with the type of an incoming object token and the type of an output parameter of the transformation must be compatible with the type of an object token that is expected by a target node.

9. *Is it the case that input Parameter of a selection defined for an edge shall have a type which is compatible with the type of source ObjectNode of that edge?*
   As mentioned before, in case both a transformation and a selection are defined for an edge, the transformation is executed first and a result of this execution (which

can be, for example, an attribute of passed object) is passed as an input Parameter to the selection. As described on page 428 of [9], "a selection Behavior has one input Parameter and one output Parameter. The input Parameter must have the same as or a supertype of the type of the source ObjectNode".

Consider the Activity shown in Fig. 4. The ObjectFlow has both the transformation and the selection. The first action completes and a token holding object of type T, which was produced by this action, is offered to the outgoing edge of OutputPin. This token is passed to the input Parameter of the transformation. The transformation returns a bag of objects of type S, but, according to the previous quote, the selection is expecting an object of type T.

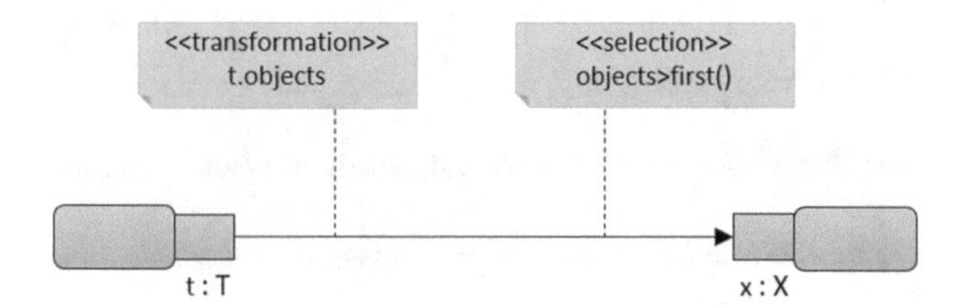

Fig. 4. An illustration of the problem with compatibility of objects types on an ObjectFlow.

10. *What is the order of evaluation guard, weight, selection and transformation?* If a transformation and a selection are both defined for an ObjectFlow, then the transformation is executed first and the result of this Behavior is set as an input Parameter of the selection. If the ObjectFlow has also a guard, then it is evaluated for every new token offered to this edge, and if a number of tokens for which the guard was evaluated to true is less than a weight value, none of them can traverse this edge. It implies that weight is verified after evaluation of guard, However, there is no information in which order guard, weight, selection and transformation are evaluated. For instance, is a weight verified after the evaluation of a selection or before that?
11. *What happens with a token after evaluation of a transformation Behavior?* As described on page 376 of [9], a result of a transformation Behavior "is put in an object token that is offered to the target ActivityNode instead of the original object token"; if the transformation returned no value, an empty token is offered. But in case that this Behavior returns multiple values, each of them is "put in a separate object token, all of which are passed to the target ActivityNode". Thus, is it really the case that token holding a result of a transformation is offered to the target node when the transformation returned none or one value, and passed otherwise?
12. *Is it really the case that a result of a selection Behavior is passed to a target node?* After a selection Behavior is evaluated, its "output value is put in an object token and passed to the target ActivityNode" (page 377 of [9]). Consider the Activity shown in Fig. 5. An object token is offered to the outgoing ObjectFlow of the OutputPin and the selection is evaluated. It returns an object which is put into a token and passed to the target node of the ObjectFlow, which is the DecisionNode. But what happens if this token will not be accepted by any of the outgoing edges

of the DecisionNode? It was passed to this node so it seems that it cannot return to the OutputPin, but it also cannot rest at the DecisionNode, as it is a ControlNode (and tokens cannot rest at ControlNodes).

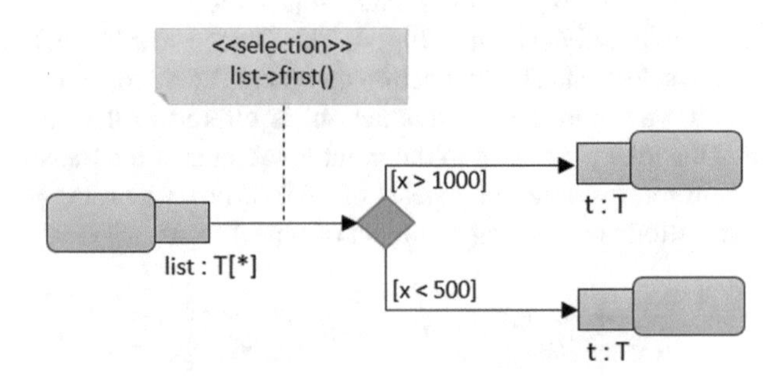

Fig. 5. An illustration of the problem with passing tokens after an evaluation of a selection.

13. *In case an ObjectFlow has both a transformation and a selection, which one is evaluated when an offer is withdrawn?*
Selection Behavior is evaluated whenever a new token is offered to the ObjectFlow, or an offer is withdrawn (page 377 of [9]). If an edge has also a transformation Behavior, then it is invoked first when a new token is offered to the ObjectFlow and the resulting value is used in the invocation of the selection behavior (page 377 of [9]); there is no information about withdrawing an offer. Thus, if an edge has both a transformation and a selection, what happens when an offer is withdrawn? Is it selection the only Behavior that evaluates, or none of them is invoked?
14. *What happens with a token resting on or traversing an edge between nodes contained in an InterruptibleActivityRegion?*
According to pages 407–408 of [9], "when a token offered along an interruptingEdge is accepted and traverses that edge, then the execution of all containedNodes of the region is terminated and all tokens are removed from them. However, the token traversing the interruptingEdge still arrives at its target and, further, any accepted tokens traversing non-interrupting edges from a source node within the region to a target node outside the region also still arrive at the target nodes, even if the interruption occurs during the traversal."
If a token reaches an interruptingEdge, tokens which traverse other interruptingEdges or edges leading from a source node contained in the InterruptibleActivityRegion to a target outside the region are still traversing – those flows are not interrupted. However, there is no information about tokens resting on or traversing edges between nodes contained in an InterruptibleActivityRegion.
15. *What happens with a token resting on or traversing edge between source outside an InterruptibleActivityRegionand target contained in the region?*
A similar problem concerns interrupting a flow of tokens traversing edges whose source node is outside the region and target node is inside the region. Does this flow terminates after termination of the region?

4 Conclusions

The UML is widely adopted academic and industry standard, implemented in many open-source and commercial tools. Despite that it has been subject of much research for almost 20 years, there are not many studies on the UML semantics correctness, although as we indicate in the paper, it still has many semantics incorrectness.

In the papers [1, 3, 13], the authors focus on verifying and validating OCL statements of the UML specification. In [3] the OCL statements of the UML 1.4 are checked for syntactical and static semantics correctness, by the way of a model checker tool development. Found issues (around 450) are classified into three different groups: non-accessible elements, empty names, miscellanea. In order to solve the issues, authors suggested, for example, to always check whether names are empty prior to their comparison.

The similar check is made in [1], however for OCL statements of the UML 2.0, which results in about 360 deficiencies being such commons errors like type casting problem, statement syntax issues, etc.

Finally, the authors of the [13] investigate the UML 2.3. As a result of found issues, they suggest changes (simplifications) in the OCL language in order to improve quality of future UML specification. Moreover, they propose changes to the UML specification process.

Because in the previous articles [1, 3, 13] there was pointed that some of the OCL constraint and operations defined on UML metamodel are erroneous, we verified the OCL statements of the UML 2.5.1 for their syntax correctness; however, we did not find any syntax errors.

There are several work related to the correctness of the semantics of the UML.

In [2] authors present a list of problems and deficiencies of UML 2.0 behavioral state machines, encountered during the attempt to formalize state machine semantics. The deficiencies are classified into appropriate categories (ambiguities, inconsistencies, unnecessary strong restrictions, and forgotten restrictions) and illustrated on state machines, legal according to the UML specification, however, having a problematic meaning. The suggestions for the problems improvement are discussed.

Whereas in [4] authors discuss inconsistencies and ambiguities of the UML 1.3 standard, particularly relating to system decomposition and use case models. The deficiencies are illustrated using the TRMCS (Teleservices and Remote Medical Care) case study. Authors analyze the nature of the deficiencies and investigate whether and how they can be overcome.

Ambiguities, inconsistencies, and deficiencies concerning UML 2.0 Activity are widely discussed in [11, 12]. The author in his doctoral thesis was involved in the formalization of an activity. He defined the token flow semantics of UML 2.0 Activity using Abstract State Machines. However, in order to do that, the author had to consider correctness of the activities' semantics. In results, several problems were identified in the field of considerations. For most of them, the author proposed solutions/unambiguous interpretations in his approach. The author considered signals semantics, which is omitted in this paper, but he did not take into account the flow of objects and related with them concepts such as selection, transformation. Moreover, the author's works refer to the UML 2.0 specification, therefore some of the identified problems are not actual and with the new revision of the language (UML 2.5.1), new ambiguities have emerged.

Problems listed in this paper concern only a subset of elements related to Activities, so the future work is to take into consideration elements that were ignored. Moreover, we are planning to report identified unclarities to [5]. Before they will be fixed, it is useful to propose solutions and clarifications for them.

References

1. Bauerdick, H., Gogolla, M., Gutsche, F.: Detecting OCL traps in the UML 2.0 superstructure: an experience report. In: Baar, T., Strohmeier, A., Moreira, A., Mellor, S.J. (eds.) UML 2004. Lecture Notes in Computer Science, vol. 3273, pp. 188–196. Springer, Heidelberg (2004)
2. Fecher, H., Schönborn, J., Kyas, M., de Roever, W.-P.: 29 new unclarities in the semantics of UML 2.0 state machines. In: Lau, K.K., Banach, R. (eds.) Formal Methods and Software Engineering. ICFEM 2005. Lecture Notes in Computer Science, vol. 3785. Springer, Heidelberg (2005)
3. Fuentes, J.M., Quintana, V., Llorens, J., Génova, G., Prieto-Díaz, R.: Errors in the UML metamodel? ACM SIGSOFT Softw. Eng. Notes **28**(6) (2003)
4. Glinz, M.: Problems and deficiencies of UML as a requirements specification language. In Proceedings of the 10th International Workshop on Software Specification and Design, IWSSD 2000, pp. 11–22, IEEE Computer Society, Washington, DC (2000)
5. OMG Issues. http://issues.omg.org/issues/spec/UML/2.5. Accessed: 15 May 2018
6. OMG Meta Object Facility Core Specification 2.5.1. https://www.omg.org/spec/MOF/2.5.1/PDF. Accessed 21 May 2018
7. OMG Model Driven Architecture Guide rev. 2.0. https://www.omg.org/cgi-bin/doc?ormsc/14-06-01.pdf. Accessed: 21 May 2018
8. OMG Object Constraint Language 2.4. https://www.omg.org/spec/OCL/2.4/PDF. Accessed 21 May 2018
9. OMG Unified Modeling Language 2.5.1. http://www.omg.org/spec/UML/2.5.1/PDF. Accessed 21 May 2018
10. Rączkowska, K.: Operational semantics of activities. MSc thesis, Wroclaw University of Science and Technology (2017)
11. Sarstedt, S.: Overcoming the limitations of signal handling when simulating UML 2 activity charts. In: Proceedings of the 2005 European Simulation and Modelling Conference (ESM 2005), pp. 61–65 (2005)
12. Sarstedt, S.: Semantic foundation and tool support for model-driven development with UML 2 activity diagrams. Ph.D. thesis, University of Ulm (2006)
13. Wilke, C., Demuth, B.: UML is still inconsistent! How to improve OCL constraints in the UML 2.3 superstructure. In: Proceedings of the Workshop on OCL and Textual Modelling (OCL 2011). Electronic Communications of the EASST, vol. 44 (2011)

Enhancements of Detecting Gang-of-Four Design Patterns in C# Programs

Anna Derezińska(✉) and Mateusz Byczkowski

Institute of Computer Science, Warsaw University of Technology, Nowowiejska 15/19, 00-665 Warsaw, Poland
A.Derezinska@ii.pw.edu.pl

Abstract. Gang-of-Four design patterns are valuable architectural artefacts in object-oriented design and implementation. Detection of design patterns in an existing code takes an important role in software evolution and maintenance. A lot of work has been devoted to development of methods and tools that support automatic detection of design patterns. There have been scarcely any attempts to detect design patterns in C# programs. We have focused on the refinement and extension of the approach of A. Nagy and B. Kovari. In this paper we discuss the rules for mining of a subset of GoF design patterns in C# applications. These rules have been used to enhance the program that detects design patterns in C# applications. The mining results of both tools were compared.

Keywords: Software maintenance · Design patterns · Design pattern detection Gang of four · C#

1 Introduction

Gang-of-Four design patterns belong to the mostly used pattern category in software development [1]. These general architectural ideas have successfully been applied while implementing programs in different notations, including the C# language [2].

Detection of design patterns in an existing code is an important issue of software evolution and maintenance [3, 4]. Knowledge of patterns used in a program might support comprehension of its design, especially if we are not in touch with an author of the code, we need to maintain a legacy software or one prepared by external collaborators via outsourcing, etc. Information about patterns pointed up could also be applicable in software refactoring or in quality software evaluation [5, 6].

Many approaches to design pattern detection have been studies and implemented in various software engineering tools [7–10]. However, there is a gap in mining and evaluation design patterns in C# programs.

The work of Nagy and Kovari [11] has been focused on a language-independent solution. They employed general structure-based criteria. One of implementations of this approach was the only one used to evaluate C# programs. Based on this solution, we have proposed similar criteria that were refined in case of several design patterns. Moreover, the list of considered patterns was extended with five additional ones.

L. Borzemski et al. (Eds.): ISAT 2018, AISC 852, pp. 277–286, 2019.
https://doi.org/10.1007/978-3-319-99981-4_26

In this paper, we have presented the structural-based criteria to design pattern detection in C# programs. The proposed criteria were implemented in a tool used for static analysis of C# programs and applied in some case studies. According to experiment results, it can be seen that even the minor refinements could have a significant effect on the design pattern detection.

In the paper *Design Patterns* will be abbreviated as DP. This paper is organized as follows. The next section summarises approaches to DP detection and other related work to DP in C# applications. In Sect. 3 criteria of DP detection used during static analysis of C# programs are presented. Section 4 describes briefly tool support to analysis of C# programs and comparison of results. Finally, Sect. 5 concludes the paper.

2 Background and Other Related Work

In general, design pattern detection approaches can be divided into two main types: those focused on the static program analysis, and the second that take into account some behavioral program features. Static program analysis is mainly based on the structural analysis, in which inter-relations between classes are examined. Structural approaches use direct code analysis, usually in some intermediate form of the code as AST (Abstract Syntax Tree), or other graph representations based on the reverse engineering techniques. In behavioral analysis, constraints could be satisfied by a candidate instance to be classified as a "valid" design pattern.

Yu et al. [9] used a code representations of Class-Relationship Directed Graphs in which sub-patterns are discovered and jointed to compare with DP templates and method signatures. Combination of static analysis with subsequent dynamic verification can be found in ePAD^{evo} [12]. Guéhéneuc et al. [13] combined former experiences with constraint programming and numeric signatures in design motifs to deal with complete and incomplete occurrences of DP.

One of the problems is existence of many variants of a pattern. Stencel and Wegrzynowicz used static analysis and SQL to find many implementations of DP in Java code [14]. Multiple variants were treated in [15] by machine learning.

Design pattern identification could be proceeded by a filtering step in order to improve performance and precision rate. A metric-based approach to preliminary selection of pattern candidates and their further evaluation was described in [16].

Many different tools have been developed to support design pattern mining in a source code. They can mainly detect patterns in Java or C++ programs, but also in Eiffel, Smalltalk, UML or XML.

Rasool et al. [7] have evaluated different tools that recognize design patterns in source code. However, neither of six compared tools, nor 30 only reviewed in the paper, were detecting patterns in C# programs. Moreover, it has been noted that many tools give only results about the number of recovered patterns but do not show an exact location in the source code. In a PhD thesis by Zanoni [8], many pattern detection tools have been described. Though, there were no candidates to assist mining of C# programs. More recent list of design pattern detecting tools could be found in [10]. But also in this survey, no tools dealing with C# have been mentioned.

Design patterns could be developed in the C# programming language as in other object-oriented ones. The basic discussion of pattern implementation and their variants related to C# has been presented by Metsker [2]. However, not much work has been published that considered research or experiments on C# programs.

Gatrell et al. [17] examined a C# system in order to confirm a hypothesis that pattern-based classes were less change prone than other classes. In this work detecting of 13 GoF DP was not automated but based on manual code inspection.

Rasool et al. [18] reported on development of a customizable approach to feature-based pattern recognition. Apart from parsers supporting Java and C++, a module for C# was also implement. However, experiments were dealt only with two first languages and not with any C# programs.

To the best of our knowledge, the only tool practically used for mining DP in C# programs was one of implementations of the approach described in [11]. Nagy and Kovari employed static program analysis of an AST form and structure-based criteria for DP. The approach was not bounded by developer intentions, and could also reveal patterns that were not deliberately designed. Our work follows their method.

3 Criteria of Pattern Detection in C# Programs

3.1 Preliminaries

We have modified the approach of [11] focusing only on the variant for the C# language. Nagy and Kovari intended to have a programming language independent solution, hence, language-independent criteria. In their tool, there were separate modules for processing of a source code, C# and Java code in particular. A data model from one of those modules was transmitted to a common criterion matching algorithm that used a list of criteria to design pattern analysis, and returned DP instances.

Based on the evaluation of many papers presented by Budgen [19], the mostly studied patterns have been *Composite*, *Observer* and *Visitor*. *Observer* and *Composite* appeared also to be the mostly useful patterns according to the conducted survey. Results of *Visitor* were mixed, as it was useful for limited purposes. Therefore, selecting new patterns to extend the tool, we have added *Observer* and *Visitor*, as *Composite* was already one of the patterns implemented in the tool [11]. Besides, two structural patterns were supplemented, namely *Adapter* and *Proxy*, that were not counted to a group of "patterns of little real use". In result, we propose the refined detection criteria that cover the list of DP supported in the original tool plus the additional five patterns mentioned above.

3.2 Criteria of Design Pattern Detection

The presented DP detection criteria are based on static analysis, i.e. general object-oriented structural program features, but taking into account objectives of C# as a source program. In general, classes that are components in a pattern are recognized by interrelations they meet and additional conditions of their fields and methods. In the following criteria, classes will be denoted by capital letters: A, B, etc. If two classes are named by different letters, it is assumed that they are different, e.g. $A \neq B$.

A rule of each DP is illustrated by a class diagram. The diagrams do not define the rules but improve their readability. A diagram presents only one structural variant of a pattern, if a rule implies existence of two possible relations, e.g. an inheritance or realization of an interface. Usually only the solution with a base class is shown.

The class diagrams (Figs. 1, 2, 3, 4, 5, 6, 7, 8, 9 and 10) do not replicate exactly the known original models that explain DP at the modelling level [1]. They should be interpreted as a UML reverse engineering design at the implementation level. Therefore, generalizations reflecting code inheritance relations are shown, but simple associations or aggregations are not present. In some cases, class fields are presented that are checked in the rule. They are mainly code level realizations of associations of various kind.

Elements in the diagrams are additionally annotated with stereotypes « .. » that relate to their roles in a pattern. It should be noted that they are not a part of the code under concern, but could only easy comprehension of particular design pattern components. They could be added to elements recognized as design pattern members in their comments or code annotations, if source code modifications were considered.

1. *Singleton*
   (a) There exists class A that has only private and protected constructors.
   (b) Class A includes exactly one non-public and static field which is of the same type as the class A.
   (c) Class A includes exactly one public static method that returns an object of type A.

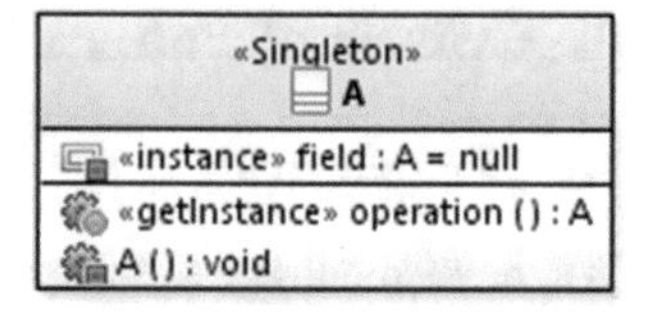

Fig. 1. Singleton

2. *Factory method*
   (a) There are classes A and B such as B inherits from A or class B implements an interface A.
   (b) Class B includes a method *operation()* that overrides the method of class A.
   (c) There exist also classes C and D such as D inherits from C or D implements an interface C.
   (d) According to its signature the method *operation()* returns an object of type C. In its body in B, the method returns an object of type D.

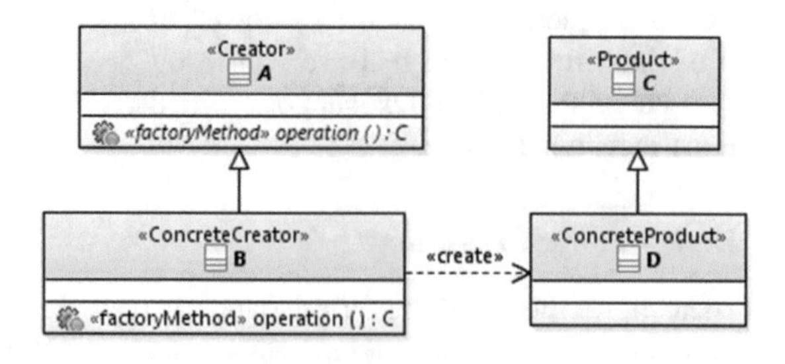

Fig. 2. Factory method

3. *Proxy* (Fig. 3)
(a) There are classes A, B and C such as B and C inherit from A or classes B, C implement an interface A.
(b) Class C includes a field that stores a reference to an object of type B.
(c) Class A includes a method *operation()*. The method is overridden in B and C.
(d) The method *operation()* from class C calls the method *operation()* from class B.
4. *Decorator* (Fig. 4)
(a) There exist classes A, B and C such as B and C inherit from A, or classes B, C implement an interface A.
(b) Class C includes a field that stores a reference to an object of type A.
(c) Class C includes a constructor C that has a parameter – an object of type A.
(d) Class A includes a public, protected or default method *operation()* that is overridden in class B and in class C.
(e) The method *operation()* from class C calls the method *operation()* from class A.

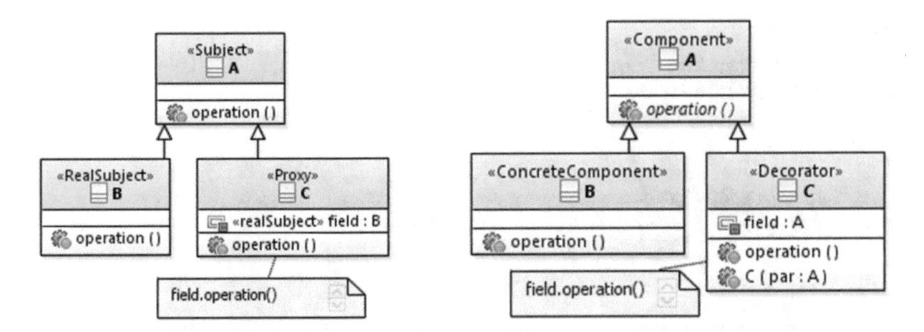

Fig. 3. Proxy

Fig. 4. Decorator

5. *Composite* (Fig. 5)
(a) There exist classes A, B and C such as B and C inherits from A.
(b) Class C cannot inherit from B.
(c) Class C includes a collection of elements of type A.
(d) Class A includes a method *operation()* that is overridden in B and C.
(e) Inside the method *operation()* in class C, method *operation()* of the class A is called for elements of type A.
6. *Adapter* (Fig. 6)
(a) There exists classes A, B and C such as C inherits from A and implements an interface B.
(b) Class A does not implement interface B.
(c) Class A has its public method *specificOperation()*. Interface B has a public method *operation()* and class C has a public method *operation()* such as it overrides the method from interface B. The method *specificOperation()* from class A is called inside the method *operation()* from class C.

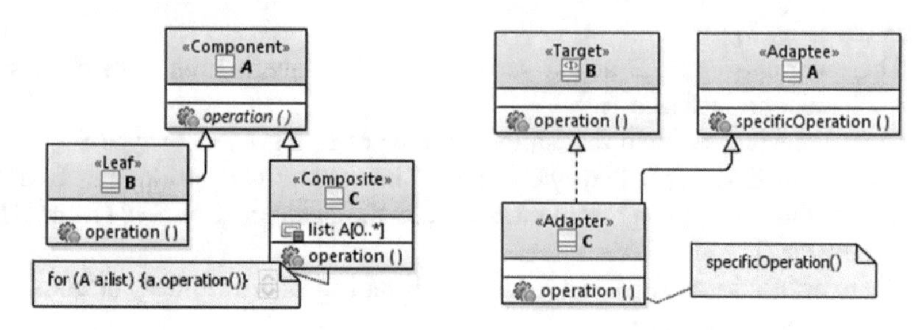

Fig. 5. Composite

Fig. 6. Adapter

7. *Mediator (a variant with two communicating classes).*
(a) There exists classes A, B and C such as B and C inherit from A.
(b) There exists classes D and E such as E inherits from E.
(c) Each of classes A, B, and C has at least one field that stores a reference to an object of type D.
(d) Class E includes one field storing a reference to an object of type B, and one field storing a reference to an object of type C.

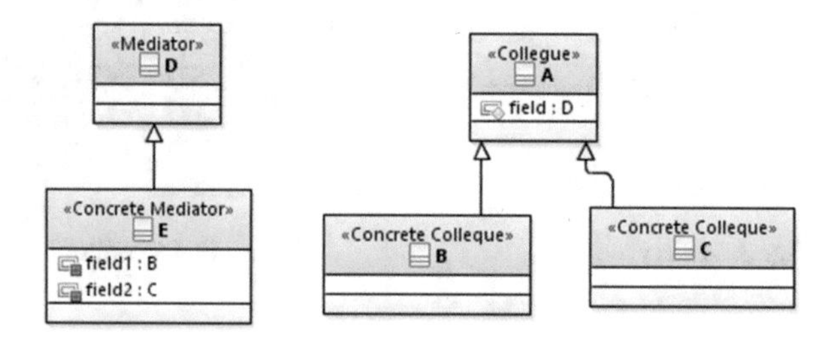

Fig. 7. Mediator

8. *Chain of responsibility*
(a) There exist at least two classes B and C that inherit from class A.
(b) Classes A, B, and C includes at least one field that has a reference to class A.
(c) Class A has a public, non-static method *operation()* that is overridden in class B and in class C.
(d) Methods from classes B and C call the method *operation()* for an object o type A.

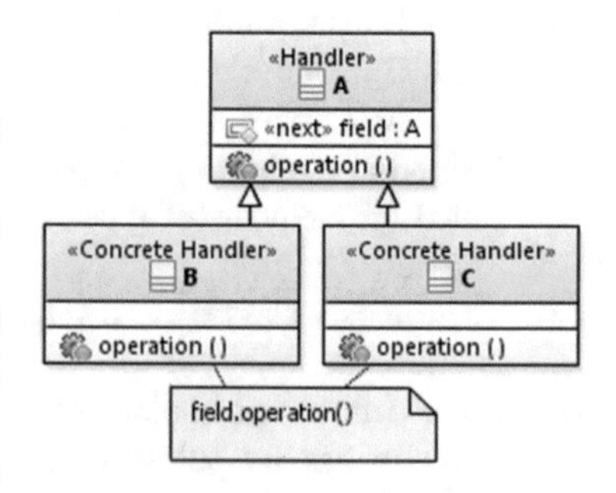

Fig. 8. Chain of responsibility

9. *Observer* In the case of this design pattern, there are omitted conditions that assume existence of a base class for the observer.

(a) There are classes A and B.
(b) Class A includes a collection of elements of type B.
(c) Class A has a method that calls another method included in class B for any component of the collection.
(d) The method in class B has a parameter that is an object of type A or one of base class of class A.

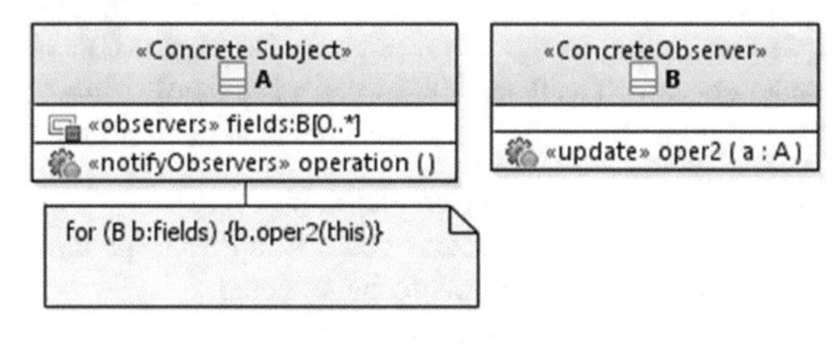

Fig. 9. Observer

10. *Visitator*

(a) There exist classes A, B, C and D such as B inherits from A or implements an interface A, and D inherits from C or implements an interface C.
(b) Class A includes a method *operation()* that has a parameter – an object of type C.
(c) Class B includes a method that overrides the method operation() from class A.
(d) Class C includes a method *oper2()* that has a parameter – an object of type B.
(e) Class D includes a method that overrides the method *oper2()* from class C.
(f) The method from class B has a parameter of an object of type C, and calls the method from class C.

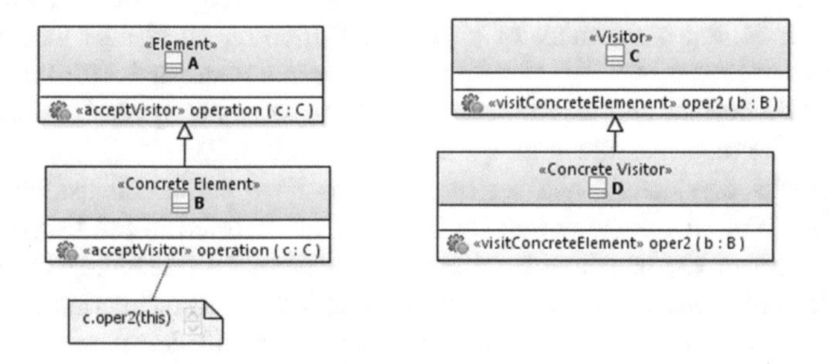

Fig. 10. Visitator

4 Design Pattern Detecting with C# Analyzer

C# Analyzer has been developed to integrate different functionalities basing on static analysis of C# programs. The mail goals of the tool are the following:

- Identification of classes and methods that implement selected design patterns.
- Calculation of software metrics.
- Support of a project comprehension.
- Support of software quality analysis based on the above results.

The tool was designed as an extension of the Visual Studio environment. A significant part of C# Analyzer is a module responsible for detection of DP. It inputs code of a program under analysis, and creates a data model, i.e. its syntax tree. The tree is generated using.NET platform compiler - Roslyn. Then, the source program is analyzed according to the given criteria. The tool points at classes that participate in DP under concern. The detection program is a modified version of the tool implemented for C# programs by [11]. The main revisions have referred to the following areas:

1. mechanism of creating model data,
2. criteria to qualify a code extract as a design pattern,
3. addition of new patterns to be detected,
4. identification of all class components in a pattern.

A modification within the first area has been taking into account relations to interfaces in the data model, which can be used in C# programs. Another adaptation implies usability of fields and methods that are inherited from a base class. Based on a recurrent approach, all base classes of a given class are visited starting from the top level class in its inheritance tree. During return routine, the corresponding methods are substituted by overriding methods, and the fields accordingly. In result, all fields and methods accessible due to public or protected modifiers are recognized in the data model of a descendant class.

The program has been extended with detection facility for additional DP (*Adapter*, *Composite*, *Observer*, *Proxy* and *Visitor*), and refinement of other DP detection criteria. The criteria for detection of the design pattern set supported by the current tool are given in Sect. 3.2.

The former detector has identified and outputted only a main class that was accepted as a central component of a pattern. Currently, all classes that are pattern components are recognized. In an output file, they are annotated with the name of a pattern being detected, and a role served in the pattern, i.e. symbols A, B, C, D,... according to the diagrams illustrating the criteria (Sect. 3.2).

Experiments conducted on a set of programs have confirmed usefulness of the solution [20]. Not all DP supported by the detector occurred in the tested programs. Therefore, a set of C# benchmark programs was prepared illustrating application of different DP. All kinds of the patterns were identified in these experiments.

In comparison to the former approach [11], we could detect new patterns and pattern instances corresponding to more structures of C#. On the other hand, some potential patterns were excluded due to criteria refined with regard to the C# language.

In Table 1. we have shown the numbers of different patterns detected in the Log4net C# library. It was used in experiments both with the previous tool [11] and C# Analyzer. Therefore, we could compare DP detection abilities of the tools. In the experiments, 22 pattern instances were detected by the old tool whereas 28 by the new one.

Only in case of *Singleton*, both tools recognized exactly the same pattern instances. Diversities in other cases followed from the differences in the pattern recognition criteria. In evaluation of *Decorator*, the former tool only used limited criteria 4.a and 4.c. While taking into account additional criteria (4.b, 4.d, 4.e) two cases were not classified as *Decorator*. The similar situation refers to *Composite*. All four pattern instances were not accepted due to an extra condition (5.d) about a collection of elements of type A implemented in class B. An opposite situation can be experienced in case of *Factory Method*. This pattern specification takes into account two additional relations, namely: detection of interface realization and inheritance of fields and methods. Therefore, 3 additional occurrences of *Factory Method* were recognized apart from 10 instances previously accepted. Finally, *Proxy* was not addressed before but was detected by the current tool.

Table 1. Numbers of design patterns detected by tools in the Log4net C# library

Tool	Design pattern occurrences					
	Singleton	Decorator	Composite	Factory method	Proxy	Sum
Tool [11]	5	3	4	10	–	22
C# Analyzer	5	1	0	13	9	28

5 Conclusions

In this paper we have presented a design pattern detection approach based on static code analysis towards structural feature mining. The approach was adopted to the C# programing language and implemented in the enhanced tool. The tool was combined within the C# Analyzer system that was integrated with the Visual Studio. Other components of the system refer to software metrics. Evaluation of results of pattern detection in the relation to the software metrics is described in [20].

Introduced improvements have resulted in higher number of recognized pattern instances mainly due to additional types of pattern detected. However, more refined criteria have eliminated some of pattern instances accepted formerly, as they counted now being false positive. It should be noted, that the former tool applied to C# programs was an instance of the language-neutral concept of pattern detection. All in all, it could be questionable whether it is not better to have language-specific solutions based on a general one, as proposed in this paper.

The presented tool could further be extended with detection of remaining GoF design patterns and other patterns. Moreover, other criteria to design pattern recognition could be added, taking into account e.g. naming conventions, comments, or metadata. Open issue remains also investigation of different pattern variants in respect to a general structure and specific features of the C# language, performance evaluation of detection facilities, as well as thoroughly estimation of detection precision.

References

1. Gamma, E., Helm, R., Johnson, R., Vlissides, J.: Design patterns: elements of reusable object-oriented software. Addison-Wesley, Boston (1995)
2. Metsker, S.J.: Design patterns in C#. Addison-Wesley, Boston (2004)
3. Ampatzoglou, A., Charalampidou, S., Stamelos, I.: Research state of the art on gof design patterns: a mapping study. J. Syst. Softw. **86**(7), 1945–1964 (2013). https://doi.org/10.1016/j.jss.2013.03.063

4. Mayvan, B.B., Rasoolzadegan, A., Yazdi, Z.G.: The state of the art on design patterns: a systematic mapping of the literature. J. Syst. Softw. **125**, 93–118 (2017). https://doi.org/10.1016/j.jss.2016.11.030
5. Ali, M., Elish, M.O.: A comparative literature survey of design patterns impact on software quality. In: Proceeding of International Conference on Information Science and Applications (ICISA). IEEE, pp 1–7 (2013). https://doi.org/10.1109/icisa.2013.6579460
6. Pradhan, P., Dwivedi, A.K., Rath, S.K.: Impact of design patterns on quantitative assessment of quality parameters. In: Second International Conference on Advances in Computing and Communication Engineering (ICACCE), pp 577–582. IEEE (2015). https://doi.org/10.1109/icacce.2015.102
7. Rasoola, G., Maedera, P., Philippow, I.: Evaluation of design pattern recovery tools. Procedia Comput. Sci. **3**, 813–819 (2011). https://doi.org/10.1016/j.procs.2010.12.134
8. Zanoni, M.: Data mining techniques for design pattern detection. Ph.D. thesis, Universita degli studi di Milano Bicocca (2012)
9. Yu, D., Zhan, Y., Chen, Z.: A comprehensive approach to the recovery of design pattern instances based on sub-patterns and method signatures. J. Syst. Softw. **103**, 1–16 (2015). https://doi.org/10.1016/j.jss.2015.01.019
10. Al-Obeidallah, M.G., Petridis, M., Kapetanakis, S.: A survey on design pattern detection approaches. Int. J. Softw. Eng. (IJSE) **7**(3), 41–59 (2016)
11. Nagy, A., Kovari, B.: Programming language neutral design pattern detection. In: Proceedings of 16th IEEE International Symposium on Computational Intelligence and Informatics (CINTI), pp. 215–219. IEEE (2015). https://doi.org/10.1109/cinti.2015.7382925
12. De Lucia, A., Deufemia, V., Gravino, C., Risi, M., Pirolli, C.: ePadEvo: a tool for the detection of behavioral design patterns. In: Proceedings of ICSME, pp 327–329. IEEE (2015). https://doi.org/10.1109/icsm.2015.7332480
13. Guéhéneuc, Y.-G., Guyomarc'h, J.-Y., Sahraoui, H.: Improving design-pattern identification: a new approach and an exploratory study. Softw. Qual. J. **18**(1), 145–174 (2010). https://doi.org/10.1007/s11219-009-9082-y
14. Stencel, K., Węgrzynowicz, P.: Implementation variants of the Singleton design pattern. In: Meerrsman, R., Tari, Z., Herrero, P. (eds.) On the Move to Meaningful Internet Systems: OTM 2008 Workshops, LNCS vol. 5332, pp 396–406, Springer, Berlin Heidelberg (2008). https://doi.org/10.1007/978-3-540-88875-8_61
15. Zanoni, F., Fontana, A., Stella, F.: On applying machine learning techniques for design pattern detection. J. Syst. Softw. **103**, 102–117 (2015). doi:0.1016/j.jss.2015.01.037
16. Issaoui, I., Bouassida, N., Ben-Abdallah, H.: Using metric-based filtering to improve design pattern detection approaches. Innovations Syst. Softw. Eng. **11**, 39–53 (2015). https://doi.org/10.1007/s11334-014-0241-3
17. Gatrell, M., Counsell, S., Hall, T.: Design patterns and change proneness: a replication using proprietary C# software. In: Working Conference on Reverse Engineering (WCRE), pp 160-164 (2009)
18. Rasoola, G., Mader, P.: A customizable approach to design patterns recognition based on feature types. Arab. J. Sci. Eng. **39**, 8851–8873 (2014). https://doi.org/10.1007/s13369-014-1449-0
19. Budgen, D.: Design patterns: magic or myth? IEEE Softw. **30**(2), 87–90 (2013). https://doi.org/10.1109/MS.2013.26
20. Derezinska, A., Byczkowski, M.: Evaluation of design pattern utilization and software metrics in C# programs, unpublished

The Performance Analysis of Distributed Storage Systems Used in Scalable Web Systems

Dominik Oleś[1] and Ziemowit Nowak[2(✉)]

[1] Tieto Czech s.r.o., 28. října 3346/91, 702 00 Ostrava, Czech Republic
[2] Faculty of Computer Science and Management, Wrocław University of Science and Technology, Wybrzeże Wyspiańskiego 27, 50-370 Wrocław, Poland
ziemowit.nowak@pwr.edu.pl

Abstract. Scalable web systems are directly related to distributed storage systems used to process large amounts of data (*big data*). An example of such a system is Hadoop with its many extensions supporting data storage such as SQL-on-Hadoop systems and the "Parquet" file format. Another kind of systems for storing and processing big data are NoSQL databases, such as HBase, which are used in applications requiring fast random access. The Kudu system was created to combine the advantages of Hadoop and HBase and enable both effective data set analysis and fast random access. As subject of the research, performance analysis of the mentioned systems was performed. The experiment was conducted in the Amazon Web Services public cloud environment, where the cluster of nine virtual machines was configured. For research purpose, containing about billion rows fragment of "Wikipedia Page Traffic Statistics" public dataset was used. The results of the measurements confirm that the Kudu system is a promising alternative to the commonly used technologies.

Keywords: Big Data · Hadoop · HBase · Kudu

1 Introduction

Every day a huge number of new websites are created. Some of them achieve market success and systematically acquire new users. As a result, an increase in the number of requests to the web system can be seen. To keep up with requests handling, the number of servers is increased, and when that is not enough – the number of whole data centres. From the technical perspective, it is possible, among others by balancing the load of application servers. There are a number of issues to be solved, though, including, among others, distributed web systems performance forecasting [3].

Copying data generated by users between servers is not the best solution, hence much more often the data storage layer is separated from the application server layer. Data present in a central location means that the rest of the environment can be scaled, and the user always has access to current and identical data regardless of the application server to which they connected. Unfortunately, the data collected in a central location quickly reaches sizes going beyond the capabilities of a single system. In this context, the concept of *big data* is encountered [7].

L. Borzemski et al. (Eds.): ISAT 2018, AISC 852, pp. 287–298, 2019.
https://doi.org/10.1007/978-3-319-99981-4_27

To deal with the issue of *big data* infrastructure is required that can process and store large amounts of real-time data of different structure [7]. Traditional relational databases turned out to be insufficient for this purpose. The computing power of the system can be scaled vertically, by using more efficient equipment, or horizontally, by adding consecutive system nodes working in parallel. The conditions of relational databases, resulting from their specificity (ACID), allow for practically vertical scaling only, which is a considerable limitation. That is why in recent years we have observed the growing interest in Open Source technologies capable of collecting large amounts of data in a scalable and at the same time economical way [4, 6, 10].

Due to their way of operation on the data, the two types of *big data* processing systems can be distinguished. Operational *big data* are systems focused on interaction (saving and reading small portions of data) in a reasonable time (real time or close to it). Examples of such systems are, for example, NoSQL databases. In turn, systems used to carry out advanced analyses of larger data sets are referred to as analytical *big data*. It is interesting that both types complement each other and are often implemented simultaneously [2, 5].

2 Motivation

The market of solutions dedicated to *big data* processing is rich in many technologies from different manufacturers. Based on the publication made available by Google which describes the concept of the architecture of their system, the Hadoop system was created and is developed on an Open Source basis. The dynamic development of this project, as well as the involvement of many companies and environments, have made the Hadoop "ecosystem" the dominant system on the *big data* market.

In the case of the Hadoop system, static data are generally stored in the HDFS file system (Hadoop Distributed File System) in the form of text or binary files (e.g. Parquet). However, the mentioned system and file formats do not allow editing individual lines, and do not allow effective random access, either. Therefore, variable data sets are usually stored in semi-structured databases, such as HBase or Cassandra. These systems enable read and write operations on individual records in real time, but they offer much lower performance in the case of sequential readings of more extensive data sets (e.g. when analysing an entire data set) [11].

There developed a gap between the higher analytical performance of static files collected in HDFS, and the capabilities of HBase and Cassandra databases in the field of free access at the level of the line in real time. If a single application is to be used for both purposes, it is necessary to build a more complex architecture and connect systems. In practice, the problem is often solved by using the HBase database, the contents of which are periodically exported to Parquet files for later analysis [5]. Such an approach, however, has several disadvantages:

- system designers must manage synchronization between systems (data export),
- there is a need to manage and support multiple systems (making backups, managing permissions),
- there may be a significant gap between the appearance of data in the HBase database and their availability for analysis (export time to HDFS),

- sometimes there arises the need to modify/delete data that have already been exported to HDFS (incorrect data, privacy protection). It requires rewriting and replacing whole fragments of the data set.

In order to solve the problem described, the Kudu project was started - a new data storage system to fill the gap between HBase and static files in HDFS, combining the advantages of both systems. Although HBase and HDFS will still have an advantage in certain applications, Kudu is an alternative that can greatly simplify system architecture in many cases [5].

The authors decided to perform a performance analysis of scalable systems that constitute the data storage layer for web applications, and compare the results obtained with the results of other research in this area. The subject of the research was to refer to the performance of the Kudu system and verify whether it is a compromise between the HDFS and HBase systems in the chosen case of use. An analysis of the impact of partitioning and data compression on system performance was also conducted. As a platform for the experiment, the authors used the Amazon Web Services public cloud environment.

3 Research Stand

The comparison of Hadoop, HBase and Kudu systems was carried out using the Impala system, which handles data collected in the three technologies mentioned above. In addition, HBase system tests were also carried out without the intervention of the Impala system, using Apache Phoenix, due to the fact that it does not require any additional services – it is a tool optimizing the use of coprocessors and scanners, which are typically offered in HBase, and enabling the use of SQL language in operations on data collected in this system. In the case of the Hadoop system, there arose the issue of selecting the file format in which the data will be stored. Due to the fact that this system serves mainly as a reference point and not a fully functional alternative in comparative analysis, the Apache Parquet file format was used to carry out the research, which is one of the most efficient formats for HDFS, especially in the case of sequential scanning operations of large ranges of records.

The experiment was carried out in the Amazon Web Services public cloud environment, more specifically Amazon Elastic Compute Cloud, which is the foundation of infrastructure for many scalable web systems. Thanks to the use of the Cloudera Data Hub trial version, which is adapted to manage various Hadoop services, the process of starting and parameterizing the environment was significantly improved.

The research platform contained a cluster of nine virtual machines, one of which (the Master node) served as the coordinator for individual cluster services (including NameNode for HDFS, MasterServer for HBase). Seven machines constituted the scalable part of the environment (Worker nodes) – DataNode roles for HDFS, RegionServer, Impala Daemon, and Tablet Server were launched on them. These cluster nodes were responsible for storing and processing data. The last instance, independent of the cluster, assumed the role of the client for the compared systems. The technical specification of the machines described is illustrated in Table 1.

Table 1. Specification of virtual machines used for the purposes of the experiment.

Node function	Master node	Worker node	Client
Number of nodes	1	7	1
Type of Amazon EC2 instance	m4.large	d2.xlarge	m4.large
Number of vCPU	2	4	2
RAM capacity	8 GB	32 GB	8 GB
Disk type and capacity	SSD (30 GB)	SSD for OS (50 GB), HDD for data (3 × 2 TB)	SSD (30 GB)
Operating system	CentOS 7 64-bit (HVM)	CentOS 7 64-bit (HVM)	CentOS 7 64-bit (HVM)
Running roles	Name Node (HDFS) Master (HBase) Master (Kudu) Impala Statestore Zookeper Server	Data Node (HDFS) Region Server (HBase) Tablet Server (Kudu) Impala Daemon	–

4 Data Set Used for Research

Data collected and presented in web systems have a variety of forms. Most often, however, those are collections of many elements of the same type (messages on social networks, discussion forums, statistics). Oftentimes they are formed into records, uniquely identified by one or several fields, as is the case in relational databases. All analyses that the data are subject to focus mainly on the processing of the value of a specific field for a given range of records, or the calculation of the number of records meeting the given criteria.

The publicly available "Wikipedia Page Traffic Statistics" data set which includes hourly statistics of visits to Wikipedia articles for 7 months - from October 1, 2008 to April 30, 2009 was used for the research [8]. In the course of the research only the part collected during the first 10 days of April (nearly one billion records) was used. It is a typical example of analytical *big data*, while data with similar structure and characteristics are also often used in real-time systems. Messages in messengers, or data from forms and online surveys are also characterized by a large number of relatively short records.

The data set used consists of several hundred compressed files (a separate file for each hour), with the name containing the date and time of the statistics collection, e.g. page counts 20090430–230000.gz. The files do not contain headers and each record contains 4 fields - language, page name, number of visits, and page size (in bytes). Such a record format, however, does not allow conducting the intended tests (the time of collecting statistics is in the file name), therefore the data were processed accordingly. A separate column was added containing the date and time, separated from the file name. Another modification (this time resulting from the adopted assumptions, not technological limitations), is the change of the main key elements. In the original data

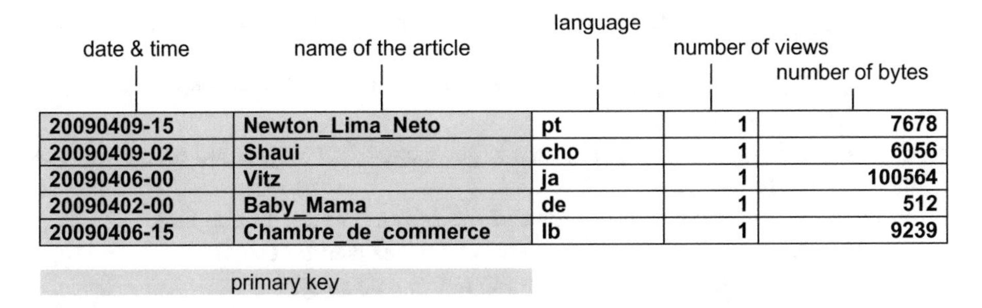

Fig. 1. The structure of rows in the table used during experiments.

set, the record is uniquely identified by the name of the file in which it is located (time of statistics collection), the name of the article, and the language of the article. In order to avoid a situation in which three of the five columns are part of the master key, two columns were adopted as the unique row identifier: with the time and date of collecting statistics, and the name of the article. The final table structure is shown in Fig. 1.

5 Description of Experiments and Obtained Results

The selection of analytical queries (operating on the ranges of records) takes into account potential use cases for the surveyed data set, e.g. returning the sum of views of a specific article on a given day or time, or calculation of the transfer generated by specific articles on a given day (the product of the number of views and the size of the article). It also allows partial reference of results to the results of tests carried out so far on another platform and using data sets with different characteristics [2, 5, 9].

The research was based solely on performing SELECT queries in several different variants. Due to the fact that the data were not modified, there was no need to reload them to the system before each query execution cycle. An essential procedure, however, was the deletion of the data cached by the operating system, as well as the systems involved in the experiment. It was automated by using scripts. The Kudu system did not offer the possibility of clearing the data buffered in the memory at the time of conducting the experiment. Therefore, between the subsequent query cycles, the servers acting as TabletServer were restart.

The queries were divided due to their specificity into four groups:

- Experiment 1 – queries operating on individual records:
  - query 1.1 – by the number of views of the selected article at a given time,
  - query 1.2 – by the product of the number of views and the size of the selected article at a given time,
- Experiment 2 – queries operating on a larger range of records, based on the value of the first element of the primary key:
  - query 2.1 – by the number of all records in the table for the selected time,
  - query 2.2 – for the sum of the products of the number of views and the size of all articles at a specific time,

- query 2.3 – for the sum of views of articles in the selected language at a specific time,
- query 2.4 – for the sum of the products of the number of views and the size of all articles on a given day.
- Experiment 3 – queries operating on a larger range of records, based on the value of the second key element:
  - query 3.1 – for the sum of views of the selected article in one day,
  - query 3.2 – for the sum of all views of the selected article.
- Experiment 4 – queries operating on the entire data range:
  - query 4.1 – the product of the number of views and sizes for all records.

The results of the measurements of each completed query cycle for a specific system were redirected to separate text files. To eliminate the impact of the so-called "cold start" [9] phenomenon on the measurement results, before the right series of queries a query was made, the measurements of which were not taken into account. The data from the logs created in this way were collected and the median was determined from the obtained values.

The interpretation of the results for each group of queries was carried out separately, due to different characteristics of the operations carried out. Two variants of the study were carried out for the Kudu system and Parquet format files. In the case of the Kudu system in the first variant, the number of partitions corresponded to the number of regions in HBase (28), the other variant was a table partitioned four times less frequently, where the number of partitions corresponded to the number of TabletServer nodes. In the case of Parquet format files, the first variant means a non-partitioned table, while the second variant means a partitioned table based on the values of the column containing the date and time of log collection. In addition, all the technologies compared (except for Parquet files) was examined in three variants of coding and compression (c&c). The results of the experiments were collected in Table 2. Time is expressed in seconds.

In almost all of the measurements carried out, the Kudu compressed tables were characterized by higher efficiency than the non-compressed tables. The compression with the use of the Snappy codec offers a slightly worse degree of compression than e.g. GZip, yet it causes less CPU load during compression and decompression.

The results of the tests described in [2] allow to conclude that Snappy compression allows the reduction of the volume of the data set even several times, which has a positive effect on the system's performance. Considering the results of the measurements carried out for the purpose of this analysis, this theory also finds support – tables using Snappy compression are characterized by higher efficiency, but this is not the rule (e.g. full scan operations on tables in Impala-HBase or Parquet files). This can be directly related to the characteristics of the testing platform, as the selected virtual machine instance type is more optimized for high disk performance than computing power. It is worth noting that in the case of the Kudu system, the compression never caused a drop in performance, and given some of the queries, it even allowed a double increase.

Table 2. The results of the experiments. Time is expressed in seconds.

	Q 1.1	Q 1.2	Q 2.1	Q 2.2	Q 2.3	Q 2.4	Q 3.1	Q 3.2	Q 4.1
Kudu 28-part, u&u	0.130	0.240	0.645	1.440	1.240	13.11	10.62	56.47	24.39
Kudu 28-part, c&u	0.130	0.240	0.230	0.630	1.340	3.710	9.640	71.71	10.61
Kudu 28-part, c&c	0.130	0.240	0.230	0.630	1.140	3.350	6.020	36.15	9.100
Kudu 7-part, u&u	0.130	0.230	0.540	1.740	1.540	17.62	12.12	103.8	44.43
Kudu 7-part, c&u	0.130	0.130	0.230	0.740	1.240	3.565	6.575	71.73	16.87
Kudu 7-part, c&c	0.130	0.130	0.230	0.530	1.240	3.050	5.460	29.89	12.11
Parquet part, c&u	1.590	1.285	0.230	0.430	0.330	0.955	4.420	20.69	2.690
Parquet part, c&c	0.955	1.205	0.180	0.430	0.430	0.740	2.960	16.90	3.080
Parquet no-part, c&u	3.155	3.855	1.150	1.860	2.260	2.515	3.375	3.375	1.250
Parquet no-part, c&c	6.520	7.195	1.140	2.050	2.405	2.655	6.845	6.790	1.495
Impala-HBase, u&u	0.530	0.240	14.87	33.17	7.160	280.1	–	–	1068
Impala-HBase, c&u	0.330	0.285	13.87	33.79	8.260	296.0	–	–	1309
Impala-HBase, c&c	0.330	0.230	15.62	34.32	9.090	498.3	–	–	1484
Phoenix-HBase, u&u	0.261	0.193	6.667	9.088	7.183	84.29	1.326	2.359	241.6
Phoenix-HBase, c&u	0.107	0.117	6.945	10.26	7.756	95.55	1.379	1.961	229.9
Phoenix-HBase, c&c	0.058	0.086	7.563	11.12	8.881	104.6	1.254	1.599	261.4

The other factor examined in the Kudu system and Parquet files – table partitioning had a significant impact on the measurement value. The Kudu system tables were partitioned into two variants, in proportion to the number of servers. A table with more partitions was generally more efficient, especially for operations on a larger range of rows. The analysis of query measurements for tables in the form of Parquet files leads to different conclusions. Non-partitioned tables were characterized by lower performance for operations on smaller ranges of records. In the case of full scans, the reverse situation took place – such operations turned out to be more than twice as slow on partitioned tables. It should be noted, however, that the effect mentioned may result from the method of partitioning tables stored in Parquet files. Partitioning is carried out based on the column indicated when creating the table and for each unique value of this column a separate partition is created. Due to the fact that the table used for the experiment was divided based on the column values over time, it had more than 200 partitions of similar size. Considering the amount of disk space occupied by tables in Parquet format files, performance drops for full scan operations can be associated with too small a partition size (operations on small files in HDFS are not effective).

6 Interpretation of Results and Conclusions

The analysis of the measurement results, as well as side observations, led to a number of conclusions about the performance of scalable systems in a given use case. Most of them coincide with the results of other studies, but a few differences can be noted that most probably result from a different characteristic of the data set used, and another platform used to conduct the research.

6.1 Assessment of Kudu System Performance

The conducted research showed that, as expected, the Kudu system is a compromise between the performance of the HBase base in operations on individual records, and the effectiveness of scanning large numbers of rows in Parquet tables. Interpretation of the results was carried out separately for each experiment. It allows to conclude which system showed the best performance in the case of operations on individual records (experiment 1), operations on the ranges of records based on the first (experiment 2), and the second (experiment 3) element of the main key, as well as operations on the whole data set (experiment 4). Table 3 illustrates how many times faster or slower the Kudu system is than other technologies when applying default parameters for tables (coding and compression).

Table 3. Comparison of the Kudu system performance with respect to the rest of the systems tested – default coding and compression settings.

System	Parquet (with partitions)	Parquet	Impala-HBase	Phoenix-HBase
Experiment 1	5x ÷ 7x faster	30x ÷ 50x faster	1.2x ÷ 2.5x faster	1.2x ÷ 2x slower
Experiment 2	1.3x ÷ 5x slower	1.4x slower ÷ 5x faster	6x ÷ 80x faster	6x ÷ 30x faster
Experiment 3	3.3x slower	1.4x ÷ 13x slower	n/a	7x ÷ 36x slower
Experiment 4	3.4x slower	7x slower	123x faster	22x faster

It should be noted that compared to the non-partitioned table in the Parquet format, the Kudu system offers more than 30x performance gain in queries for a single record, at the expense of 7 times slower full scan operations. In comparison with Phoenix-HBase, Kudu offers more than 16x times gain in analytical queries, at the expense of twice as slow operations on individual records. However, several anomalies should be pointed out. Experiment 3 revealed that in some analytical applications, the Phoenix system can be many times faster than tables in Parquet files. Further important conclusions can be drawn after analysing the measurements of query 2.3, in which the Kudu system was more efficient than Phoenix-HBase and Impala-HBase, while the difference in efficiency in favour of Kudu was only 6 times. The comparison of results in the case of optimal tables for a given query is presented in Table 4.

The relationship between the systems is presented in the same way as for tables with the default coding and compression format. Performance differences between tables stored in pure text, encoded and compressed are significant when considering results for a specific technology, e.g. compressed and uncompressed Phoenix tables.

Table 4. Comparison of the Kudu system performance with respect to the rest of the tested systems – optimal coding and compression settings.

System	Parquet (with partitions)	Parquet	Impala-HBase	Phoenix-HBase
Experiment 1	5x ÷ 7x faster	16x ÷ 24x faster	equal ÷ 2.5x faster	2.3x ÷ 2.8x slower
Experiment 2	1.3x ÷ 4.5x slower	1.3x slower ÷ 5x faster	6x ÷ 60x faster	6x ÷ 30x faster
Experiment 3	1.7x ÷ 2x slower	1.8x ÷ 17x slower	n/a	5x ÷ 23x slower
Experiment 4	3.4x slower	7x slower	117x faster	25x faster

Gains or losses usually reach a dozen or several dozen per cent. In the case of selection of the system itself (e.g. Kudu or Phoenix), the differences in performance reach the order of multiple. Experiment 3 showed, however, that in certain cases the highest efficiency gain (over 100 times) can be noted by properly harnessing the full potential of a given system in a strictly defined case of use. The conducted research also confirmed that the right decisions at the stage of determining the partitioning scheme and the structure of the primary key have the decisive impact on the system's performance. For Kudu system tables it is important as the system, unlike Phoenix-HBase, does not offer the possibility to create additional indexes (the so-called secondary index) [1].

Of the technologies listed, it is impossible to determine the optimal system regardless of the case of use. Therefore, the decision on the choice of the system should each time consider the characteristics of operations in the system (frequency and type of inquiries performed), and specify the selection criteria. In some cases a 30 times slower data set analysis is less of an obstacle than a few milliseconds of overhead for operations at the level of a single record. It should be noted that the technologies listed were created for various purposes. The HBase system was not created to analyze the data collected, just as the HDFS concept did not provide for operations at the level of a single record. The purpose of the Kudu system was to fill the gap between HDFS and HBase systems and offer a certain compromise. The results of the research carried out so far show that the Kudu system meets this assumption. On the other hand, it should be noted that the project is currently at a relatively early stage of development, which results in some functional limitations of the system, which was also mentioned in [2, 9].

6.2 Evaluation of the Performance of SQL Systems for HBase

The analysis of HBase performance measurements confirms that in all types of operations, the Phoenix-HBase system is more efficient than HBase queried via the Impala system. In the case of operations at the record level, the difference may be due to the temporary overhead caused by planning and coordination of the query by the Impala system, which was noted in the course of other studies [2, 5]. However, the measurement results indicate that the Phoenix system is several times more efficient than the Impala HBase also for long-term analytical or full table scans.

The Phoenix system, in addition to efficiency improvements (coprocessor implementation, use of filters), also offers functionalities relevant to SQL tables – it enables the creation of the master key consisting of several columns (in a way which is transparent for HBase) or additional indexes. Attention should be paid, however, to the assumptions of both projects. The Impala system was created to conduct data analysis in the HDFS file system. The possibility of referring to HBase tables should be treated as an additional functionality, useful if there is a need to integrate data from various systems (e.g. table merging operations in Parquet and HBase files). The Phoenix project, in turn, is focused on improving performance and developing HBase functionality in applications where this system – as the NoSQL database – was not originally designed.

6.3 Evaluation of Disk Storage Usage

Due to the fact that the technologies being compared are to store large amounts of data, an important factor is the amount of disk space occupied by the tables. Figure 2 illustrates the amount of space occupied by the data set depending on the system, coding and compression. The uncoded table stored in Parquet files was not used in the experiment, therefore the amount of space occupied in this case was not demonstrated.

As expected, the uncompressed HBase tables occupied the most space, which results from the organization of data in rows, and the structure of the record. The following research led to similar conclusions [2]. However, the amount of space occupied by the Phoenix system table after applying coding and compression unexpectedly turned out to be the lowest among the compared technologies.

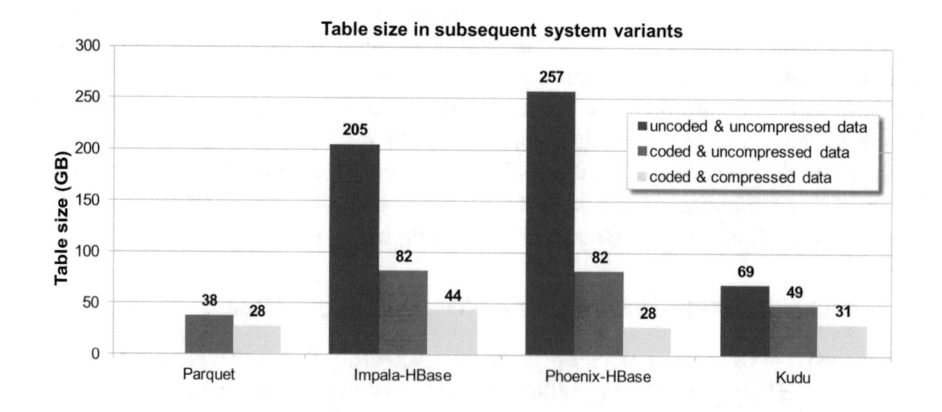

Fig. 2. The chart illustrating the amount of disk space occupied by tables of the compared systems. The table stored in Parquet files, in the uncoded form, was not used in the research.

6.4 Evaluation of Public Cloud Infrastructure

An important point of the research was the reference of the results of previous research conducted on clusters of efficient physical servers [2, 5, 9], to the results achieved in the increasingly popular type of platform for Web systems – the public cloud. It should be noted that general observations confirm the conclusions drawn from other studies: the Kudu system offers a compromise between HBase and HDFS. However, it is worth noting the issues arising from the applied platform. In the case of instances characterized by high disk efficiency (recommended for the Hadoop system), the "bottleneck" may be the computing power of virtual machine processors. It may play an important role especially in the case of operations on compressed tables. This is justified by the measurement results, which in some cases turned out to be worse for compressed tables. In studies carried out by CERN for the needs of the ATLAS system [2], the use of Snappy compression allowed for a two-fold increase in performance in the case of the Kudu system and Parquet files. The results of measurements carried out for the purposes of this analysis lead to different conclusions. For the Kudu system, compression in each case resulted in an increase in efficiency, while in terms of the Impala-HBase system and Parquet files, the reverse situation can be observed. The use of a stronger compression format (e.g. GZip) could further enhance the performance degradation. The authors presume that this might be related to the fact that Kudu employs modern hardware technologies (it is written in C++) and therefore better manages the resources of processor.

An important issue was also to check the stability of the platform. Virtual machine instances in public clouds generally share physical hardware resources with dozens of other virtual machines, and each of them may temporarily affect the performance of others. The virtualization layer ensures isolation, but there is no guarantee that the performance parameters will remain unchanged over a longer period of time. Still the analysis of variances and amplitudes of the measured values leads to an assumption that the hardware platform behaves stably – all the more so that there was more than a two-hour interval between subsequent measurements of a given query for a given table. The total measurement time was over 20 h. With almost identical values of the majority of measurements, it can be concluded that the isolation of resources in AWS EC2 cloud services is at a good level and there is no widespread effect of reduced efficiency in the IT infrastructure depending on the time of day.

6.5 Further Research

The potential direction of further research is the analysis of the efficiency of said systems in the case of simultaneous operations. This is particularly important in the context of the comparison of HBase and Kudu systems as these systems allow quick operations at the level of a single record, which is often characterized by systems operating in real-time. This type of research, however, requires a different platform than the Impala system since in the case of operations performed on individual records, it is not as effective as performed directly by the program's programming interface (API). An interesting aspect of such comparisons, especially for the Kudu system, is to examine how the implementation of complex data set analyses influences the performance of operations on individual records.

References

1. Apache Kudu. Documentation (2018). https://kudu.apache.org/docs/
2. Baranowski, Z., Canali, L., Toebbicke, R., Hrivnac, J., Barberis, D.: A study of data representation in Hadoop to optimize data storage and search performance for the ATLAS EventIndex. J. Phys: Conf. Ser. **898**, 062020 (2017). https://doi.org/10.1088/1742-6596/898/6/062020
3. Borzemski, L., Kamińska-Chuchmała, A.: Distributed web systems performance forecasting using turning bands method. IEEE Trans. Ind. Inf. **9**(1), 254–261 (2013). https://doi.org/10.1109/TII.2012.2198664
4. Lakhe, B.: Practical Hadoop Migration - How to Integrate Your RDBMS with the Hadoop Ecosystem and Re-Architect Relational Applications to NoSQL. Apress, New York (2016)
5. Lipcon, T., Alves, D., Burkert, D., Cryans, J.D., Dembo, A., Percy, M., Rus, S. Wang, D., Bertozzi, M., McCabe, C.P., Wang, A.: Kudu - Storage for Fast Analytics on Fast Data. Cloudera, Inc. (2015). https://kudu.apache.org/kudu.pdf
6. Marz, N., Warren, J.: Big Data - Principles and Best Practices of Scalable Realtime Data Systems. Manning Publications, New York (2015)
7. Press, G.: A Very Short History of Big Data. Forbes, 9 May 2013. https://www.forbes.com/sites/gilpress/2013/05/09/a-very-short-history-of-big-data/
8. Skomoroch, P.N.: Wikipedia Page Traffic Statistics - 7 months of hourly pageview statistics for all articles in Wikipedia. Amazon Web Services (2015). https://aws.amazon.com/datasets/wikipedia-page-traffic-statistics/
9. Tyukin, B.: Benchmarking Impala on Kudu vs Parquet. Blog about Big Data, Business Intelligence, Data Warehousing and ETL, 5 January 2018. https://boristyukin.com/benchmarking-apache-kudu-vs-apache-impala/
10. Vohra, D.: Practical Hadoop Ecosystem A Definitive Guide to Hadoop-Related Frameworks and Tools. Apress, New York (2016)
11. Yegulalp, S.: Cloudera's Kudu: Like HDFS and HBase in one. InfoWorld Tech Watch, 28 September 2015. https://www.infoworld.com/article/2986675/hadoop/cloudera-kudu-hdfs-hbase-in-one.html

E-Business Systems, Mobile Systems and Applications, Internet of Things

Phonetic String Matching for Languages with Cyrillic Alphabet

Viacheslav Paramonov[1,2(✉)], Alexey Shigarov[1,2], Gennady Ruzhnikov[1], and Evgeny Cherkashin[1,2,3]

[1] Matrosov Institute for System Dynamics and Control Theory of SB RAS, Irkutsk, Russia
{slv,shigarov,rugnikov,eugeneai}@icc.ru
[2] Institute of Mathematics Economics and Informatics, Irkutsk State University, Irkutsk, Russia
[3] National Research Irkutsk State Technical University, Irkutsk, Russia

Abstract. The usage of phonetic similarity in comparison of textual strings and elimination of misprints is one of significant issues in philology. It is widely used in automatic text checking. Nowadays most of phonetic algorithms are designed for English language words processing. The quality of comparison may be decreased for non-English languages especially for languages, which have rich morphology and use non-Latin alphabet symbols, e.g. East Slavic languages with Cyrillic letters. We propose an approach to phonetic comparison of Russian language words. It is based on detection letters and letter sequences that have similar pronunciation according to rules of the language. The resultant phonetic representation of the words are coded by prime numbers. The efficiency of the reviewed algorithm is considered in the paper. The algorithm was adopted for Mongolian language phonetic processing.

Keywords: Natural language processing · Phonetic algorithms
String comparison · Cyrillic letters

1 Introduction

Integration of heterogeneous documents which have different formats of representation allows one to accumulate information datasets and use them for analysis by different criteria. Textual data collected by persons may have different formats of presentation and may contain fallacies. It is a serious obstacle to data processing and integration [1].

Collected data very often presented in spreadsheets. Fallacies in data arise as a result of misprints, spelling mistakes, invalid usage of date or currency delimiters, faults of text recognition etc. In this regard, data cleanse is required before the integration. It is one of important steps in data integration processes. It embodies many aspects such as detection and automatic correction of spelling mistakes, incorrect values, logical inconsistencies, and missing data.

L. Borzemski et al. (Eds.): ISAT 2018, AISC 852, pp. 301–311, 2019.
https://doi.org/10.1007/978-3-319-99981-4_28

Some of integrable data is possible to juxtapose with dictionaries and classifiers. Example of such data are types and habitat places of vegetation and spiders (ticks) in a region, proper names etc. One of strings comparison methods is based on their phonetic codes comparison. This paper considers Russian language words detection and correction by applying methods of phonetic comparison.

The idea of phonetics algorithms is based on word identification by seeking similarly pronounced words [3]. The typical application of phonetic algorithms intended for nouns as names and surnames [4] identification. Changing of noun case and form results in e.g. an inefficient use of phonetic algorithms. Such changes do not affect the quality of results. Therefore, phonetic algorithms are most suitable for word comparison with directories, dictionaries and classifiers.

An algorithm adoption for the Mongolian language is proposed in the paper also.

2 Data Anomalies

The Russian language belongs to the East–Slavic language group. It uses by more than 250 million speakers [2]. The Russian language alphabet contains 33 Cyrillic letters. It has 10 vowels, 22 consonants and two special letters "ъ" ("hard sign"), "ъ" ("soft sign").

In this research, we consider scientific data to be processed in circumstances when there is no special software and electronic dictionaries that can be refer to. This kinds of source data may contains some mistakes, called as anomalies [5]. The anomalies are often arisen in processes of manual data input. The rate of these mistakes is about 5% and depends on many circumstance [5].

3 Usage Phonetic Algorithms for Russian Text Cleansing

3.1 Phonetic Spelling Anomalies

Types of spelling anomalies depend on language properties. For example, the Russian language words spelling anomalies is a subject to following classes [6]:

- **morphological** due to uniform graphic symbol of morphemes of a letter, i.e. a person tries to write all audible sounds by letters [7];
- **phonologic**, i.e. preservation of a writing of phonemes spelling regardless the word of change;
- **phonetic**, i.e. words are written as they heard;
- **traditional** due to historic or traditions style of old times or as in a language this word was borrowed from.

Most of spelling errors in Russian are associated with language phonetic norms. The type of mistakes generally depends on person's education level [8].

In this paper, Russian language words will be accompanied with transliteration forms for reader convenience. Transliteration is made in accordance to the standard GOST R 52535.1-2006 [9] and shown in [ ] brackets. This designation will help to facilitate the understanding of text for persons who are not aware of Cyrillic characters.

3.2 Phonetic Algorithms

General idea of phonetic algorithms is based on word comparison according their pronunciation (phonetic forms). It does not depend on orthographic rules. In this approach, words are considered to be phonetically similar if their codes are matching.

Phonetic algorithms allow one to figure out typos related to changing places of two adjacent letters and typos based on pronunciation similarities. Well-known phonetic algorithms are based on English-words coding. There are some algorithm modifications for Turkish, Spanish languages. They described in [11]. However this languages are based on Latin alphabet symbols. There are some approaches of phonetic coding for non-Latin languages. This approach is usually based on symbols transliteration and on Soundex phonetic algorithm [12] application.

In order to use well-known phonetic algorithms for Russian texts, the texts have to be transliterated into the Latin characters coding. However transliteration method in Russia is not defined precisely due to absence of standards. There are transliteration rules in International Civil Aviation Organization, Ministry of Internal Affairs of Russia, The Federal Migration Service are differ that rules in [9].

4 Russian Language Phonetic String Matching

Transliteration allows us to get Cyrillic text in Latin letters representation. Misspellings in East–Slavic languages with Cyrillic letters generally differ from these in English or German texts. The reason is the different rules of pronunciation and writing. The same letter sequence can be represented differently in different languages.

In paper [10] we suggested algorithm Polyphon that uses word transformation with regard to the rules of the Russian language and according to its phonetic properties. It allows to get more correct phonetic code for conformable strings. The stages of algorithm are as follows:

1. substitution of Latin letters which are similar to Russian with Russian ones;
2. removal of all non-Russian alphabet characters from the string;
3. modification of letters before dividers, i.e., special letters as “ъ” and “ь” ;
4. transformation of doubled characters into one;
5. reduction of a row of the same letters into one letter;
6. transformation of specially defined character sequences.

Let us consider algorithm steps in details. Some letters in the Russian alphabet have equivalent Latin ones. These are letters а **[a]**~ a, е **[e]** ~ e, о **[o]** ~ o, с **[es]** ~ c, х **[kha]** ~ x. Some letters are equal in capital form only: В **[ve]** ~ B, М **[em]** ~ M, Н **[en]** ~ H . Sometimes these letters are substituted (incidentally or purposely) in text typing. If this kind of Latin letters are presented in the text they are exchanged to Cyrillic ones.

Any others characters that do not belong to the Russian alphabet are removed from text.

Special letters “ь” and “ъ” do not have pronunciation at all. They are used for giving softness or hardness for consonants respectively. These characters are deleted.

The next stage is the transformation of similar letters following each other into one e.g. “xx” to “x”. It was done because it is not always possible to define double letters in hearing. Therefore, we carry out these transformations for rule of generalization.

Thus, Polyphon uses coding letters by sounds, which are heard. The Table 1 provided ways of different pronunciations. The aim of the proposed phonetic algorithm is construction of a general letter sequences describing sound combinations [13]. The reason of the generalization is that some sounds from a letter and letter sequences depends on a stress position. Such norm deviations are occurred in social and territorial dialects in Russia [14]. A reduction of vowels occurs in Russian when word has three or more syllables. Vowels at the beginning and the end of a word remain unchanged. Thus, if a word contains three and more syllables, we remove all vowels in the middle of the word. The basis of splitting into syllables is the number of vowels in the word. However, some vowels may be put in the word by a mistake. We assume that if there are more than 4 consonants they presented at least in 2 syllables.

Table 1. Ways of different pronunciation

Correct writing form	Ways of pronouncing
интерпретация [interpretatsiya]	интерпритация [interpritatsiya]
	инт**э**рпр**и**тация [int**e**rpr**i**tatsiya]
	инт**э**рпретация [int**e**rpretatsiya]
	интерпретац**ы**я [interpretats**i**ya]
	инт**э**рпр**и**та**тсы**я [int**e**rpr**i**ta**tsy**ya]

The next stage is the substitution of the sequences of letters taking into account the changes made in Table 2. Examples of letters sequence substitution are presented in Table 3. Letter combinations sometimes lead to different sound. The review from [13] was used as a basis for these combinations.

If a transformation took place, we should restart the stage of doubled character searching.

Table 2. Replacement of letters

Letters	А, Е, Ё, И, Й, О, Ы, Э, Я	Б	В	Г	Д	З	Щ	Ж	М	Ю
Modification result	А	П	Ф	К	Т	С	Ш	Ш	Н	У

Table 3. Letters sequence conversion

Sequence	АКА	АН	ЗЧ	ЛНЦ	ЛФСТФ	НАТ	НТЦ	НТ	НТА	НТК	НТС
Result	АФА	Н	Ш	НЦ	ЛСТФ	Н	НЦ	Н	НА	НК	НС
Sequence	НТСК	НТШ	ОКО	ПАЛ	РТЧ	РТЦ	СП	ТСЯ	СТЛ	СТН	СЧ
Result	НСК	НШ	ОФО	ПЛ	РЧ	РЦ	СФ	Ц	СЛ	СН	Ш
Sequence	СШ	ТАТ	ТСА	ТАФ	ТС	ТЦ	ТЧ	ФАК	ФСТФ	ШЧ	
Result	Ш	Т	Ц	ТФ	ТЦ	Ц	Ч	ФК	СТФ	Ш	

The result of word transformation is a phonetic code. Let us present examples of transformations of the same original word: "Эхирит-Булагатский" [**eherit-bulagatskij**] (the name of district in Irkutsk Region of Russia) $\rightarrow$ "ахаратпулакатска" [**aharatpulakatska**]. This coding **is admit** recognition of different spelling errors related with word pronunciation like "эх**ерид**-була**к**атски" [**eherid-bulakatskij**], "**ехирит** булага**т**ский" [**eherit bulagatskij**] etc. So when phonetic codes are found, software is able to juxtapose a manually input word with a predefined classifier.

It is possible to improve Polyphon algorithm to use fuzzy phonetic comparison. All repeating letters are deleted from the string in this case. Thus, a letter is present in the string only once. Each letter has a prime numerical code according to Table 4. The resulting code is the sum of primes. Usage sum of primes guarantees that strings with different letters will have different codes.

Table 4. Coding letters by primes

letter	А	П	К	Л	М	Н	Р	С	Т	У	Ф	Х	Ц	Ч	Щ	Э	Я
code	2	3	5	7	11	13	17	19	23	29	31	37	41	43	47	53	59

5 Experimental Testing of the Algorithm

We performed the following experiment for estimation the efficiency of the proposed algorithm. The experiment consists of several stages:

- data preparation is a generation a data set of words with specially amended mistakes;
- testing the phonetic algorithm;
- comparison with existing algorithms.

The basis for testing is words from Ozhegov's explanatory dictionary [15]. The descriptions of words were not used in the experiment. Some words were removed from review because they are homonym. A method for error generation was proposed. The errors, which expressed in words, reflect the phonetic phenomena and processes of the Russian language. The error generation is based on

- position changes, a phonetic rule applied at the a word (devocalization of a paired consonant), and qualitative reduction that substitute e.g. "о" **[o]** and "и" **[i]**, "е" **[e]** to "и" **[i]** , etc. in a weak position;
- assimilation and deassimilation processes, i.e. devocalization and vocalization of concordats in the word. The phenomenon of assimilation is based on the similarity of sounds, i.e. (ножка **шк** [nozhka **shk**], отдать **дд** [otdat' **dd**], сдоба **зд** [sdoba **zd**], косьба **зьб** [kos'ba **z'b**]) [16].

We did not use already prepared data set of erroneous words for our experiment due the absence of such set. Some of services such as https://wordstat.yandex.ru/[1] provides very limited number of **queries**. So we used a positional voicing of consonants in accordance with [6,16] description. Voiced pair located at the end of a word and before voiceless consonants are stunned. (мозг **ск**[mozg **sk**] , пароход **т** [parahod **t**]) . Voiceless consonants are converted to voiced consonants in the case when their location is before voiced (сдать **здать** [sdat' **zdat'**]) . The exception is the unpaired voiced consonants and "в" character. In the diaeresis process a sound is removed out and a different sound appears (сердце **с'эрцъ** [serdtse **s'ertc'**], солнце **сонцэ** [solntse **solntce**]) . The fusion process is a merging of consonants (жарится – моется [zharitsya – moetsya] – жарит(**ц**)а [zharit(**c**)ya], мыться [myt'sya] – мы(**ц**)а [my(**tc**)a]).

The basis for the third category errors is orthography errors in unstressed vowels and "ь", "ъ" . Wrong writing of "ь" for assimilation softness of consonants in combinations зд(*) [zd], -ст(*) [st],-зн(*) [zn],-тн(*) [tn],-сн(*) [ch],-ст(*) [st],-нн(*) [nn], -нч(*) [nch], -нщ(*) [nsh], - нт(*) [nt], -дн(*) [dn] , where (*) is vowel е [e], ё [jo], ю [yu], я [ya], и [i] . (гвозьди [gvozdi], есьть [es't'] жизьнь [zhizh'n'], защитьник [zash'itnik], лисьтья [list'ya], раньнего [ran'ego], сеньтябрь [sentyabr'], утреньнюю [utren'yuyu], шерсьть [shers't'], коньчились [konchilis'], опусьтели [opus'teli], отьнес [otnes], песьня [pes'nya], полдьню [pold'nyu]) . The substitution of "ь" instead of "ъ" in words with "ъ" before vowels "е", "ё", "ю", "я", "и" — (бьют [b'yut] – бъют [byut]) and (сьезд [s'ezd]) and the contrary is considered as well.

Anomalies of incorrect "не" [ne] and "ни" [ni] writing were not considered in the paper. Errors of letters mixing and their shift were not generated also.

The software use all words and tries entering errors of each type into the word if it is possible. The number of generated words, which contain errors, is 261977.

[1] Yandex keyword statistics service.

The resulting document has two columns: the original "correct" word and the same word with phonetic errors. Examples of words with mistakes are listed in Table 5.

Table 5. Example of words with generated mistakes

Original word	Word with mistake(es)
АВАНЗАЛ [AVANZAL]	АВВАНЗАЛ [AVVANZAL]
	АВАН**ЗЗ**АЛ [AVAN**ZZ**AL]
	ЕВАНЗАЛ [**E**VANZAL]
	АВАНЗА**ЛЛ** [AVANZA**LL**]
	АВ**АА**НЗАЛ [AV**AA**NZAL]
	АВА**НН**ЗАЛ [AVA**NN**ZAL]
	ААВАНЗАЛ [**AA**VANZAL]
	АВАНЗ**АА**Л [AVANZ**AA**L]
	АВАН**САА**Л [AVAN**SAA**L]

5.1 Algorithm Testing and Comparison

The proposed algorithm was applied to the prepared set of test data. As a result we had an accurate verification of correct words and words with phonetic mistakes added. The accuracy of Polyphon data matching is higher than in [10]. It is the result of a repetitive application of the algorithm stages resulting in a transformation. The increase in computational time is not significant in this case.

Table 6. Algorithms comparison results

Algorithm	Matches of phonetic codes, %	Time, ms
Proposed algorithm	97.02	3756
Proposed algorithm (fuzzy phonetic comparison)	99.2	1720
Soundex	90.24	1096
Metaphone	90.29	870
Double Metaphone	96.15	1451
Caverphone	90.41	9770
NYSIIS	75.97	1517
DaitchMokotoffSoundex	96.84	1763

We compares phonetic algorithms Soundex, Metaphone, Caverphone, Daitch-Mokotoff Soundex. These algorithms use the English alphabet characters only. The standard of transliteration GOST R 52535.1–2006 was used [9].

The results of testing obtained by the proposed algorithm, Double Mataphone, Caverphone and Daitch–Mokotoff Soundex are shown in Table 6. It should be noted that string which were shown as different ones in Double Mataphone, Caverphone and Daitch–Mokotoff Soundex are displayed as one in proposed algorithm.

It should also be noted that all algorithms have been tested on single words rather than on sentences.

5.2 Description of Test Results

The results of testing demonstrate that application of prime coding is the most effective for comparison. However, most of information about a word might be lost in application Daitch-Mokotoff Soundex algorithm and similar ones. It can lead to wrong comparisons. Word transformations by means of the suggested approach allows one to compare a word according to its possible phonetic transformation. The suggested approach allows one to compare words more accurately. Unrecognized words are words with several types of mistakes, including reduction.

6 Extending the Algorithm for Other Languages

Authors also made attempts to apply Polyphon to the Mongolian language [17], which belongs to the Mongolic language family, but it is greatly differ from the languages of East–Slavic group. However modern Mongolian alphabet is based on Russian alphabet. It consists of 35 letters. 20 letters among them are vowels and 12 are consonants, letter й **[j]** and two special letters "ъ", "ь" .

It is possible to use of Mongolian language phonetic rules for transformation of word letters. The shortness of the vowels in the letter is made up by writing a single letters: о **[o]**, а **[a]**, э **[e]** etc.

The repetition of vowels in the written form is made up by doubling the writing of the corresponding letters, for example, оо **[oo]**, аа **[aa]**, үү **[uu]** , etc. However, a number of vowel sounds are characterized by a special pronunciation and in some cases does not adhere to the graphic writing of letters.

Among the phonetic features of the Mongolian language, diphthongs are also singled out. It is a combination of two sounds. In total there are 5 diphthongs in the language: ай **[aj]**, ой **[oj]**, уй **[juj]**, үй **[uj]**, эй **[ej]** . One of the basic rules of the phonetics of the Mongolian language is synharmonism – the law of harmony of vowels. In addition to the synharmonicity of the Mongolian phonetics, a complex syllabic structure is also inherent, which makes possible the presence of up to three consonants at the end of the syllable. It should also be noted that there is no grammatical gender in the Mongolian language. A noun in the form of a stem can perform the syntactic functions of the subject, the definition, the complement, and the nominal part of the compound predicate. Stress in words is not used, because it is always on the first syllable.

So in this case it is need to transform diphthong to one letter, for example ай **aj**, ой **oj**, эй **ej** → **e**; **uj**, **uj** → **u** .

Shortness and longness vowels in the modern Mongolian language a meaning-determining function. Concerning this, more than two consecutive letters are converted into two: ааа – аа, ооо – оо, ууу – уу etc., so that misprint is possible here. Moreover transformation of doubled characters to one is not produced. This action may affect word semantic interpretation: цас **tcas** – цаас **tcaas**, ул **ul** -– уул **uul** и т.п.

Letter replacements refer to both vowel and consonant. Features of the syllabic structure are not taken into account in this case. Further, the analysis of the word is carried out taking into account the possible substitutions that have been made earlier. As a result, a phonetic comparison of words can be used to improve the efficiency of comparing the user text of information contained in various classifiers. Unlike the Russian language, the algorithm can be used for verification not only nouns.

7 Conclusion

In this paper we presented the algorithm for Russian words phonetic comparison including fuzzy phonetic comparison option. This algorithm can be used not only for surnames but for establishing correspondence of the word to the qualifier entry. Moreover developing of suggested approach may be used for text rhythmicity analysis [18]. The described approach might be useful for data integration process. The proposed methods can be applied in data cleanup tools for Russian text processing. Accurate data cleanup may be useful for integrating data from different sources.

The proposed approach is based on Russian language phonetic rules. Phonetic coding is more exact in comparison with the algorithms using transliteration. We suggest using algorithms of such kind traditionally for last names and for establishing a correspondence of a word to a denotation in classifiers. We also adopted algorithm for Mongolian. Letters transformation rules allow to be used for Cyrillic alphabet languages such as Russian, Belorussian, Ukrainian, Serbian, Mongolian.

Acknowledgement. The reported study was supported in part by RFBR (grants 18-07-00758, 17-57-44006, 16-07-00411), RFBR and Government of Irkutsk Region – grant 17-47-380007. Experiments were performed on the resources of the Shared Equipment Centre of Integrated information and computing network of Irkutsk Research and Educational Complex http://net.icc.ru.

References

1. Storeya, V.C., Songb, I.-Y.: Big data technologies and management: what conceptual modelling can do. Data Know. Eng. **108**, 50–67 (2017)
2. Cubberley Russian, P.: A Linguistic Introduction, 396 p. Cambridge Press (2002)
3. Parmar, V.P., Kumbharana, C.K.: Study existing various phonetic algorithms and designing and development of a working model for the new developed algorithm and comparison by implementing it with existing algorithm(s). Int. J. Comput. Appl. **98**(19), 45–49 (2014). (0975 — 8887)
4. Zahoranský, D., Polasek, I.: Text search of surnames in some slavic and other morphologically rich languages using rule based phonetic algorithms. IEEE/ACM Trans. Audio Speech Lang. Proces. (T–ASL), 553–563. IEEE (2015)
5. Orr, K.: Data quality and systems theory. Commun. ACM **41**(2), 66–71 (1998)
6. Skripnik, Y.N., Smolenskaya, T.M.: Phonetics of modern Russian Language (Фонетика современного русского языка: Учебное пособие). Skripnik, Y.N. (ed.) Stavropol — VoSIGI (2010). 152 p. (in Russian)
7. Valgina, N.S., Rozental', D.E., Fomina, M.I.: Modern Russian Language: Textbook (Современный русский язык: Учебник), 6th edn. In: Valgina, N.S. (ed.) . Moscow Logos (2002). 528 p. (in Russian)
8. Parubchenko, L.B.: Hypercorrection errors (Ошибки гиперкоррекции). Russian Literature 4, 23–27 (2005). (in Russian)
9. GOST R 52535.1-2006. Identification cards. Machine readable travel documents. Part 1 Machine Readable Passports. National Standard of the Russian Federation (ГОСТ Р 52535.1-2006. Карты идентификационные Машиносчитываемые дорожные документы. Часть 1. Машиносчитываемые паспорта. Национальный стандарт Российской Федерации). Moscow, Russia (2006). 18 p. (in Russian)
10. Paramonov, V.V., Shigarov, A.O., Ruzhnikov, G.M., Belykh, P.V.: Polyphon: an algorithm for phonetic string matching in russian language. In: Proceeding of the 22nd International Conference Information and Software Thechnologies, ICTIST 2016. Communications in Computer Science, vol. 639, pp. 568–579 (2016)
11. Alotaibi, Y., Meftah, A.: Review of distinctive phonetic features and the Arabic share in related modern research. Turk. J. Electr. Eng. Comput. Sci. **21**(5), 1426–1439 (2013)
12. The Soundex Indexing System. National archives. http://www.archives.gov/research/census/soundex.html
13. Ivanova, T.F.: New orthoepic dictionary of Russian. Pronunciation. Accent. Grammatical forms (Новый орфоэпический словарь русского языка. Произношение. Ударение. Грамматические формы), 2nd edn. Russian language-Media (2005). 893 p. (in Russian)
14. Zhirmunsky, V.: National Language and social dialects (Национальный язык и социальные диалекты). The State Publisher of Fiction, Moscow (1936). 300 p. (in Russian)
15. Ozhegov, S.I.: Dictionary of Russian language. About 53000 words (Словарь русского языка: Ок. 53 000 слов). In: Skvortsova L.I. (ed.) 24 edn. Oniks, World and Education, Moscow (2007). 1200 p. (in Russian)
16. Kasatkin, L.L.: Modern Russian dialectics and literary phonetics as a source for the history of the Russian language (Современная Русская диалектика и литературная фонетика как источник для истории русского языка). Nauka, Moscow (1999). 528 p. (in Russian)

17. Budnjam, S., Paramonov, V.V., Ruzhnikov, G.M.: Phonetic strings comparison with particularities of the Mongolian language. Scientific Notes of the University of Science of Mongolia (Фонетическое сравнение строк с учетом особенностей монгольского языка. Эрдмийн Сургуулийн эрдмийн бичиг), Ulaanbaatar, N 1 , pp. 40–47 (2017). (in Russian)
18. Damaševičius, R., Kapociute-Dzikine, J., Wozniak, M.: Towards Rhythmicity analysis of text using empirical mode decomposition. In: Proceeding of the 9th International Joint Conference on Knowledge Discovery, Knowledge Engineering and Knowledge Management (IC3K 2017), vol. 1, pp. 310–317. KDIR (2017)

Conversion Rate Gain with Web Performance Optimization. A Case Study

Kamil Szalek and Leszek Borzemski(✉)

Faculty of Computer Science and Management,
Department of Computer Science,
Wrocław University of Science and Technology, 50-370 Wrocław, Poland
leszek.borzemski@pwr.edu.pl

Abstract. In this paper, we show how e-business performance metrics gain from Web Performance Optimization (WPO). We study the application of WPO to improve the conversion and bounce rates for the credible opinions form of one of the biggest Polish opinion web portals. We evaluated and optimized two types of opinion forms, namely the company and product opinion forms. We applied modern WPO techniques suggested in the literature. The conversion of the company opinion form was the primary aim of our optimization. We observed and analyzed the effects using Google Analytics tools connected to the application and internal analytic panel. We collected the conversion and bounce rates before and after the optimization. The result for the company opinion form is auspicious, showing its high conversion increase from 6.35% to 14.30% within six months of observation nevertheless the product opinion form got the significant decrease, from 3.69% to 1.68%. We discuss and explain the possible reasons for such a situation. We used WebPageTest to test the performance of both opinion forms, before and after the optimization.

Keywords: Conversion rate · Bounce rate · Mobile first
Web Performance Optimization

1 Introduction and Motivation

Performance is one of the main aspects of modern websites that affects conversion rate. The performance aim is to decrease load time. How Google's research shows, 75% of the people will abandon a website if load time takes more than 5 s. Moreover, the long page load time is also one of the main reason for the frustration of websites users. Moreover, 60% of them want to have web page loaded faster than 3 s. The Amazon company shows that every 100 ms of latency cost them 1% of revenue. Also, every 1 s of improvement increases 2% in conversion rate [7]. Web performance can be considered within two contexts – backend and frontend performance.

The purpose of this research is to identify and examine how the website's performance influences on conversion and bounce rate. This issue has been touched and studied in the context of an opinion collector system of one of the leading opinion portals in Poland.

L. Borzemski et al. (Eds.): ISAT 2018, AISC 852, pp. 312–323, 2019.
https://doi.org/10.1007/978-3-319-99981-4_29

The main aims of the audit and developer works were to increase the efficiency of opinion service, decrease the time necessary to fill out the forms, gain usability and user experience. We got reliable results of taken work that effects with a more than a double number of opinions issued. WPO applies the modern web solutions – separation of server side and client side technologies and communication by created by usage of REST API. The application was written with the usage of ReactJS – JavaScript library built by the Facebook corporation.

Another aspect of modern web services and its performance is Responsive Web Design trend. Mobile devices have been more popular during the last years with increasing market share of mobile phones and tablets. Every day is new devices developed with different resolutions and sizes. The point is to realize websites and applications that could support hundreds of possible screen resolutions. Flexible layouts are already not a luxury feature but 'must have' to meet the requirements of the users [6]. A layout has to be intuitive, easy to use and readable on high-resolution monitors and small mobile phones. Images have to be adjusted automatically, text content fills all possible space, and unnecessary information should be hidden on smaller devices. It is a huge challenge that developers and designers must meet. Responsive web design is important because of user experience, and also because of Google websites indexing. Google introduced in 2015 – "Starting April 21, we will be expanding our use of mobile-friendliness as a ranking signal. This change will affect mobile searches in all languages worldwide and will have a significant impact on our search results" [10]. That states that mobile and responsive websites are mandatory to show in Google's mobile searches and increases SEO position in internet search engine.

Understanding website complexity, doing measurements, using specific web and performance metrics is a challenging process. Especially when applied to the real-life businesses to identify its specificity and answer to the question of how web performance may correlate to specific business performance [3, 5, 14]. We used available in the Internet performance analysis tools including PageSpeed Insight [11], WebPageTest [16], Google Analytics, and Chrome Developer Tools.

The paper is organized as follows. First, we introduce some aspects of performance that matters showing their change while the Internet is developing decade by decade. Next, we present the transformation of our web application following introduced web performance optimization (WPO). After there, we show the conversion and bounce rates changes with WPO. In conclusion, we claim that to improve conversion and bounce rates it is not a simple task. We show and discuss as in a short message [13] that commonly accepted optimization options could be not always optimal therefore we need all-inclusive and innovative WPO techniques which among others would take into account not only the front-end issues but also on the back-end as well as transmission problems, especially with database oriented solutions.

2 Aspects of Performance that Matters

2.1 Performance Definition Before

Web performance theory changed over the years from early web existence until now. The first theory of web performance was introduced in 2002 by Patric Killelea [9]. He presented ten key activities to improve and optimize internet websites:

1. Check for standard compliance by using Weblint or other HTML checking tool
2. Minimize the usage of JavaScript and style sheets
3. Turn off reverse DNS lookups in the web server
4. Try out free analysis tools
5. Use simple servlets or CGI
6. Get more memory
7. Index database tables well
8. Make fewer database queries
9. Look for packet loss and retransmission
10. Set up monitoring and automated graphic on website performance

How we could see in 2002, web optimization was focused mainly on hardware limitations that completely changes over the years. Code optimization is still importing and but not that dramatically as before. Today's computers have the most significant computing power ever, every two years since 1990 computer speed is doubled. For this reason, most of this optimization techniques became obsolete.

2.2 Performance Definition Now

With comparison to Patric Killelea's web performance view, it is hard to imagine a modern internet without JavaScript, style sheets and images, which are primary elements of websites, which have to satisfy a need for clean, intuitive and eye-friendly layouts. The market for mobile devices started to multiply in 2010 and each year increasing its position in use for web browsing when desktop usage begins to decrease slowly. Finally, they overtake the desktop first time in 2016. We can now observe from many sources that at the end of 2015, usage of mobile devices gain 51% on the market where there were only 41% of desktop users. That gives a clear sign that websites need to be fully optimized for mobile users. It also started a revolution in website design process that became focused on the mobile first challenge.

If the website project will be built on Google's assumptions, there is a high chance that it will deal great with Google's search engine. But not only, but the most common search engines also have the same assumptions. In the 2009 Google gained 65% of a search result from the whole web, where second search engine – Yahoo gained only 19.6% and on third place Microsoft with 8.4%. In Europe Google is even stronger with 80% of search results [8].

Google's index has the main impact on our position in Google's search engine. So the performance of the websites is grounded factor when Google's boot is crawling pages and could directly influence the crawling budget and page visitors. Web page loading speed is also a ranking factor of Google search engine results. Google is giving rules for proper web optimization like Patric Killelea in 2002. These main points of interest for Google are the following:

1. Avoid landing page redirects
2. Enable compression
3. Improve server response time
4. Leverage browser caching
5. Minify resources
6. Optimize images
7. Optimize CSS delivery
8. Prioritize visible content
9. Remove render-blocking JavaScript
10. Focus on responsive web design
11. Deliver images responsively

2.3 Mobile First

It is not easy to deliver pretty, useful websites for all kind of devices. To achieve this goal, we need to combine two main design factors which are – responsive design and responsible practices. It needs to stay in mind, that responsive design is to make toward easy usability not only on desktops where a user is operating with a mouse. Developers have to care about things like keyboard, mouse, touch, and others. They have to care about the main points of design [7]:

1. Usability – Way how the user interface is presented to the end user, with design and interactions
2. Access – Ability to access webpage or application on all kind of devices and browsers
3. Sustainability – Ability to work with devices now and in the future
4. Performance – Speed how the page is loading and presenting to the user

When it comes to responsive web design, designers have to think how to submit visible content, how to prioritize what should be visible at the beginning of the loading process. It is important to plan how features also react to different input mechanism and how to simplify the interface to be understandable and easy to use. Vital is also to fit all of the necessary content and do not create noisy layout's which could reduce conversion. Luke Wroblewski also had the suggestion, that instead, it is better to hide inconvenient content than use it on smaller device's screens [15].

2.4 Mobile Networks

Going deeper it is visible that mobile devices are essential to market share. Devices are usually using mobile networks which are generally slower than regular networks. There is also a huge possibility to break the connection. Mobile networks force new requirements for the development process and changes in presentation layer [4]. Applications designed for mobile devices introduce other aspects of performance like battery life which is straight correlated with requests latency. Having this aspect in mind, we could not elaborate presentation layer too much. Reducing battery consumption is concert involved for every person from the group of application developers, device designers, manufacturers and end users. Battery consumption is inherently linked with the efficiency of the mobile network.

Wireless mobile networks are gaining first place when looking for most battery consumption's aspects of mobile devices. There are even consuming more battery power than high resolution's mobile screens. For example, [13] shows the main elements of the battery energy consumption in mobile phones. They took on research two most popular mobile devices available on the market and detected what parts of applications took most battery power. In this research, we have developed a similar approach.

2.5 Predictive Web Performance Analytics Needed

Modern web systems increasingly require the prediction of their performance. This feature may be necessary, for example, in a case when the application has to decide on the choice of the server with which it will work best in the future regarding performance. The results of such studies show, among others, papers [1, 2] that the web performance prediction may be an excellent way to improve available web performance.

2.6 Takeaways

Looking at the provided research main conclusion is to reduce the number of requests and page size to a minimum. Many types of research show that modern internet changed a lot during the last fifteen years. Already average page size has been doubled since 2011. So there is a tremendous field of optimization needed to reduce file sizes. The principal fact is that mobile devices market growth significantly and it is still growing up. Going this way it is crucial to prioritize mobile users. Mobile devices should have priority for optimization. Mobile networks have higher latency than conventional networks, and with every second of delay part of the users is going to abandon the page. Web page size has also impact on mobile devices battery power duration. Not only performance matters but also good prioritizing of the content for mobile. Web performance prediction may be an excellent technique to improve obtainable web performance.

3 Opinion's Form Transformation

3.1 General Transformations Overview

The subject of this research is a credible opinions form of opinion portal in Poland. Old form introduced ten years ago has an unintuitive layout and could be upgraded with modern UX trends to simplify UI. Transformation also anticipated creating responsive form during a significant increase in usage of web devices on the Internet, where the old form was designed only for devices with the resolution not higher than 1024 pixels.

The most important part of the edition was to move functionality and speed up user's choices. The filling of the form should be intuitive and encourage entirely. The goal was achieved by usage of vote buttons that decreased the time of selecting the right option. Also flow of filling the form has changed. Before optimization all forms were displayed on one page, already all of them are separated into different ones. Changes have been made on multiple fields:

- New form flow
- Responsive design
- Usage of ReactJS frontend JavaScript framework with Redux pattern
- Usage of asynchronous HTTP requests that prevent reloading the pages

New opinions form has been created with all best practices for responsive web design to scale between different devices and resolutions. Applications have been developed with *mobile first* design pattern – it forces to design and think what is the most important on the web page and prioritize it.

3.2 ReactJS Framework and Asynchronous HTTP Requests

Two fundamental modern web technologies were used in our WPO, namely ReactJS and Ajax. ReactJS used in the development process for credible opinions form is JavaScript library design to build user interfaces with the usage of JavaScript. It is created especially for single web page applications that aim at application speed and simplicity. Solution based on components give a developer tools to increase the scalability of the application. It was introduced by Jordan Walke in the Facebook company in March 2013. It has a one-way data flow, that means properties are passed from parent component to child components by its properties. After any property changes, the state of application component will automatically apply in it render function. It also provides the virtualDOM feature, that creates the virtual state in application stored in memory and recognizes and compute changes and uses them to components which are already affected by them. It gave a possibility to separate backend and fronted part of the application. The code could be easier organized with a more effective way, and it is easy to debug and find where errors could appear.

Ajax (Asynchronous JavaScript and XML) is used to create web applications without reloading webpage. It gives a possibility to make HTTP requests on the fly and make applications more dynamic than with usage of regular HTTP requests. They could be done when a page is already loaded. Main technique of doing this is a usage of XMLHttpRequest JavaScript class created by Microsoft Corporation. First used in Mozilla Firefox web browser. It could be used not only for XML data formats but also for plain texts, JSON, and others.

4 Optimization Results

We observed and analyzed the effects using Google Analytics tools connected to the application and internal analytic panel. After introducing a new opinion form, we get a noticeable boost in credible company opinions. Conversion of the form gained twice from 6.35% in September 2017 to 12.45% in December 2017 and resulted in 14.30% in February 2018. New form performed well with the increased number of filled to more than 130000 opinions for every month. Noticeable is also fall of product opinions. Conversion became on the level of 3.69% in September 2017, and after implementation of the new form, it changed its level to 1.68% in October 2017.

Before form transformation (03.09.2017–03.10.2017):
- Sent fill forms requests: 1009802
- Filled company forms: 64057
- Conversion: 6.35%

After form transformation (10.10.2017–10.11.2017):
- Sent fill forms requests: 1099413
- Filled company forms: 136891
- Conversion: 12.45%

After form transformation (02.02.2018–02.03.2018):
- Sent fill forms requests: 1066881
- Filled company forms: 152669
- Conversion: 14.30%

Before form transformation (03.09.2017–03.10.2017):
- Sent fill forms requests: 1009802
- Filled product forms: 37339
- Conversion: 3.69%

After form transformation (10.10.2017–10.11.2017):
- Sent fill forms requests: 1099413
- Filled product forms: 18477
- Conversion: 1.68%

The optimization options could not always be optimal [12]. The fall of products opinions is not a negative result but could be performed better. It is essential to understand that a new form product is suggesting a mechanism and its difference to the old one. In the old form, it was possible to choose by product category for a given producer, and find it on the list or write an opinion about the product that was not in the portal database. Every user could create a new product with its custom name. For example, it was possible to create a product with the name "My new phone" and add an opinion about it. It caused many problems. Of course, there was a mechanism of filtering out such opinions. This solution gave a lot of worthless products in the database however they increased the number of opinions that were added to the system.

The primary goal of the introduction of the new form was not only to increase numbers of forms filled but also to increase the quality of opinions stored in a system. During this process, new credible opinions form had implemented a new suggest mechanism. Because of technical debt and a significant amount of already stored products in a database, this search mechanism was not perfect.

Device Category	Browser	Acquisition			Behavior		
		Users	New Users	Sessions	Bounce Rate	Pages / Session	Avg. Session Duration
		1,085,461 % of Total: 99.35% (1,092,559)	1,092,268 % of Total: 100.10% (1,091,142)	1,515,511 % of Total: 100.00% (1,515,513)	13.26% Avg for View: 13.32% (-0.48%)	7.61 Avg for View: 7.58 (0.40%)	00:01:56 Avg for View: 00:01:56 (0.21%)
1. mobile	Chrome	301,479 (27.56%)	299,351 (27.41%)	524,347 (34.60%)	20.86%	10.85	00:01:56
2. desktop	Chrome	291,709 (26.67%)	293,373 (26.86%)	339,059 (22.37%)	1.75%	5.45	00:02:00
3. desktop	Firefox	196,029 (17.92%)	195,853 (17.93%)	231,550 (15.28%)	5.64%	4.91	00:02:14
4. mobile	Safari	90,384 (8.26%)	90,714 (8.31%)	115,749 (7.64%)	5.11%	5.36	00:01:20
5. mobile	Samsung Internet	48,033 (4.39%)	47,375 (4.34%)	94,871 (6.26%)	29.31%	11.27	00:02:01
6. desktop	Internet Explorer	35,894 (3.28%)	35,266 (3.23%)	45,468 (3.00%)	52.81%	1.90	00:01:00
7. desktop	Opera	35,846 (3.28%)	35,918 (3.29%)	42,451 (2.80%)	4.62%	5.24	00:02:03
8. desktop	Edge	31,726 (2.90%)	31,759 (2.91%)	37,316 (2.46%)	3.64%	5.26	00:02:08
9. tablet	Chrome	14,394 (1.32%)	14,342 (1.31%)	18,899 (1.25%)	8.49%	9.23	00:02:25
10. desktop	Safari	11,456 (1.05%)	11,447 (1.05%)	13,172 (0.87%)	2.90%	5.28	00:01:41

Fig. 1. Detailed information for device categories and web browsers.

Figure 1 shows specific information for device categories and web browsers. Mobile device users have a significant impact on the results of gained opinions and conversion of the form. How it was introduced before, responsive web design is a must-have in the modern internet. The results of launching a new form to confirm that trend. 42.05% of the users use mobile devices to fill credible opinions form where 55.31% gains desktop browsers and 2.64% of users using tablets.

Looking at results, we may see how important is optimization and performance for mobile devices. With another kind of devices also changes the bounce race. It is highest for mobile devices where scores 19.68%. The lowest bounce rate could be observed for desktop browsers where achieved 6.60%. This results could describe that application could be better optimized for mobile devices.

Going deeper it is also possible to discover how pages convert for specific resolutions. Research clearly shows that there are only three most recent used resolutions, but important is to focus minimum on two of them – 360 × 640px and 1366 × 768px. These two resolutions are using more than fifty percent of the users. Ranking of most three used resolutions: 27.80% – 340 × 640px – Typical mobile resolution, 19.74%– 1366 × 768px – Typical laptop resolution, 8.71% – 1920 × 1080px – FullHD screen resolution.

For all the devices, the most popular browser is Google Chrome. Looking on the session time, bounce rate and pages on session it is visible that application has good performance. Also, there are no troubles for Mozilla Firefox. Problems are evident for Internet Explorer for desktops and Samsung Internet browser for mobile devices. Bounce rate increased dramatically. The reason could be a fact, that application has not been tested using this two browsers, and opinion form could perform with problems.

Another juxtaposition also shows the usage of different mobile devices. Here it is also visible that less popular operating systems have a higher bounce rate. It is correlated with the fact, that developers do not have these devices during the development process for tests and optimization. Windows phone and Blackberry – two operating systems that have the highest bounce rate are gaining only half percent each of total page views. It is hard to interpret Android and iOS systems, where the former has a higher bounce rate but also this users are more active when looking at the number of pages per session.

Both forms have been tested using WebPageTest (Figs. 2 and 3). The most important performance features discovered using WebPageTest before and after optimization are presented in Fig. 4. The comparison of loading times gives similar results in loading time and visually complete document. After lean of the form, it has to download fewer assets and images than before. It is easy to see in requests difference. New form reduces the number of requests by 625% and reduces the number of downloaded bytes from 659 kB to 333 kB what gives 195% boost even after the change to microservices architecture. Both form documents are complete with a small difference, but here is visible render blocking JavaScripts that increases the time of fully loaded webpage for old form. After fully loaded content, some scripts are reloaded for next 3.5 s what is visible in the difference between document complete and fully loaded.

								Document Complete			Fully Loaded		
	Load Time	First Byte	Start Render	Visually Complete	Speed Index	First Interactive (beta)	Result (error code)	Time	Requests	Bytes In	Time	Requests	Bytes In
First View (Run 1)	4.983s	1.287s	2.200s	3.200s	2446	> 4.465s	0	4.983s	87	659 KB	8.579s	100	677 KB

First Interactive (beta)	Colordepth	RUM First Paint	domInteractive	domContentLoaded	loadEvent
> 4.465s	24	2.153s	2.321s	2.321s - 2.418s (0.097s)	4.983s - 4.987s (0.004s)

Fig. 2. Old form download times with WebPageTest.

								Document Complete			Fully Loaded		
	Load Time	First Byte	Start Render	Visually Complete	Speed Index	First Interactive (beta)	Result (error code)	Time	Requests	Bytes In	Time	Requests	Bytes In
First View (Run 1)	4.893s	1.041s	2.500s	3.100s	2635	> 2.429s	0	4.893s	15	333 KB	5.204s	16	347 KB

First Interactive (beta)	Colordepth	RUM First Paint	domInteractive	domContentLoaded	loadEvent
> 2.429s	24	2.429s	2.232s	2.233s - 2.234s (0.001s)	4.892s - 4.892s (0.000s)

Fig. 3. New form download times with WebPageTest.

	Metric	Old form	New form	Result boost [%]
	Load time [s]	4.983	4.893	101.84
	First byte [s]	1.287	1.041	123.63
	Start render [s]	2.2	2.5	88
	Visually complete [s]	3.2	3.1	103.23
	Speed Index	2446	2635	92.83
Document complete	Time [s]	4.983	4.893	101.84
	Requests	87	15	580
	Bytes in [kB]	659	333	197.90
Fully loaded	Time [s]	8.579	5.204	164.85
	Requests	100	16	625
	Bytes in [kB]	677	347	195.10

Fig. 4. WebPageTest performance metrics comparison before and after optimization.

5 Conclusions

Web developers have to take into account many aspects for web development application. There are many problems that they have to solve during the development process. They have to take a look not only on front end side solutions but also on the back-end side. In the modern internet, they do not only take care of strict performance side but also on a comparison between design and performance which are equally important. There are many points where optimization could happen.

It should be the main point where performance optimization should start. How our research shows, mobile devices are gaining 50% of the internet users market. More than 40% of users visiting credible opinions form was using mobile devices with high priority for mobile phones. Tablets are not as active, but resolutions of applications should also take care for this device category.

Conversion of credible opinions form grows double after introducing new solutions, and the main reason is a responsive layout which is highly improving the user experience. The main resolutions where developers should focus are 360 × 640px and 1366 × 768px. Laptops and mobile phones typically use these resolutions. Moreover, here optimizations are most important. Also, resolutions with high definition monitors are next in the queue. These three resolutions are gaining more than 54% of visitors, so they are worth attention.

Modern web technologies give a lot of solutions and improvements. They are making applications more dynamic and increase user experience. With technologies like ReactJS and Redux, building applications are more comfortable, faster and more effective. They are easy to build and debug problems and create solutions for secure applications improvement. Usage of tools like webpack or sass should increase the time needed for development and increase web performance by automatic code minification. This decrease highly files sizes and then also decreases server response times. It is visible how performance could be improved by reducing the number of files and reducing their size during web page test. The number of requests has been reduced from 87 to 15, and file sizes from 659 kB to 333 kB. Page has been fully rendered with the same amount of time. However, the same full page render time does not mean that performance will be the same. The mobile user is often could visit our pages using low internet connections, and here the number of requests and file sizes matters. This way is improving not only compression and user fillings, but also decrease server traffics.

Website optimization and compatibility of the webpage could be the essential factor for page bounce rate. Wrong optimization and low compatibility with different browsers could result in the impossibility to show part of the page, problems with receiving data or errors that wouldn't allow showing the page. Our application has been written with the usage of ReactJS – JavaScript library created by a Facebook corporation. That means that the server forwards JavaScript file which is responsible for rendering webpage. Any error in the JavaScript file could crash rendering. This possibility forces web developers to test web pages on multiple devices with different browsers to protect any oversights.

References

1. Borzemski, L., Kamińska-Chuchmała, A.: Distributed web systems performance forecasting using turning bands method. IEEE Trans. Industr. Inform. **9**(1), 254–261 (2013)
2. Borzemski, L.: Data mining in evaluation of internet path performance. In: Innovations in Applied Artificial Intelligence. LNCS, vol. 3029, pp. 643–652. Springer, Heidelberg (2004)
3. Butkiewicz, M., Madhyastha, H.V., Sekar, V.: Understanding website complexity: measurements, metrics, and implications. In: Proceeding of the 2011 ACM SIGCOMM Conference on Internet Measurement Conference, pp. 313–328. ACM, New York (2011)
4. Grigorik, I.: High Performance Browser Networking: What Every Web Developer Should Know About Networking and Web Performance. O'Reilly Media Inc., Sebastopol (2013)
5. Fork, R.: How Walmart.com correlates web performance to business performance. https://rigor.com/blog/2012/03/how-walmart-com-correlates-web-performance-to-business-performance. Accessed 21 Apr 2018
6. Hogan, L.C.: Designing for Performance, Weighing Aesthetics and Speed. O'Reilly Media Inc., Sebastopol (2014)
7. Jehl, S.: Responsible Responsive Design. The Book Apart, New York (2014)
8. Jerkovic, J.: SEO Warrior: Essential Techniques for Increasing Web Visibility. O'Reilly Media Inc., Sebastopol (2009)
9. Killelea, P.: Web Performance Tuning. O'Reilly Media Inc., Sebastopol (2002)
10. Makino, T., Jung, Ch., Phan, D.: Finding more mobile-friendly search results. https://webmasters.googleblog.com/2015/02/finding-more-mobile-friendly-search.html. Accessed 21 Apr 2018
11. PageSpeed insights. https://developers.google.com/speed/docs/insights/about. Accessed 21 Apr 2018
12. Robledo, A.: Optimization options not always optimal. https://rigor.com/blog/2011/09/optimization-options-not-always-optimal. Accessed 21 Apr 2018

13. Tawalbeh, M., Eardley, A., Tawalbehb, L.E.: Studying the energy consumption in mobile devices. Procedia Comput. Sci. **94**, 183–189 (2016)
14. Vihervaara, J., Loula, P., Tuominen, T.: Performance gains from web performance optimization – case including the optimization of webpage resources in a comprehensive way. In: Proceedings of the International Conference on Web Information Systems and Technologies, vol. 1, pp. 188–193. Scitepress, Setúbal (2016)
15. Wroblewski, L.: Mobile First. The Book Apart, New York (2011)
16. Web Page Test. https://webpagetest.org. Accessed 21 Apr 2018

A Prototype of Evacuation Support Systems Based on the Ant Colony Optimization Algorithm

Yasushi Kambayashi[1(✉)], Kota Konishi[1], Rikiya Sato[1], Kohei Azechi[1], and Munehiro Takimoto[2]

[1] Department of Computer and Information Engineering, Nippon Institute of Technology, Miyashiro, Japan
yasushi@nit.ac.jp, {c1145217,c1145230,c1145104}@cstu.nit.ac.jp
[2] Department of Information Sciences, Tokyo University of Science, Noda, Japan
mune@is.noda.tus.ac.jp

Abstract. We have proposed and implemented a system that supports evacuation after a large-scale disaster. When a large-scale disaster such as earth-quake or conflagration occurs, it may not be possible to pursue predefined evacuation route due to collapsed buildings or fire. The refugees have to select an optimal evacuation route according to circumstances. In such situations, however, it is almost impossible for refugees to grasp the precise circumstance and find the correct evacuation route. In order to mitigate this situation we have proposed a system based on smart phones and server/client system, and implemented it. We make the server side perform the basic processing for evaluating the dynamic situation based on the information collected from refugees' smartphones using crowd-sourcing technique so that the system configuration is flexible. The evaluation is performed based on the idea of the ant colony optimization (ACO) algorithm on the server side. We have implemented the client side of the evacuation route guiding system on both Android OS and iOS, and the server side on Linux system. We have achieved to construct a practical system applicable for real world assuming network infrastructure is intact.

Keywords: Disaster mitigation · Ant colony optimization
Route recommendation

1 Introduction

When the Great East Japan Earthquake occurred in 2011, we observed the tsunami following the earthquake that caused enormous damage to the coastal areas. The damage that people suffer from such a wide-area disaster can be classified into three patterns: the direct disaster casualties, the damage caused by collapsing of structures and fire during evacuation, and the damage caused by the disastrous tsunami. In general, these damages do not occur simultaneously. In most cases, people saved from the first damage have to escape from the third damage while avoiding the second damage. For example, people had to complete the evacuation before the tsunami strikes. Cracks and liquefaction on the ground, however, invalidate predefined evacuation routes. As another

L. Borzemski et al. (Eds.): ISAT 2018, AISC 852, pp. 324–333, 2019.
https://doi.org/10.1007/978-3-319-99981-4_30

example, in the case of Itoigawa conflagration in 2016, we observed many flying sparks and fire leaps that suddenly shutdown the valid evacuation passes and forced people to find a new evacuation route. Therefore, it is extremely important to dynamically find safe evacuation routes depending on the situation where a wide-area disaster occurs.

In this paper we propose an evacuation support system that mitigates the difficulties people face such situations. The system consists of smart phones and server/client system. We make the server side perform the basic processing for evaluating the dynamic situation using ant colony optimization (ACO) approach and the information collected from refugees' smartphones. Such configuration is flexible and easy to implement. We have implemented the client side of the evacuation route guiding system on both Android OS and iOS, and the server side on Linux system. We have achieved to construct a practical system applicable for real world assuming network infrastructure is intact.

The structure of the balance of this paper is as follows. The second section describes the background and discusses the related works. The third section describes the design and implementation of the evacuation support system we have implemented. In order to demonstrate the feasibility of the system, we have conducted experiments using popular smartphones that runs both Android OS and iOS. We report the results and discuss problems we have found through the experiments in the fourth section. We discuss the future research directions and conclude our discussion in the fifth section.

2 Background

Asakura et al. investigated the calculation of the evacuation routes after a large-scale disaster. They have proposed a method that uses Ant Colony Optimization (ACO), and have shown it is useful in a simulator [1].

Avilés et al. investigated how to support people escaping to the emergency exit from the building [2]. In their study, they assumed some of evacuees use portable device based guidance system like in our study. They have also adopted MANET and the ACO in order to derive escape routes.

We have proposed several evacuation support systems based on multi-agent systems [2–6]. Although, we have implemented a simulation system for each proposed system, we have been aware of the insufficiency as feasibility studies. Also, all of the previous four systems are based on ad hoc systems. Theoretically, ad-hoc system should be robust when a natural disaster occurs and communication infrastructure has got damaged. We have found, however, from the experiences of the Great East Japan Earthquake, current infrastructures are robust enough when they are struck by a disaster. Even though voice communication are hindered by the congestion, communication through packets was possible due to the redundant network routes.

On the other hand, we have found disaster recovery of the database and servers is important. Therefore we stress the database replication and switching of servers for our evacuation support system. In this paper, we rely on two technologies; one is ant colony optimization (ACO) for adapting dynamic situation changes, and the other is distributed system with microservices for disaster recovery.

ACO is an optimization algorithm that mimics the foraging behavior of ants. Ants go back and forth between the feeding grounds and nest in order to bring food from feeding areas to nests [7]. At that time, ants put down volatile chemical substance called pheromone to the routes they went. Other ants that are back and forth feeding grounds and nest follow the pheromone and replenish drop the pheromone. By these actions, the long paths to the feeding grounds lose their pheromone by evaporation before the pheromone replenished. On the other hand, other ants strengthen shorter path pheromone before evaporation. As the result, ants derive the optimum route to feeding area. We call methods that are using this characteristic ACO [8]. Goodwin et al. took advantages of ACO for planning safe escape routes in a chaotic crisis situation [9], and Baharmand and Comes used ACO for deciding optimized locations of shelters [10].

We have composed our system as a distributed system. A distributed system is a system that consists of many computer nodes [11]. The computers that comprise the system need not be installed at the same location, but they can be geographically distributed. Recently, microservice architecture is gaining popularity in server configuration [12]. In the microservice architecture, each service is divided into relatively small servers so that it can be extended easily and thus provides scalability. It is known a system constructed in a microservice architecture is easier to expand than one constructed in monolithic architecture. In a context of disaster mitigation, microservice architecture is an effective way to provide redundancy and high-availability.

Our system is built based on the microservice architecture, and our service is divided into several servers. Servers are redundantly placed in geographically different locations with replicated database. We employ Container technology to packaging an application server with runtime environment to geographically deploy microservice.

3 Evacuation Support System

3.1 The Design

Figure 1 shows the system configuration and the flow of data. We have installed the servers in several different regions that are geographically distant enough so that even when a disaster struck, the system can provide continuous services. In order to achieve this purpose, we have made the region configuration into micro-services, i.e. the configuration has many small servers and each has only one function. Micro-services make the servers easily replicated. In this system, we have made three servers, namely (1) way point generation server, (2) pheromone database server, and (3) shelter server. The way point generation server calculates transit points for the evacuation route so that the application server can revise the evacuation route for each user. The pheromone database server is a supporting server for the way point generation server and accumulate the pheromone value for each intersection in the evacuation area. The shelter server is a database that has the location of shelters in the area.

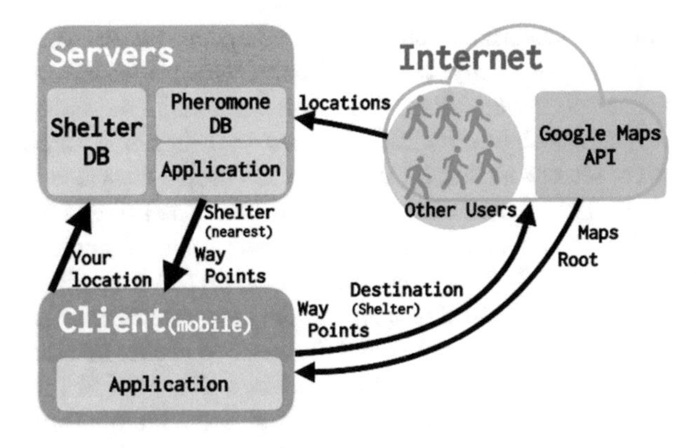

Fig. 1. The system configuration and the flow of data.

The application uses the map of Japan provided by ZENRIN DataCom Co., Ltd through Google Maps API, so that the application server can use the latest map data of Japan. We have employed Apache as the middleware and MariaDB as the RDBMS so that we can easily deploy in different regions with different environments, because it is compatible to the de facto standard, MySQL. The framework of the system is developed in an object-oriented programming language PHP because of its high compatibility with CGI environments. Since we have needed to perform many numerical computation, we have partially employed Python programming language for that purpose.

3.2 The Implementation

As the baseline of the system, we have to provide application users map data with certain accuracy. For this purpose, we use Google Maps. We post various data to Google Maps API, and receive a route from a source to a destination by way of several transit points as well as detailed map. We have used this facility to revise evacuation routes.

The system needs to identify each user. It is not feasible to request the users to register because the users use system when a disaster occurs. MAC address is the ideal identifier, but it is not possible to acquire MAC address from iOS devices. Therefore we decided to use UUID for the user identifier [13].

The system first provides each user "initial evacuation route" and then provides "revised evacuation route" that is calculated based on ACO periodically. The steps the system takes are as follows:

1. Acquires the location of each user (latitude and longitude, hereafter L/L).
2. Searches and acquires the nearest shelter.
3. Generates the initial evacuation route.
4. Acquires the location of each user (L/L).
5. Acquires the transit points for each user.
6. Generate the revised evacuation route.

The location of user means the location of the smartphone that each user has, precisely speaking. Therefore the system only counts the specific refugees that have smartphones and installed the client of this system. Steps 1 through 3 are initial procedures, and then the system repeat steps 4 through 6 periodically to revise the evacuation routes.

First, the system select a shelter location for each user. The system simply chooses the one closest to each user based on the L/L of the user. The L/L of a user is calculated based on GPS. In fact the calculation can be performed through Google Maps Place API and Google Maps Geocoding API. When the application server received each user's L/L, it send the information to the database server, and the server returns the closest shelter location. Upon receiving user's location and its specified shelter location, they are sent to the client, and then the client invokes Google Maps Direction API to receive and to display the route. Google Maps Direction API returns the route information in JSON format, which is a sequence of transit points. JSON is a format for data exchange based on JavaScript [14]. It is an ordered list of pairs that consist of name and value, and can be parsed more easily than XML. The initial route is returned from Google Maps Direction API that uses Dijkstra's algorithm to calculate the shortest path [15].

In order to revise the evacuation route, the system also employs Google Maps Direction API with a new sequence of transit points. The system generates the new sequence of transit points by using ACO. In order to generate an evacuation route, we have made a list of nodes based on the intersections of the real map. Traditionally, a map is modeled as a graph and pheromones are associated with arcs instead of nodes. Our implementation, however, employs microservice and we needed to make the application server as small as possible. We could not afford a graph. Instead, we modeled a map into just a list of nodes, and associate pheromone to each node. This implementation trick saves memory space greatly. We delegate the calculation of the shortest path to Google Maps Direction entirely. We have set the initial value of each of the selected node to a half of the maximum value, and zero for each unselected node. When a user visits a node, the system increments the pheromone by very small value (0.03). This small value is calculated based on the average number of refugees for the target shelter. The Cabinet Office of Japan defines that each shelter contains three hundred people in maximum, and at each intersection a user has three directions to choose. Therefore we set the incremental value of pheromone three divided by one hundred tentatively. In preliminary experiments in simulation, this value works well. In our experiment using real servers and clients, it also works well too.

As a pedestrian user passes an intersection, the pheromone value of the corresponding node increases, and the system revises the evacuation route based on the new pheromone values. The revision is performed every thirty seconds. In the early stages, information about unpassable routes is not collected and is not yet recorded in the system. Therefore the revised route is just a trace of the route that the majorities of the users actually took. In some situation, however, our system produces wrong pheromone values and misguides pedestrian users to wrong ways. The reason why such a situation occurs is that, unlike standard ACO, the source points of user are distributed over the evacuation field, and the system has to generate different optimal evacuation route for each user. Especially, at an early stage of the system operation, since many users try their own imaginary optimal routes and the number of samples is small, the pheromone values are not always reflect the right evacuation routes. In order to mitigate this situation, at the early stage, the route is selected in order to come close to the destination (shelter).

On the other hand, the choice of majority does not guarantee the safe evacuation route. Therefore the system performs corrections, as sample data have accumulated. When the system receives the signals that is submitted by the users who have successfully reached to the shelter, the system increases the pheromone values along that routes so that other refugees can follow the success routes. Also when an authority, i.e. police or fire department, defines unpassable points, and notifies the location of unpassable points to the server, the system eliminates the pheromone values of those points. Current implementation simply makes the pheromone value zero at such points, we can implement minus pheromone values for seriously dangerous points.

4 Experiments

4.1 Evacuation Route Generation

We have performed a series of feasibility studies using the following scenario: (1) one particular user starts from Nippon Institute of Technology, (2) the destination is an elementary school nearby, (3) nine other users are around (4) after certain time, an unpassable point occurs. Figure 2 shows the evacuation route that the system recommended initially. When the system invokes Google Maps Direction API with source and

Fig. 2. The initially recommended evacuation route.

destination points, it returns the shortest route. The current location mark indicates the starting point of the particular user, and the flag indicates the destination.

Figure 3 presents the visualized pheromone values in the route that the other nine user took to reach the destination. Since pheromone values are set at the nodes corresponding intersections, the pheromone occurrences represent transit points they took (orange circles). Those points are sent to the client, and the client invokes Google Maps Directions API to display a new evacuation route as shown in Fig. 4.

Fig. 3. Visualized pheromone values deposited by the pedestrian users.

When this new evacuation route was displayed on the particular user's smartphone, three of other nine users had successfully reached to the destination. Then our scenario engendered unpassable area along the recommended route. The area is shown as a green square in Fig. 4. Such information should be provided from authorities, and our server is ready to accept such notification. Then the system generates deodorant pheromone to erase the accumulated pheromones so that the Google Maps Direction API does not choose routes in such unpassable area. After a few revisions of evacuation route, the system provide a new route that bypasses the unpassable area as shown in Fig. 5.

Fig. 4. Unpassable area advents (green area).

Fig. 5. A new route that bypasses the unpassable area is generated.

4.2 Disaster Recovering

We have examined recovery experiments as well as evacuation experiments. The scenario is as follows: (1) a server in region 1 (IP: xxx.xxx.xxx.16) is down, (2) the surveillance software on a surveillance server for the service (IP: yyy.yyy.yyy.200) sends a message to the service manager by an e-mail, (3) service manager sends a message to the DNS that tells the server has been down, and (4) interacts with DNS so that it can communicate with a backup server, then (5) rewrite the IP address so that the backup server can communicate with the rest of the system. In the experiments, we set up a backup server on a backup machine (IP: yyy.yyy.yyy.16) as a hot-standby and test whether the switching of the servers is smooth, and database replication is correctly performed.

We have confirmed the switching is smooth, but we have found database replication is not perfect. Since we set the synchronization period of the database every two minutes due to the resource restriction, we have found that the data generated in thirty seconds immediately before the server down are lost. The down time for the service is less than one minute and this loss of service produces revision error two times for evacuation route generation. This does not seriously affect the evacuation route recommendation on client's machine.

5 Conclusion

We have proposed and implemented an evacuation support system with real servers on Linux machines and real clients on both Android machines and iOS machines. We have confirmed evacuation support systems we have proposed and demonstrated on simulators are feasible in a real world environment as long as Internet infrastructure is intact. We also confirmed the disaster recovery can be achieved through distributing server packages in different regions by using Containers. On the other hand, we found the surveillance server is the single point of failure (SPOF). When we develop our prototype system into practical system, we have to introduce more redundancy in the server configuration. Even though we are aware of shortcomings in our prototype system, the microservice base architecture that uses Containers to deploy micro-servers in geographically different locations certainly contributes robustness of the system. This is an important finding for especially disaster mitigation system such as ours.

We have also found the revision of the evacuation route largely depends on the number of users. Since the current system is just a prototype, we could not provide enough client machine. We have to perform a series of large scale experiments before deploying the system in practice. Our utilization of ACO should be also improved. The parameter setting is largely ad-hoc in the current system. We need to go back to simulator to estimate pheromone values in various environments. On the other hand, when we simply increase the number of user in the current system, congestions on the evacuation routes may occur, because the system leads people to the route the majority took. We need to take consideration of the velocity of pedestrian users so that congested routes get to less pheromone and comfortable routes get more pheromone. Distributing refugees to the right shelters is another optimization problem. We are planning to integrate another pheromone with minus value to prevent refugees from concentrating on a few shelters.

Acknowledgements. This work is partially supported by Japan Society for Promotion of Science (JSPS), with the basic research program (C) (No. 17K01304 and 17K01342), Grant-in-Aid for Scientific Research (KAKENHI).

References

1. Asakura, K., Fukaya, K., Watanabe, T.: Construction of navigational maps for evacuees in disaster areas based on ant colony systems. Int. J. Knowl. Web Intell. **4**, 300–313 (2013)
2. Avilés, A., Takimoto, M., Kambayashi, Y.: Distributed evacuation route planning using mobile agents. In: Transaction on Computational Collective Intelligence XVII. LNCS, vol. 8790, pp. 128–144. Springer, Heidelberg (2014)
3. Ohta, A., Goto, H., Matsuzawa, T., Takimoto, M., Kambayashi, Y., Takeda, M.: An improved evacuation guidance system based on ant colony optimization. In: Lavangnananda, K., Phon-Amnuaisuk, S., Engchuan, W., Chan, J. (eds.) 19th Asia Pacific Symposium on Intelligent and Evolutionary Systems, pp. 15–27. Springer, Heidelberg (2015)
4. Taga, S., Matsuzawa, T., Takimoto, M., Kambayashi, Y.: Multi-agent approach for return route support system simulation. In: Proceeding of the Eighth International Conference on Agents and Artificial Intelligence, vol. 1, pp. 269–274. INSTICC, Rome (2016)
5. Goto, H., Ohta, A., Matsuzawa, T., Takimoto, M., Kambayashi, Y., Takeda, M.: A guidance system for wide-area complex disaster evacuation based on ant colony optimization. In: Proceeding of the Eighth International Conference on Agents and Artificial Intelligence, vol. 1, pp. 262–268. INSTICC, Rome (2016)
6. Taga, S., Matsuzawa, T., Takimoto, M., Kambayashi, Y.: Multi-agent approach for evacuation support system. In: Proceeding of the Ninth International Conference on Agents and Artificial Intelligence, vol. 1, pp. 220–227. INSTICC, Porto (2017)
7. Beckers, R., Deneubourg, J.L., Goss, S., Pasteels, J.M.: Collective decision making through food recruitment. Insectes Soc. **37**, 258–267 (1990)
8. Dorigo, M., Maniezzo, V., Colorni, A.: Ant system: optimization by a colony of cooperating agents. IEEE Trans. Syst. **26**(1), 29–41 (1996)
9. Goodwin, M., Granmo, O., Radianti, J.: Escape planning in realistic fire scenarios with ant colony optimisation. Appl. Intell. **42**(1), 24–35 (2015)
10. Baharmand, H., Comes, T.: A framework for shelter location decisions by ant colony optimization. In: Proceeding of the 12th International Conference on Information Systems for Crisis Response and Management, Kristiansand (2015)
11. Tanenbaum, A.S., Van Steen, M.: Distributed Systems: Principles and Paradigms, 2nd edn. Createspace Independent, North Charleston (2016)
12. Newman, S.: Microservices: Designing Fine-Grained Systems. O'Reilly Media, Cambridge (2015)
13. The Internet Engineering Task Force (IETF): A Universally Unique Identifier (UUID) URN Namespace, IETF RFC 4122 (2005). https://tools.ietf.org/html/rfc4122. Accessed 10 May 2018
14. ECMA International: The JSON Data Interchange Syntax, Standard ECMA-404 (2017). http://www.ecma-international.org/publications/files/ECMA-ST/ECMA-404.pdf. Accessed 10 May 2018
15. Dijkstra, E.W.: A note on two problems in connexion with graphs. Numer. Math. **1**, 269–271 (1959)

Understanding Mobile Purchase Intentions in Poland: Extension of the Technology Acceptance Model

Mariusz Trojanowski[1(✉)] and Jacek Kułak[2]

[1] University of Warsaw, Warsaw, Poland
trojan@wz.uw.edu.pl
[2] LSC Communications Europe, Kraków, Poland
jacekulak@gmail.com

Abstract. The purpose of this article was to identify factors influencing *intention to use mobile device during purchasing process*. To achieve this goal Technology Acceptance Model extended with additional variables (*social influence* and *perceived enjoyment*) was chosen as the research model. There were 500 respondents from 5 Polish cities – Warsaw, Poznań, Gdańsk, Wrocław and Białystok. Data was collected with paper questionnaire and analyzed with partial least square path modeling (PLS-SEM) method in Smart PLS 3 software. Model explained over 50% of variance of endogenous variable. According to results perceived usefulness has the strongest influence on *intention* (0.389; $p < 0.01$). Other statistically significant predictors of *intention* were *perceived ease of use* (0,222; $p < 0.01$) and *perceived enjoyment* (0,172; $p < 0.01$), while *social influence* turned out to be the sole insignificant factor in the model (0.050; $p > 0.05$). Outcomes of the study suggest that sellers should focus on the utilitarian aspects of the mobile sales in the first place (such as providing free delivery of products, offering multitude of payment options and increasing overall performance of their websites), and then also on the facets related to simplifying of the mobile shopping (production of video tutorials, clear and plain explanations regarding purchasing process on the website) and pleasure of the mobile buying process (creation of additional features in the sellers' applications e.g. quizzes, games or downloadable materials).

Keywords: Mobile commerce · Poland · Technology Acceptance Model

1 Introduction

1.1 The Importance of Mobile Commerce in Poland

In 2017 according to Euromonitor International mobile Internet retailing was accounted for less than 1% of total retailing in Poland, which may seem unimpressive. However, looking deeper into the numbers, mobile commerce was worth 3.74 billion dollars[1]

[1] Without value added tax.

L. Borzemski et al. (Eds.): ISAT 2018, AISC 852, pp. 334–344, 2019.
https://doi.org/10.1007/978-3-319-99981-4_31

(Fig. 1) in the same year, also the growth dynamic was by far the highest from all types of sales in Poland with the average of 49% in years 2012–2017 [5] (Fig. 2). It is also very popular topic among researchers all over the world, as number of articles[2] grew from 140 in 2000 to more than 650 in each of the last 5 years [10]. Yet, at the same time, m-commerce research is still in very early stage of development, especially in the Eastern European countries.

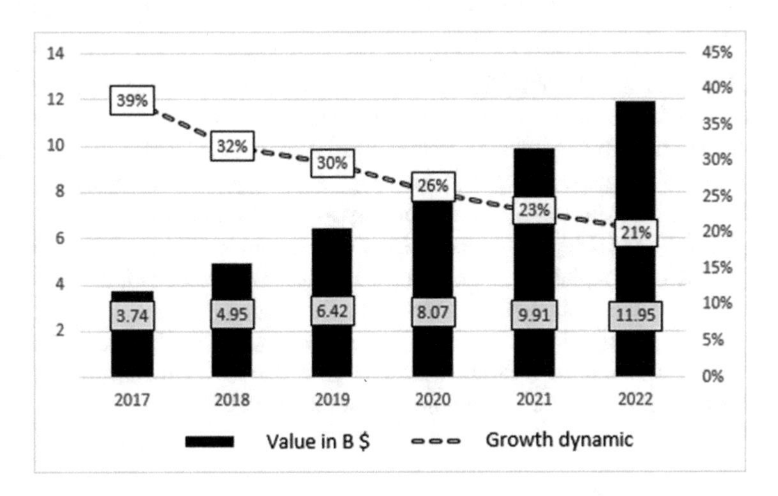

Fig. 1. Value (without VAT) and growth dynamic of mobile commerce in years 2017–2022 [5]

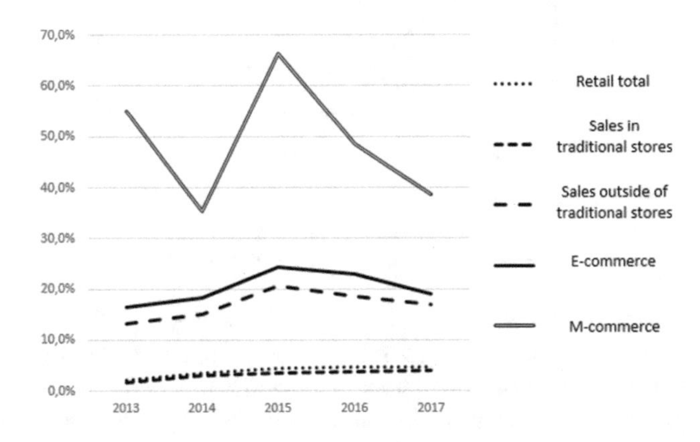

Fig. 2. Growth dynamic for various type of sales in Poland in years 2012–2017 [5]

[2] Phrase “mobile commerce” or “m-commerce” used in article title, abstract or keywords.

2 Theory, Research Model and Hypotheses

2.1 Technology Acceptance Models

The aim of this article is to identify factors influencing *intention to use mobile device during purchasing process* for Polish consumers. In literature such phenomenon are usually explained by using one of the technology acceptance theories. The most popular ones over the years were:

- Theory of Reasoned Action (TRA), which was originally designed as general theory. It posits that *intention* to use a device for particular purpose depends on individual's attitude toward the behavior and perceived social pressure related to that behavior [1].
- Technology Acceptance Model (TAM, Fig. 3), which was originally designed to predict acceptance and usage of technology in the work environment. Technology Acceptance Model (Fig. 3) was used as the base model in this research. It posits that two variables are explaining *intention* – *perceived usefulness* (PU in short), which was defined by the author of the theory as "The degree to which a person believes that using a particular system [device] would enhance his or her performance" [4] and *perceived ease of use* (PEU), which is "The degree to which a person believes that using particular system would be free of effort" [4].

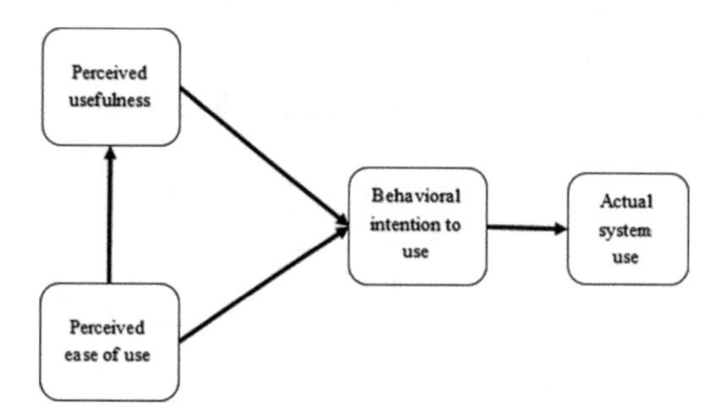

Fig. 3. Technology Acceptance Model [4]

- Unified Theory of Acceptance and Use of Technology (UTAUT), which was designed to consolidate constructs from prior prominent technology acceptance theories. It posits that main predictors of *intention* to use a device for particular purpose are variables associated with performance, effort, social influence and available support [15].
- Unified Theory of Acceptance and Use of Technology 2, which adapted UTAUT theory to consumer context by inclusion of variables related to price, pleasure and habit [16].

All of the above theories supports underlying concept first proposed in TRA, which is critical and most important role of *intention* as predictor of behavior.

2.2 Research Model and Hypotheses

Technology Acceptance Model was chosen as the model for this research due to its parsimony, efficiency, popularity and ease of application. However, the TAM model was often critiqued for being too simplistic and not including predictors related to opinions of important others and pleasure connected with using particular technology or device [3, 14]. Due to above, *social influence* (SI) and *perceived enjoyment* (PE) variables were added into the research model (Fig. 4).

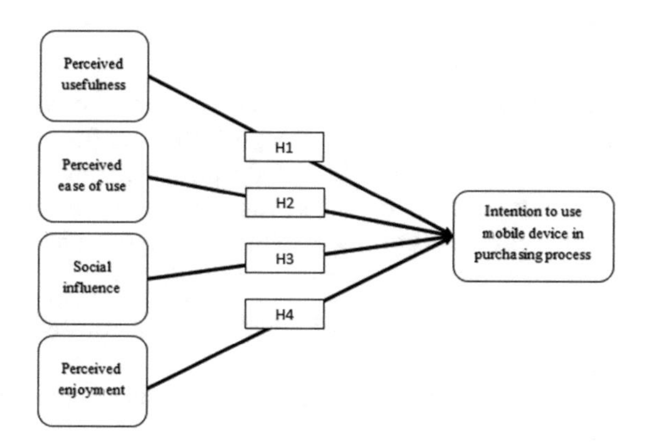

Fig. 4. Research model

Literature review and preliminary empirical research were used for formulation of the following hypotheses:

H1: *Perceived usefulness* is the strongest predictor of the *intention to use mobile device in purchasing process.*

The primary function of a mobile device during purchasing process is to allow consumers to research, compare, order and buy the product they need at any time of the day regardless of the place. This, and much larger selection of the goods available, are the main advantages of m-commerce in comparison to traditional shopping. Variables related to usefulness were found to be the most important antecedents of *intention* in other studies [12, 15]. The variable was measured with the following items:

1. I think that using mobile device during purchasing process is useful.
2. Using mobile device during purchasing process increase probability of buying things that are important to me.
3. Using mobile device during purchasing process allows to buy products faster.
4. Using mobile device during purchasing process increase frequency of purchases.

H2: *Perceived ease of use* is the *intention to use mobile device in purchasing process.*

The easier the process, the greater the chance that it will be completed and repeated. People tend to abandon complicated, hard to understand procedures. Using mobile device in purchasing process is very intuitive for more and more people, especially the younger ones, for whom it's often hard to imagine a day without tablet or mobile phone in their pockets. Positive relationship of *perceived ease of use* and *intention* was confirmed in articles [2, 16] Statements below were used to measure PEU:

1. It was easy for me to learn how to use mobile device in purchasing process.
2. Using mobile device in purchasing process is clear and understandable for me.
3. I think that using mobile device in purchasing process is easy.
4. Using mobile device in purchasing process is complicated [reverse coding]

H3: *Social influence* is statistically significant predictor of the *intention to use mobile device in purchasing process.*

Most purchasing decisions aren't made without asking for opinion or advice, whether it comes from family, friends, coworkers or anonymous people from an Internet forum. It's easier to take action if there is support for it. Judgement of important others regarding m-commerce may greatly affect individual's propensity to use mobile device during purchasing process. Social factors positively influenced *intention* in works of [15, 16]. Following statements were formed to measure *social influence:*

5. People, who are important to me think, that it's good to use mobile device in purchasing process.
6. People, who have influence on my behavior, accept that I use mobile device in purchasing process.
7. People, whom opinion I value, approve that I use mobile device in purchasing process.

H4: *Perceived enjoyment* is statistically significant predictor of the *intention to use mobile device in purchasing process.*

People like activities that are enjoyable. Some of them may find pleasure in shopping with mobile device. This way they can effortlessly read about newest products, communicate with others fans of particular brand, use coupons or interact with the seller. Enjoyment factors were statistically significant in articles of [2, 12]. *Perceived enjoyment* in this research was measured with items below:

8. Using mobile device in purchasing process is fun.
9. Using mobile device in purchasing process is pleasure.
10. Using mobile device in purchasing process is entertaining.

In addition, moderating influence of three variables were checked in the model – age, gender and experience in using the mobile Internet.

3 Research Methodology

3.1 Research Participants

There were 500 respondents from 5 Polish cities Warsaw, Poznań, Wrocław, Gdańsk and Białystok. All of them were students (BA, MA and postgraduate), mostly women

(66%), under 25 years old (77%), with more than 6 years of experience in using mobile Internet (55%). Summary of respondents is included in Table 1.

Table 1. Summary of respondents.

Variable	Number of respondents	% of respondents
Sex [Man]	170	34
Sex [Woman]	330	66
Age [>=25 years]	114	23
Age [<25 years]	386	77
Exp [>=6 years]	275	55
Exp [<6 years]	225	45

3.2 Research Questionnaire

Research questionnaire had 20 statements. All the statements were randomly placed to avoid pattern answering. All the statements were measured with 7-point Likert scale (strongly disagree – strongly agree). *Intention to use mobile devices in purchasing process* captured six different stages of purchasing process: research of information, comparison of offers, comparison of prices, checking opinions related to products, products ordering, and payment for products. Only respondents who have used mobile devices in purchasing process in the past were eligible.

3.3 Data Collection and Analysis

Data was collected with PAPI technique and analyzed with PLS-SEM (partial least square path modelling) method in SmartPLS 3 software. Advantages of this method include ability to handle non-normal data and maximization of the amount of explained variance [6].

4 Results

4.1 Measurement Model Results

Assessment of measurement model is done through checking composite reliability, convergent validity, indicator reliability and discriminant validity. All constructs had composite reliability values above recommended threshold of 0.7 [6]. For convergent validity all AVE values of constructs should be above 0.5 [17] which is the case in this study. Indicators with outer loadings values higher than 0.708 should be retained. Indicators with outer loadings values between 0.4–0.708 should be retained if deleting them is not increasing significantly AVE values and composite reliability values [6]. Due to above, all indicators were retained. Required threshold for reliability for each indicator is 0.4 [17]. All indicators were above this threshold. Discriminant validity is assessed through HTMT values, which should not be higher than 0.9 [6]. It was established for all of the constructs in this study.

4.2 Structural Model Results

Collinearity
Collinearity was not an issue in this study. VIF values for all of the constructs were below 5 [17].

Path Coefficients and R^2Value
Perceived usefulness has the strongest influence on *intention* (0.389; $p < 0.01$). Other statistically significant predictors of *intention* were *perceived ease of use* (0,222; $p < 0.01$) and *perceived enjoyment* (0,172; $p < 0.01$), while *social influence* turned out to be the sole insignificant factor in the model (0.050; $p > 0.05$). R^2 value was 51.1%. Structural model results are presented on the Fig. 5.

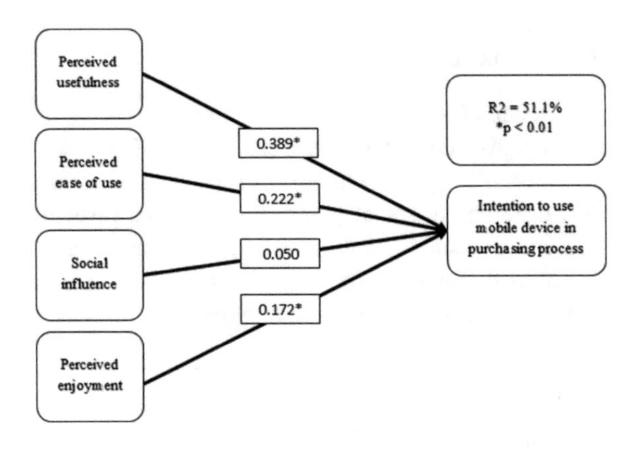

Fig. 5. Structural model results

Moderating Variables
Relation between *perceived ease of use* and *intention* was moderated by experience in using mobile Internet (for experience <6 years coefficient value was 0,144, $p < 0.05$; for experience >6 years coefficient value was 0.312, $p < 0.05$; difference was 0.168, $p < 0.05$). Other relations were not significantly moderated by any of the moderating variables. Measurement invariance was provided with MICOM procedure.

5 Summary and Conclusions

5.1 Theoretical and Managerial Contributions

This article helped to fill the gap in the mobile shopping research. This area needs further exploration as it is rapidly growing, especially in the Central and Eastern European countries, where double digits growth in m-commerce is expected until 2022 and beyond.

In the previous research many different technology acceptance theories were used to explain mobile purchase *intention*. [11] have used Technology Acceptance Model extended with perceived enjoyment and positive word of mouth intention in the context of the usage of the m-commerce website. This model explained 73% of the intention to use m-commerce and the perceived enjoyment was its most important predictor. [9] created the model based on the Social Cognitive Theory in the context of the acceptance of mobile banking by Australian youth. Results suggest that exposure to the media about mobile banking is the most important factor affecting adoption of the mobile banking. [18] modified Unified Theory of Acceptance and Use of Technology model by adding perceived credibility, perceived financial cost and perceived self-efficacy. The research was conducted in the context of the acceptance of the mobile banking. Social influence, performance expectancy, and perceived credibility had the positive influence on the intention, while the perceived financial cost had significant, negative influence on the intention. The R^2 was 60%. [8] used Unified Theory of Acceptance and Use of Technology 2 extended with general privacy, system-related privacy and perceived security to explain intention to use NFC mobile payment in hotels. Performance expectancy was the strongest exogenous variable. Based on the above results and the results of different m-commerce research, there are two main conclusions: (1) There are many effective technology acceptance theories that helps to explain mobile purchase intention. The main difference is complexity and ease of application. (2) Mobile payments and mobile banking emerged as one of the most popular topics of the research connected to m-commerce in recent years.

Extended Technology Acceptance Model used in this research turned out to be efficient model in mobile purchasing context. It explained more than 50% of *intention to use mobile device in purchasing process* using only 4 predictors, which is satisfying value. Parsimony is one of the biggest advantages of extended TAM, especially in comparison with other models, such as UTAUT2 (7 exogenous variables), TAM2 (9 exogenous variables) and TAM 3 (15 exogenous variables). As a consequence it's easier to comprehend the results and contrast with other studies, as well as duplicate this research in different country or setting. Three out of four hypotheses were confirmed (Table 2).

Table 2. Summary hypotheses

Hypothesis	Supported
H1: *Perceived usefulness* ➔ *Intention* (strongest predictor)	Yes
H2: *Perceived ease of use* ➔ *Intention*	Yes
H3: *Social influence* ➔ *Intention*	No
H4: *Perceived enjoyment* ➔ *Intention*	Yes

PAPI was effective technique of data collection, as there was very low rate of invalid questionnaires (less than 2%, due to lack of responses or patterns in answering). PLS-SEM structural equation modeling was adequate for this studies. It allowed to check differences between groups of respondents and conduct bootstrap analysis for confirming significance of relationships. It is highly recommended data analysis method for technology acceptance research in mobile context by the reason of multitude of other m-commerce studies using it (ability to compare) and ease of application. For example researcher can use Smart PLS3 software, which is user friendly, yet delivers powerful and complex statistical computations. There were 500 respondents in this study, for models with 4 exogenous variables [17] suggests to use at least 65 questionnaires to achieve stability of results.

According to the results *perceived usefulness* has the strongest influence on *intention to use mobile device in purchasing process* (0.389; $p < 0.01$). Variables related to usefulness usually are among the most important ones in the research based on technologu acceptance models [13] and PE was the strongest predictor of *intention* in the original study [4], when the model was proposed. Due to importance of *perceived usefulness*, some researchers are focusing on finding factors that explains this variable, for example in TAM3 model [14] these factors are subjective norm, image, job relevance, output quality and result demonstrability. Understanding this particular predictor may be critical, as simulated model for this research with *perceived usefulness* as sole exogenous variable explained 44% of variance of *intention*. Outcomes of the study suggest that sellers should focus on the utilitarian aspects of the mobile sales in the first place. This can be done by focusing on 3 areas of business improvements: product-related improvements (reduction of defective goods, extended warranty, friendly complaint policy), website/app improvements (optimization of the website for mobile devices, reduction of site outages and bugs in application), co-operation with external partners (offering multitude payment and delivery options).

Perceived ease of use is the second most important factor influencing *intention* (0,222; $p < 0.01$). Managers should focus on simplifying mobile shopping. Mobile website and application should be intuitive, with clear and plain explanations regarding purchasing process. They should contain video tutorials, and manuals related to the product. There should be several contact options available, preferably 24/7 support for consumers.

Third statistically significant predictor of *intention* is *perceived enjoyment* (0,172; $p < 0.01$). Fun is especially important for younger consumers, who are quickly bored. Organizations can produce quizzes, games or downloadable materials. Mobile coupons strategy may be particularly effective.

Social influence does not have statistically significant influence on *intention* (0.050; $p > 0.05$). This is congruent with [12, 13] studies. SI variable may be insignificant due to the student's research group, who like to think of themselves as independent of other's opinion and advices. Marketers in their campaigns aimed at young m-shoppers shouldn't focus on social aspects of purchases, rather show the benefits and fun associated with mobile shopping.

Age was not significant moderator for any of the relationships in the model. This may be due to the fact that respondents were students only and age variance wasn't high enough to capture any differences. However in the research of [13] age was significant moderator of relation between hedonic motivation (fun connected with the usage of the mobile device) and *intention*. This variable was statistically significant only for the younger group, which suggests that marketers should prepare specifically for this group different, more engaging mobile application for m-commerce, where you can not only buy things, but also watch videos, play games or connect with the community of fans of the brand. Using gamification elements for that group (e.g. awarding different prizes for activity in the m-commerce app) may also be beneficial for the seller. Gender didn't moderate any relationships in the model, which indicates that men and woman use mobile devices for purchases in a similar way. In contrast to these results, [15] in the context of technology usage in work found out, that gender is significant moderator for relations of variable connected with usefulness (performance expectancy, effect is stronger for men) and ease of use (effort expectancy, effect is stronger for women) with *intention*. This means that additional research is needed in this field, and that marketers may consider in their promotional communication underlining different advantages of the shopping through mobile app, for example for men they can promote a message that they can buy more things in comparison to the traditional shop, faster, with free delivery and at any time of the day, in other words that buying within the mobile app will be efficient. Marketers can persuade women in the advertisement that mobile app is very intuitive and learning will not be required to buy the needed product. Experience in using mobile Internet significantly influenced relation between *perceived ease of use* and *intention*. PEU is more important for more experienced users. As they have in general visited more websites, they have better comparison, greater expectations and they demand support more often than inexperienced users and don't want to waste time for learning how to use unintuitive websites. What's interesting, result of research [15] are opposite and indicates that ease of use is more important for less experienced users. Either way managers should focus on developing machine learning algorithms analyzing user behavior patterns, and help them in buying process, for example there can be pop-ups displayed with the additional information how to navigate the site for those who click a lot of the same tabs repeatedly (which may indicate that they are lost), or more contact options should be added, like interactive chat window available 24 h a day, which may be possible with the implementation of the chat bots (computer programs which conduct conversation with the user).

5.2 Limitations and Further Research

There were several limitations identified in this research. Respondent group is not representative for Polish population as a whole (only students from big cities were included). Actual usage of mobile devices during purchasing process was not examined. Endogenous variable was *intention*, which is usually very good predictor of actual use of device in particular purpose according to the literature [1, 15], but does not always convert to sale. In data analysis method non-linear relationships were not analyzed. There were 3 moderating variables in this study: age, experience in using mobile Internet and gender. It is very probable that other moderators are significantly altering the results. Some researchers [15] claim, that parsimonious models are not best suited for explaining very complex problems like consumer behaviors.

Future research can be focused on adding different variables to the model. Instead of *social influence*, which was not statistically significant, researchers can propose factors such as trust for the seller or trust for the mobile transactions in general, perceived risk connected with mobile transaction or cost-related variables. It's worth to analyze influence of other moderating variables as well, for example operating system (iOS, Android, Windows), income, individual innovativeness or even cultural moderators for cross-country studies (power distance, individualism, masculinity, uncertainty avoidance, long-term orientation, indulgence). There is a need to conduct longitudinal studies aimed at capturing both *intention* and actual use of mobile devices for purchases with usage of Technology Acceptance Model in Poland. Finally, researchers can consider examining non-linear relationships. Matlab, Statistica, R studio among other software, allow these type of analysis.

References

1. Ajzen, I., Fishbein, M.: Belief, Attitude, Intention and Behavior: An Introduction to Theory and Research. Addison-Wesley Publishing Company, London (1975)
2. Alalwana, A.A., Dwivedi, Y.K., Ranab, N.P.: Factors influencing adoption of mobile banking by Jordanian bank customers: extending UTAUT2 with trust. Int. J. Inf. Manage. **37**, 99–110 (2017)
3. Bagozzi, R.P.: The Legacy of the technology acceptance model and a proposal for a paradigm shift. J. Assoc. Inf. Syst. **4**(8), 244–254 (2007)
4. Davis, F.D.: Perceived usefulness, perceived ease of use, and user acceptance of information technology. MIS Q. **3**(13), 319–340 (1989)
5. Euromonitor International: Retailing in Poland (2018)
6. Hair, J.F.J., Hult, G.T.M., Ringle, C.M., Sarstedt, M.: A Primer on Partial Least Squares Structural Equation Modeling (PLS-SEM), 2nd edn. Sage Publications Inc., London (2016)
7. Lin, Y.-R., Chen, Y.F., Chung, Y.C.: Mobile learning: prediction of user behavior by means of the theory of reasoned action. W: F. Saito, red. Consumer Behavior, pp. 213–225. Nova Science, Taipei (2009)
8. Morosan, C., DeFranco, A.: It's about time: revisiting UTAUT2 to examine consumers' intentions to use NFC mobile payments in hotels. Int. J. Hospitality Manag. **53**, 17–29 (2016)
9. Ratten, V.: Social cognitive theory in mobile banking innovations. Int. J. E-Bus. Res. **7**(1), 39–51 (2011)
10. Scopus database. http://www.scopus.com. Accessed 18 May 2018
11. Song, J., Koo, C., Kim, Y.: Investigating antecedents of behavioral intentions in mobile commerce. J. Internet Commer. **6**(1), 13–34 (2007)
12. Trojanowski, M., Kułak, J.: Wpływ postrzeganego ryzyka na proces korzystania przez konsumentów z mobilnego handlu elektronicznego - rozszerzenie modelu UTAUT2. In: Handel we współczesnej gospodarce. Nowe wyzwania, pp. 255–270. Uniwersytet Ekonomiczny w Poznaniu, Warszawa (2016)
13. Trojanowski, M., Kułak, J.: The impact of moderators and trust on consumer's intention to use a mobile phone for purchases. J. Manag. Bus. Adm. Cent. Eur. **2**(25), 91–116 (2017)
14. Venkatesh, V., Bala, H.: Technology acceptance model 3 and a research agenda on interventions. Decis. Sci. **2**(39), 273–315 (2008)
15. Venkatesh, V., Morris, M.G., Davis, G.B., Davis, F.D.: User acceptance of information technology: toward a unified view. MIS Q. **3**(27), 425–447 (2003)
16. Venkatesh, V., Thong, J.Y., Xu, X.: Consumer acceptance and use of information technology: extending the unified theory of acceptance and use of technology. MIS Q. **1**(36), 157–178 (2012)
17. Wong, K.K.-K.: Partial least squares structural equation modeling (PLS-SEM) techniques using SmartPLS. Mark. Bull. **24**, 1–32 (2013)
18. Yu, C.-S.: Factors affecting individuals to adopt mobile banking: empirical evidence from the UTAUT model. J. Electron. Commer. Res. **13**(2), 104–121 (2012)

Tiny TTCN for IoT-Related Testing: That Shrinking Feeling

Krzysztof M. Brzeziński(✉)

Institute of Telecommunications, Warsaw University of Technology, Warsaw, Poland
kb@tele.pw.edu.pl

Abstract. This work presents the design rationale and implementation details of a test programming language that is meant to retain the semantics and look-and-feel of the standardized TTCN-3 test language, while being able to run on Arduino-class microcontrollers (with resources many orders of magnitude smaller than originally required by TTCN-3). This is a part of a larger project that aims at harmonizing the test approaches and test tools of distinguishable research communities that joined forces in their work on Socio-Technical Systems providing subtle health-related interventions. The tiny testers (the IoT devices themselves) are meant primarily to supervise validation experiments in such systems.

Keywords: Testing · Microcontroller · IoT · TTCN-3 test language

1 Introduction

The present work is concerned with the design and implementation of tiny passive testers programmed in a version of the standardized test language TTCN-3 (Testing and Test Control Notation) [1], and in particular – with the system-level problems with implementing the constructs of a *huge*, complex test language on a platform with extremely modest resources.

The initial publication on such tiny testers [2] mostly focused on inter-community problems of methodological and tool-related “lock-in” (reluctance to seriously consider, and to adopt, external ideas). It was argued that three distinct communities: the “T-school” (telecommunications, standards, formality...), the “I-school” (Internet, informality...), and the “S-school” (social sciences, psychologists...) could greatly benefit from sharing the testing-related expertise and tools. The T-school is the “owner” of TTCN-3, and the idea was to develop a TTCN-inspired technology that could be acceptable for use in *joint projects* of the three communities. One such project area was identified as health-related *persuasive technologies* that involve technical devices of the IoT (Internet of Things) class. Such devices are native to the “I-school”. On the other hand, the “S-school” is concerned with experimentally validating the effectiveness of subtle methods of behaviour change. Such experiments (in particular those in-the-wild) can run for a long time. It is thus important to unobtrusively “test the testing”, in order to be able to *automatically* detect and correct malfunctions

L. Borzemski et al. (Eds.): ISAT 2018, AISC 852, pp. 345–354, 2019.
https://doi.org/10.1007/978-3-319-99981-4_32

while the tests are still running. This *niche setting* was chosen as a target for the *integration project*, in which a typical IoT device (here: an Arduino-class microcontroller [3]) was turned into an autonomous passive tester [4]: a self-contained, tiny Tester thinG (μTG), programmed in a test language μTTCN that is closely inspired by TTCN-3. The whole technology of such testers was dubbed μT. TTCN-3 is a *huge* language indeed, and its development environments and run-time support systems may require hundreds of megabytes and powerful host computers to install and run. On the other hand, a typical Arduino board boasts just 2–2.5 kB of RAM and 32 kB of ROM. The disparity is so staggering that the whole project might seem absurd; indeed, we have not been able to detect any traces of similar ideas in the available literature. This made the project all the more challenging. Carrying over ideas or tools to another context is a good opportunity for stripping them of their initial (possibly – crippling) preconceptions, and "*introducing variations* [...] *that others may not realise or do not dare to consider*" (Krippendorff in [5]). This also happens to be one of the inventive methods in Design Thinking; traces of the DT mindset can hopefully be identified in the present work.

The development of the μT technology has been autonomous, conceptually dating back to [6]. *A posteriori*, only indirect links could be identified with other published works. Some of these works point to the growing availability of open-source TTCN-3 tools and to their suitability for testing in the IoT setting [7]. This observation, however, does not address the context of our work: the need to provide TTCN-like tools and experience to those communities that are unlikely to invest in the full TTCN-3 technology (including the ability to set up an operational TTCN-3 test system, which is a non-trivial development task [8], at least as complex as the development of tests themselves). The reader is referred to [2] for a brief discussion of related publications, but it is fair to note that the level of their relevance to *this* work is very modest.

As a complement to [2], the present paper deals almost exclusively with technical details of design and implementation of the system-level software for tiny testers. It is assumed that the reader is aware, at least in general terms, of TTCN-3 and of the Arduino platform (see [2] for basic information on both). In Sect. 2 we synthetically present the hardware/software architecture of a tiny tester. The whole Sect. 3 is devoted to discussing consecutive classes of μTTCN constructs – their design and implementation that "opportunistically" tries to take advantage of existing features of the Arduino platform and its C-based Sketch programming language. Section 4 concludes the paper.

2 Architecture of Tiny Testers

For completeness, in Fig. 1 we present the changes introduced to the standardized TTCN-3 architecture of a test system in order to obtain the architecture of a tiny tester μTG. A test program written in μTTCN (which is described in the sequel) is also a valid Sketch program (μTS – a Test Sketch) for the Arduino platform. In TTCN-3, a Test Executable (that which executes a Test Suite) is

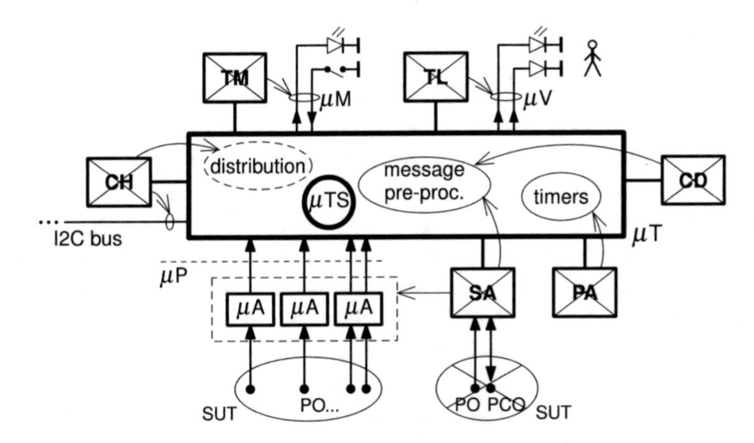

Fig. 1. Architecture of a μTG tiny tester-thing (*superfluous TTCN-3 architectural entities are crossed out, with their functionality transferred as shown by curved arrows*)

necessarily supported by conceptual entities, some of which are "external" and must be developed or configured by a test designer/programmer. In μT, all these entities are dispensed with (crossed out in the figure). A SA (SUT Adaptor) that handles communication with a System Under Test (SUT) at its Points of Control and Observation (PCO) and Points of Observation (PO), and normally contains, e.g., protocol stacks, disappears altogether. Some electrical and physical interfacing tasks are taken over by very simple tiny Adaptors (μA: a resistor, a reed relay, etc.), and message preprocessing is done by system-level μTTCN software, as described in Sect. 3.4. A μTG, being a *passive* tester, only handles POs. All communication from SUT appears at Arduino's pins (μP ports). The external Platform Adaptor (PA) is replaced by system-level software that handles timers (see Sect. 3.5). The external Coder/Decoder disappears; some of its functions (in their most simple form) are taken over by message pre-processing. The possible distribution of a tester, managed by the Component Handler (CH), is also being considered for μT; it will use the I2C communication bus of Arduino. Test Management (TM) has been replaced by rudimentary controls for starting and resetting the tester (μM). Test Logging (TL) now consists in presenting the test verdicts on μV pins of Arduino, to which rudimentary indicators may be attached (such as LEDs).

3 Tiny TTCN: Design and Implementation

Design decisions on the implementation of TTCN language constructs are not normally divulged beyond the internal documents of a tool maker (for a rare exception, see [9]), probably because they are treated as valuable scientific and commercial assets. On the other hand, these design decisions are an important part of the *essence* of our proposal. We thus describe the implementation of major parts of μTTCN (together with the rationale behind the implementation decisions), so that the examples given can be understood.

3.1 Test Case Structure

In the simplest case, the whole Test Sketch contains just a single Test Case. The program of this Test Case cannot be placed, as a whole, inside the Arduino `loop` (a main part of every Arduino Sketch), because it would then either block indefinitely, or execute to its end in one go (one loop instance). Instead, it needs to be suitably fragmented, so that it yields control to "the system" sufficiently often, and at the right places. In other words: in the "OS-less" environment, the program structure itself is largely responsible for the proper scheduling of its own execution. The way in which this is done should not be too cumbersome for a programmer, it should feel natural, and it should be "explainable" in terms of TTCN concepts and constructs. We claim that μT fulfils these requirements.

In μTTCN, a Test Case (Fig. 2) is implemented as a single outermost `switch` statement controlled by the system-level, global `tStep` variable. This syntactic form reads naturally: "*now switch to the Test Step number...*". The concept of a Test Step is unknown to TTCN-3, but it was present in TTCN-2, where it denoted "*a named subdivision of a test case, constructed from test events and/or other test steps*" [10]. The pragmatics of a Test Step is to keep together TTCN operations that can, or should, be executed together, in a single pass of the `loop`.

Names of Test Steps additionally serve as TTCN labels. With each loop execution, control is passed to one particular Test Step, according to the value of the "program counter": `tStep`. During the execution of this Test Step, `next_tStep` is set: explicitly (e.g., by the `goto` operation of TTCN, which we retained, in the form: `tgoto`), or implicitly: either to the *current* value of `tStep` (in order to repeat the same Test Step, in case it contained an `alt` statement in which no alternative was successful; see next), or to the *next* value, if a test is to continue with the next step. If there is no such *next* Test Step defined, a `default` case of the switch is executed, which causes a Test Case to stop, and to report its verdict. In μTTCN this "stop execution" statement has the form: `estop` (not to be confused with `cstop`, our form of `testcase.stop`, which is associated with the `error` verdict). One of the tasks of the explicitly called `snapshot` system function is to update the "program counter". In TTCN, snapshot is a conceptual,

```
{...}               // TTCN-related definitions (system-level functions, etc.)
void setup(){...} // Arduino part; various TTCN initializations go here
void loop(){        // Arduino part; loop will execute cyclically

snapshot();         // system-level TTCN functionality explicitly called

switch(tStep){              // an unnamed, default Test Case starts here
  case 1:  alt              // alt implemented as another switch (see fig.3)
              ...; break; // if successful: tStep++, Test Step 2 will be next
              ...; break;
                            // if no alternative successful: tStep unchanged
  case 2:  ...; break;
: case 3:  ...;
           tgoto(2);        // Test Step 2 to be executed in the next loop
           break;
  default: estop(); }}      // end of Test Case; tStep will not further change
```

Test Steps

Fig. 2. Test Case structure

semantic construct, not present at the syntactic level, and its *main* purposes (i.e., "freezing" the conditions for race-free handling of messages and timers, as explained in the sequel) are also realized in our `snapshot` function.

The `case` numbers (i.e., Test Step labels) must start with "1", and must be *consecutive* (or else the system-level "Test Step advancing" mechanism embedded in the `snapshot` function will fail). For readability purposes, however, these numbers might be defined as named constants, such as: `const byte preamble=1`.

Each Test Step (switch case) must dynamically end with a `break` statement, which yields control to system-level procedures. The list of constructs in which this statement is allowed does not include Test Steps, because TTCN-3 knows nothing about Test Steps, so our use of `break` in this context is a "creative projection".

3.2 The `alt` Statement

The most notable test-related feature of TTCN-3, not present in general-purpose programming languages, is the `alt` statement, semantically closely tied to a snapshot: "*when entering an* `alt` *statement, a snapshot is taken*" [1, Part 1]. In TTCN, `alt` is composed of a list (sequence) of branches that are "tried out" in turn, in their syntactic order. Each branch, guarded by an (optional) boolean expression, contains one of the statements that may succeed or fail, due to the configuration of messages awaiting reception, expired timers, or other conditions established (and temporarily "frozen") by a snapshot. If a branch is successful, an (optional) statement block is entered, which in TTCN may contain another `alt`. If no branch executes successfully, a new snapshot is taken and the `alt` is re-evaluated.

The original TTCN-3 version and the μT version of `alt` are shown side-by-side in Fig. 3. In μT, only the *receiving-branches* (with all sorts of receive-like statements), *timeout-branches*, and optional *else-branches* (which are always successful) have been implemented. In particular, we have not implemented the TTCN-3 mechanisms of default handling and altsteps.

```
       guard  port  template
alt{                             switch(alt){
 [g>7] p1.receive(m1)             case 1: if(receive(g>7,p1,m1))
         {...; ...}                         {...; break;}
 [] p1.receive(m2)                case 2: if(receive(p1,m2))
         {...; ...}                         {...; break;}
 [] p2.receive                    case 3: if(receive(p2))
         {...; ...}                         {...; break;}
 [] t1.timeout                    case 4: if(timeout(t1))
         {...; ...}                         {...; break;}
 [else] {...}                     case 5: telse(); ...; break;
}                                }
    a)                               b)
```

Fig. 3. The `alt` statement. (a) TTCN-3 syntax, (b) μT version

In μT, each `alt` statement must be assigned in full to a single Test Step of a Test Case. Apart from `break`, which *must* follow it, no other statements may precede or follow `alt` in a Test Step.

Unlike in the implementation of the program structure using Test Steps, in `alt` we want branches to "fall through": when the current branch is "unsuccessful", the next one should be entered. To obtain this behaviour, we use the `switch` statement controlled by the `alt` system-level constant with the value of 1. The branches are represented by numbered cases. The statement block, which is entered only if the initial statement of the branch executes successfully, must contain a `break` (which causes `alt` to be left). This `break` happens to be a valid TTCN statement, used here in its proper context. When the initial statement of a branch fails (i.e., no expected message could be received, an expected timeout did not occur, etc.), `break` is *not* executed, and the switch "falls through" to the next branch.

In μT, nesting of alts within alts is not allowed. Instead, an `alt` to be "nested" is placed in its own Test Step, and control is transferred to it with a `tgoto(new_step)`. Superficially, this is the most serious limitation of μTTCN from the programmer's point of view, but it just enforces a slightly different programming style, in which more active use is made of Test Step labels. No genuine loss in expression power is incurred. Also observe that multiple nested alts make a test program hard to understand and debug, especially for those who take TTCN programming as an incidental task.

3.3 Message Ports

In order to understand how a μTTCN program deals with signals/messages obtained from a SUT, it must first be explained what kinds of messages there are, how they differ from the "regular" notion of a TTCN message, and how they are detected/enqueued by a tester.

The observed behaviour of a μT SUT is envisaged as a sequence/partial order of events – *state change occurrences.* When a μTG detects an event, it *generates* and enqueues a respective message. A test program will then later receive the enqueued messages, e.g., using the `receive` statements. States of SUT objects are perceived by a μTG as values present at a defined subset of available Arduino pins. Accordingly, three kinds of values (and thus three sorts of events-messages) are distinguished: 1-bit (a boolean value present at a single digital pin), k-bit (a binary number present at a given set of digital pins), and analogue (a value present at a single analogue input pin, which after A/C conversion changes into a binary number). Global port-related structures are shown in Fig. 4, to which we will further refer. There is a fixed number of port descriptors (by default – 16), and a smaller, fixed pool of structures (by default – 4) for keeping multi-bit values. Referring to a port (e.g., `p1`) means pointing to a port descriptor (a row in descriptor array). This descriptor needs to be preloaded by a test designer/programmer (in the initial or `setup` part of a Test Sketch) with relevant information: the type of a port (`pType`), its assigned pin number or numbers (`pPinL`–`pPinH`), an "idle" value that will not be considered when a message is to be

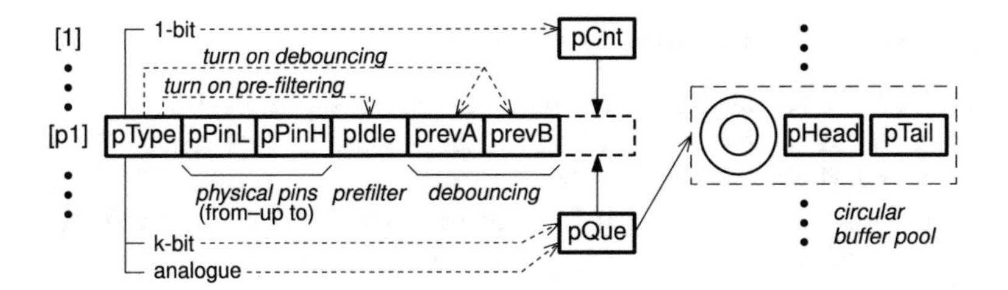

Fig. 4. Static message port structures

generated, the fields for keeping previous state reads (for "debouncing", with which Arduino programmers are plagued), and a structure to enqueue the received messages. In TTCN, each message port has its own input FIFO queue (formally – of infinite length) for messages pending reception. In μT, a FIFO queue for 1-bit ports is *just a counter* (`pCnt`), incremented when a message is enqueued, and decremented by a successful `receive`. By default, the counters are byte-long (a "queue" of length 256). Messages received (i.e., generated) by k-bit and analogue ports are kept in *circular buffers*, pointed to by `pQue`. These buffers store *bytes*, which means that multi-bit values are limited to 8 bits; for a 3-bit value, the remaining 5 bits are "wasted". All buffers have the same maximum length (by default, 16 messages), and their control with `pHead` and `pTail` is particularly frugal. Examples of handling 1-bit and k-bit ports are shown in Fig. 5. Note how "pre-filtering" (if activated for a port) allows an uninteresting, idle value to be "skipped". Also note that there is nothing that would prevent a test programmer from storing "1-bit messages" in circular buffers rather than in counters; only in this way a test program can separately react to events: "state changed to 0" and "state changed to 1" (and not just to "state changed").

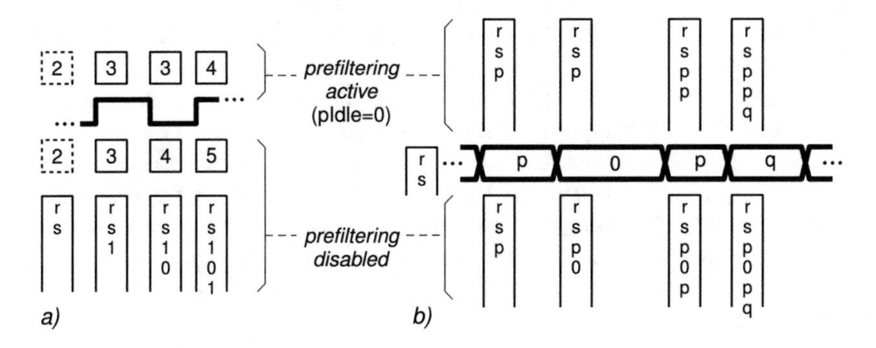

Fig. 5. Enqueueing messages. **a** for 1-bit ports, counters count "message" occurrences (*at the bottom – non-standard use with circular buffers*), **b** for multi-bit ports, buffers store message values

3.4 Receiving and Matching Messages

All three TTCN operations for the handling of messages present in port queues have been implemented in μTTCN: `receive`, `trigger`, and `check`. They differ in details, but their common feature is consulting a message at the head (top) of a port queue, and checking if it matches given conditions. We further deal only with `receive`. In TTCN, this operation "succeeds" if a message is present at the head of a specified port queue, and it matches the criteria expressed in a specified *template* (a special kind of value with "holes" and wildcards, to which a message is compared); the message is then removed from the queue. In case a template is absent, *any* present message is accepted.

In TTCN, "*a template can be thought of as being a set of instructions... to match a received message*" [1, Part 1], and a rich system of matching criteria is provided. It is presupposed that TTCN messages correspond to highly structured, complex protocol messages with fields that may be present or omitted, may have a specific value or a range of values, etc. Accordingly, the provided mechanisms are *unnecessarily rich* for our purposes. We have thus substituted a simplified, more relevant template concept and matching criteria. Note that μT templates are non-trivially used only for multi-bit ports, as for 1-bit ports the only relevant information pertains to the presence of a message: what this message is, is fixed in the definition of the port itself, and the knowledge of what this port "spies upon". Recall that any multi-bit message is represented by a single byte, so any matching operations pertain to these eight bits.

In μT, a system type `constraint` was defined ("template" is used by Arduino for other purposes, but "constraint" was present in TTCN-2 with the same meaning as "template" in TTCN-3) as a structure with three elements: `m` to *mask* out the irrelevant bits, `v` to state the expected *value* of remaining (relevant) bits, and `c` (*complement*) to state whether a message is to be accepted when its contents are consistent, or inconsistent with `v` ("anything but 7"). Templates themselves are declared in a Test Sketch as variables of type `constraint`, and "template" or "tpl" will usually be part of their name. The "set of instructions" for matching is fixed: `(message&m)^v`. The calculated result is treated as a boolean value (encoded as zero/non-zero), *true* indicating the success of operation. This value is returned directly if `c=true`, or logically inverted if `c=false`. A distinguished constant template is defined: `constraint any_message={0,0,false}`, which causes any (multi-bit) message to match. For uniformity, it is also accepted (and does nothing) for single-bit messages. Note, however, that explicitly using this catch-all template is not necessary, as omitting it altogether brings the same effect. The differences between the original receive syntax and its μT version can be seen in Fig. 3.

We have not implemented the *optional* elements of `receive`: port arrays, checking the sender of a message, as well as the possibility of receiving on *any* port – these constructs are mostly irrelevant in the envisaged context of use. We did, however, implement the possibility of manipulating the message that was just received, although in a different way than in standard TTCN: the received value is available in a single global `byte` variable: `received` (until it is overwritten by a consecutive `receive`).

We have also implemented the stand-alone versions of receive-like statements; in fact, they automatically "implemented themselves", due to the way we implemented the "scheduling mechanism" that modifies the `tStep` variable.

3.5 Timers

Timers are used, predominantly, to guard against the infinite execution of an `alt`, in which no message can be received. In μT, there is a fixed pool (array) of global timers (by default: integer values), and using a timer means just pointing to one of those (rather than creating at run-time some new structure). The syntax of timer operations slightly differs from their TTCN version. We have: `tstart(timer7, duration1)` for starting a timer, `tstop(timer7)` for stopping a particular timer or `tstop(allTimer)`(where `allTimer` is just a constant value 0), `x=tread(timer7)`, `trunning(timer7)` or `trunning(anyTimer)`, and finally `timeout(timer7)`, which can be used in an `alt` (see Fig. 3). Timers are handled in the timer part of a `snapshot`: when a given delta time has elapsed since last "timer tick" (which takes many loop executions), each timer is checked, and those running are decremented *by 2*. When a timer reaches the value (-1), it has expired, and a timeout will execute successfully. An idle (stopped) timeout has value 0. To start a timer, an *odd positive* value is written to it; the exact value is calculated by `tstart` basing on requested duration given in milliseconds. This extremely simple scheme is effective, and does everything a timer should do.

3.6 The Control Part

Control over executing multiple test cases was tool-specific in TTCN-2, and the novelty of TTCN-3 was the explicit inclusion of the Control part in a test program (which shows in the name: "...and *Test Control* Notation"). What was described herein so far, pertains to the basic (but fully usable and useful) case, in which a single μT Test Case starts executing automatically, without being "called" by any control statements, and the (optional) Control part is absent altogether. However, a μT Test Case can be packed into a function. Several such functions may be defined, and passed as parameters to the system-level `execute()` function. The Control section of a Test Sketch is implemented using the `switch` construct, similar to a Test Case and `alt`. The execution of a test case (one at a time) can be additionally guarded by the "system" timer, which corresponds to the standard TTCN-3 functionality. If Test Cases are called from the Control part, another (or the same) Test Case may be called depending on the verdict issued (returned) by the previously executed Test Case.

4 Conclusion

We have managed to re-create a useful part of semantics and look-and-feel of TTCN-3 using *unexpectedly simple methods*. The hypothesis that a *huge* testing language can "shrink" a thousand times and remain useful (and surprisingly similar to the original) was confirmed. Also the measured speed of execution of a TTCN-like test program running on a 8-bit microcontroller (from 50000 Test Steps per second, down to 5000 TS/s for complex arrangements of test ports, but insensitive to the complexity of a Test Sketch itself) turned out to be more than satisfactory. The range of TTCN-3 features implemented so far in μTTCN is consistent with the envisaged application context, and does not look overly limited. Further work will focus on streamlining the system-level operations (scheduling, port operations), and turning the common system-level parts of every Test Sketch into an Arduino Library. Note that μT is a *live* project, and particular μTTCN constructs still undergo minor (mostly syntactic) modifications.

The design of distributed ("multi-thing") tester configurations is in the preliminary stage, and as such was not addressed here. The TTCN-3 architecture takes into account the possibility of distributing a test system through the functionality assigned to CH, but the actual ways of exercising this possibility have never been considered obvious. For implementing this feature, the I2C communication facilities of Arduino were chosen (the I2C bus). They were found to operate in a way that is unobtrusive w.r.t. the assumed run-time model – a "side-effect" of the overall design, which intentionally does not rely on manipulating any low-level resources of the microcontroller.

References

1. ES 201 873: MTS: The Testing and Test Control Notation version 3. ETSI (nd)
2. Brzeziński, K.M.: Tiny TTCN-inspired testing tools for experimenting with hybrid IoT systems. In: Bujnowski, A., Kaczmarek, M., Rumiński, J. (eds.) 11th Conference on Human System Interaction (HSI 2018), pp. 261–267 (2018, in print)
3. Arduino: (nd), http://arduino.cc/
4. Brzeziński, K.M.: Towards the methodological harmonization of passive testing across ICT communities. In: Soomro, S. (ed.) Engineering the Computer Science and IT, chap. 9, pp. 143–168. In-Tech (2009). https://doi.org/10.5772/7764
5. Michel, R. (ed.): Design Research Now: Essays and Selected Projects. Birkhäuser (2007). https://doi.org/10.1007/978-3-7643-8472-2
6. Brzeziński, K.M., Mastalerz, D., Artych, R.: Practical support of testing activities: the PMM family. COST 247 WG3 report, LTiV technical report 965. Institute of Telecommunications, Warsaw University of Technology (IT P.W.) (1996)
7. Böhm, T., Csöndes, T., Réthy, T.G., Wu-Hen-Chang, A.: A flexible, multi-purpose open source test platform for IoT testing. In: 5th User Conference on Advanced Automated Testing (UCAAT 2017) (2017)
8. Sabiguero, A., Baire, A., Viho, C.: Automatic CoDec generation to reduce test engineering cost. Int. J. Softw. Tools Technol. Transf. **10**, 337–346 (2008). https://doi.org/10.1007/s10009-008-0073-2
9. Deltour, J., Gaudin, E.: TTCN-3 snapshot semantics (nd). http://www.pragmadev.com/downloads/TTCN3SnapshotImplementation.pdf
10. ISO/IEC 9646: Conformance Testing Methodology and Framework, vol. 1–7 (nd)

Industrial Internet of Things Solution for Real-Time Monitoring of the Additive Manufacturing Process

Mahmoud Salama[1], Ahmed Elkaseer[2,3(✉)], Mohamed Saied[1], Hazem Ali[4], and Steffen Scholz[2]

[1] Faculty of Engineering, Ain Shams University, Cairo, Egypt
[2] Karlsruhe Institute of Technology, Karlsruhe, Germany
ahmed.elkaseer@kit.edu
[3] Faculty of Engineering, Port Said University, Port Said, Egypt
[4] European X-Ray Free-Electron Laser Facility GmbH, Schenefeld, Germany

Abstract. Modern cyber manufacturing has been introduced into a broad range of manufacturing processes to ease their digital reconfigurability and enhance flexibility while retaining a high throughput of quality products. Such a system provides real-time data acquisition, enabling monitoring of the actual condition of the manufacturing process. The Industrial Internet of Things (IIoT) facilitates such real-time monitoring and optimization of the fabricating system, which reduces time necessary for maintenance with the possibility of almost instantaneously taking any necessary corrective measures with respect to either human to the machine/process from learned algorithms. In this research, an original IIoT approach has been proposed to monitor the process conditions, including nozzle temperature and filament breakage/runout, of the additive manufacturing process. In particular, concurrent multi-task IIoT communication was developed for a network of five nodes. This was implemented to ensure real time monitoring of the manufacturing process via multi sensors and empowered by an embedded software design. The proposed embedded software architecture offers a reliable solution to eliminate communication latency and provides real-time response to acquired information. It is worth emphasizing that the embedded software was designed so that it optimally exploits the very great potential of the hardware resources, with the ability to detect run-time issues in the nodes' performance and re-try to address such issues to maintain a high capability networking communication. The designed architecture also offers auto-scaling throughput of the data transferred to the cloud to minimize the bandwidth.

Keywords: IIoT · M2M · Industrial informatics · Industry 4.0 · AM

1 Introduction

The advent and subsequent explosive growth of advanced information- and communication-driven technologies have promoted digitalization in manufacturing industries [1], aiming at realizing the potential of cross-linking intelligent manufacturing operations with automated near-real-time data acquisition and simulation technologies [2]. This enables the so-called cyber/physical modelling of manufacturing processes, which is

L. Borzemski et al. (Eds.): ISAT 2018, AISC 852, pp. 355–365, 2019.
https://doi.org/10.1007/978-3-319-99981-4_33

considered the key enabling technology for the transformation to the next industrial revolution, Industry 4.0 [3]. The introduction of such smart cyber/physical modelling into manufacturing technologies includes integrating physical systems with cloud/edge computing, machine/deep learning algorithms, internet of things (IoT) and process simulation techniques, where cyber and physical worlds liaise profitably [4]. This approach offers fast and accurate prediction of manufacturing process outcomes, provided there is a comprehensive and seamless linkage between the physical and digital worlds. In particular, there is the possibility of instantaneous acquisition of physical problems with the almost concurrent production of necessary corrective actions, which it is expected will ultimately help optimize the performance of entire manufacturing systems [5]. However, among the aforementioned fragments of Industry 4.0, IoT is a central part that facilitates the interconnection between the intelligent manufacturing assets in a timely and effective manner. In particular, IoT, particularly Industrial IoT (IIoT) allows the manufacturing system, sub-systems, raw materials, and products to be in contentious communication sessions from the supply-chain to the final manufacturing stage. It allows for flexible and customizable production, which increases the manufacturing productivity, and flexibility, which positively reflects on the economic aspects [6].

The IIoT paradigm is presently based on two technologies: radio frequency identification (RFID) and wireless sensor networks (WSN) [7]. Whereas RFID is a microchip-based technology that interfaces with an antenna and presents identification through radio waves to define the objects, WSN is a group of interconnected intelligent sensors, which monitor a certain object for a predefined purpose.

The implementation of IIoT technology entails a series of design layers. The first is the data acquisition layer that consists of groups of intelligent sensors' nodes responsible for sensing the signals and processing these signals to collect the required data. These sensors are designed to perform the data acquisition techniques in different domains such as invasive or noninvasive techniques. The second IIoT layer is the network layer, which allows the nodes to interact/communicate with each other to exchange the acquired information. Furthermore, the network layer works to concatenate data from other networks and to establish connections between all heterogeneous systems through the internet. Thirdly, the service layer is defined as the middleware IoT technology that integrates the application and the service [8], to provide application requirements such as APIs, and to implement the control process for information storage and data management.

The design of IoT solution is facing many challenges related to the feasibility of the proposed solution to facilitate effective and seamless mass communication among the manufacturing assets, to minimize the independency of nodes to each other, which would allow for a stable solution under different working conditions [9], and to manage the throughput capacity of the transferred data.

In this context, the aim of this paper is to develop an original Industrial Internet of Things Solution to model the physical-cyber world of the additive manufacturing (AM) process to real-time monitoring of AM fabricating conditions. For demonstration purposes, the conditions of the AM process to be monitored are the temperatures of the nozzle and filament breakage. Besides, human-machine interactions have been developed based on local and remote techniques that offer monitoring of the additive manufacturing processes directly by the operator.

2 Proposed IIoT Solution Architecture

2.1 Physical Layer

The proposed IIoT network structure presents a real-time solution for monitoring and interfacing with the additive manufacturing process. The design is empowered by a convenient communication algorithm based on a wireless approach. The network structure includes sensing nodes, a HMI node, a server mirror node and a control unit node.

Sensing Nodes (two sensors): responsible for sensing, acquiring, and processing the sensing parameters of the fabrication process.

HMI Node: produces a user-friendly interface and enables the user to investigate parameters of the fabrication process to tune the actions of the fabricating conditions that might be required [10].

Server Mirror Node: facilitates acquisition of all sensed information and sends it to the cloud; it also operates as a backup for the cloud computing in case of issues with the internet connection.

Control Unit (IoT Broker): is the main handler of the entire modules, and manages mainly the information exchanges between the modules, checks the robustness of modules performance, investigates any run time issues and tries to solve them.

2.2 Communication Layer

The objective of the communication module is to manage the exchange of information through the nodes channel, and between the node network and the application network. The communication software algorithm of the broker is designed to manage multiple nodes in the network. In addition, the proposed software works to collect, split, and send the information from/to the other solution nodes. The communication between the broker and the other nodes is carried out via a convenient software algorithm to enable high-speed communication response.

2.3 Application Layer

The final layer of IIoT design is the application tier. The fabrication process state has been monetarized based on real-time HMI (Human-Machine Interaction) using an Android-based application. The Android-based application monitors and manages the AM process by using real-time commination with the broker. Also, cloud computing has been used in this design to store, manage and present the information of the AM. Cloud computing has been established based on the connection with the server mirror node, which accesses the internet, formulizes the data and then sends it to the cloud. However, if there is a problem with the internet connection, the node will automatically save the information on the local storage and send it back when the connection is reinstated.

3 Implementation of IIoT Solution

The designed IIoT solution is based on five nodes: two sensor nodes, a server mirror node, an IoT Broker node and a HMI node. The implementation of the proposed idea is as follows: an Android-based application is utilized for the HMI, and other nodes have been designed and developed based on embedded system architecture. The wireless communication between the nodes has been carried out using IEEE 802.11 communication protocol. IoT MathWorks application (ThingSpeak [11]) has been utilized as Cloud computing for the solution, as shown in Fig. 1.

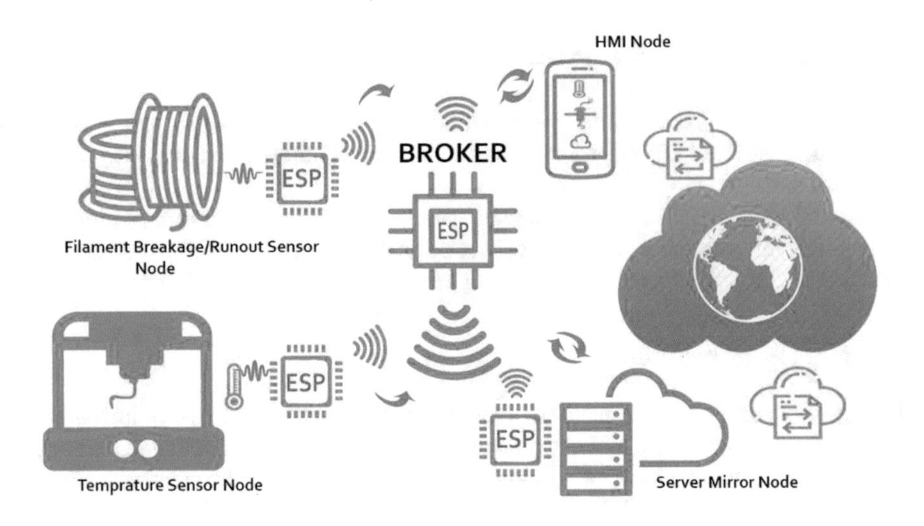

Fig. 1. Proposed design of the IIoT solution

The embedded software has been designed based on RTOS (Real-Time Operating System), due to its ability to handle multiple tasks simultaneously. RTOS categorizes the functions of the software into tasks and implements scheduling during the execution process based on demand [12]. Furthermore, RTOS facilitates the embedded software to exploit the high potential of the hardware resources. Thus, it makes all resources active concurrently which entirely eliminates the latency in the system. FreeRTOS and C Programming Language have been utilized to develop the embedded software of the nodes based on the ESP32 module, which includes Dual Core xTensa Microcontroller with a large number of peripherals, Wi-Fi and BLE (Bluetooth Low Energy).

The first sensor node has been designed to measure the temperature of the additive manufacturing nozzle. The design utilized an ESP32 module and a thermocouple K type temperature sensor. This thermocouple sensor measures the temperature up to +1024 °C. The interface between the ESP32 and the Thermocouple is performed by SPI (Serial Peripheral Interface) communication protocols through a MAX6675 Module. MAX6675 Module digitizes the analog signal from a type-K thermocouple to 12-bit resolution via internal ADC with accuracy 0.25 °C then pass it to SPI interface. The embedded software design has used the ESP32 interface with Max6675 thermocouple through SPI bus. Besides, the design allows for a communication session with the broker using the IEEE 802.11 protocol. The multiprocessing is utilized to perform two functions to avoid task blocking and communication latency since communication has been implemented using socket blocking API's. The first task of the multiprocessing is to interface with Max6675 via the SPI which is pinned to core 0 of the ESP32. The second task, a client task, is to connect to the broker and its pinned to core 1 of ESP32. The described embedded software design is shown in Fig. 2.

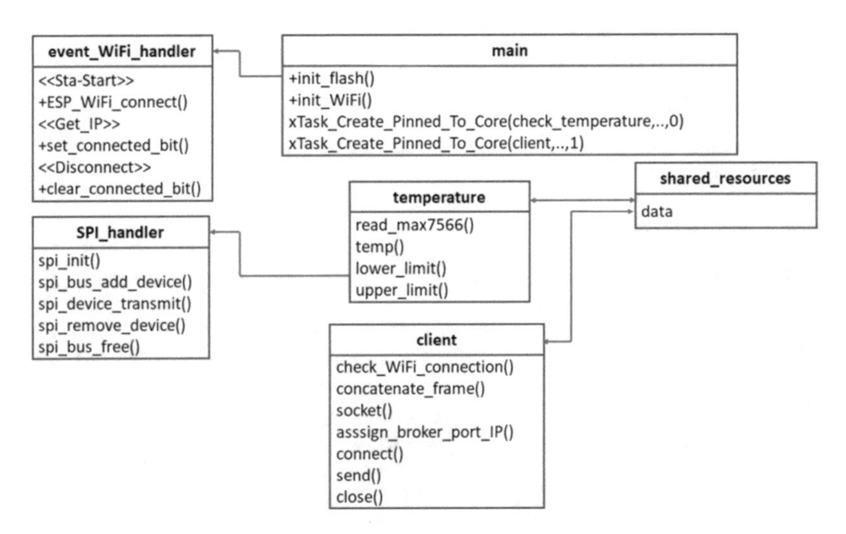

Fig. 2. Embedded software architecture of the temperature sensor node

The second sensor node is the filament breakage detection sensor node. This sensor has been devoted to detect the breakage of the 3D printer filament. In this proposed design, the sensor has been replaced with a switch to perform the same function as the sensor prototype. The embedded software is interfaced to the ESP32 with filament breakage sensor via an I/O pin. In addition, the communication between the node and the broker is performed via the IEEE 802.11 protocol. The multiprocessing is utilized to perform these node functions. Thus, core 0 of the ESP32 will perform the breakage sensor function and core 1 of the ESP32 will perform the client task to communicate with the broker, as presented in embedded software design shown in Fig. 3.

The software architecture for the broker, the main node, has been designed to perform four tasks concurrently, namely three server tasks and one client task. The software architecture is designed to handle, collect and split information via socket communication with all nodes (temperature sensor node, filament breakage sensor node, server mirror node and android-based application). Multiprocessing and multithreading has been utilized in the embedded software design for the broker. Every two tasks are pinned to one of the cores of the ESP32. In addition, every core has multi-tasks as threads. The broker software design diagram is presented in Fig. 4.

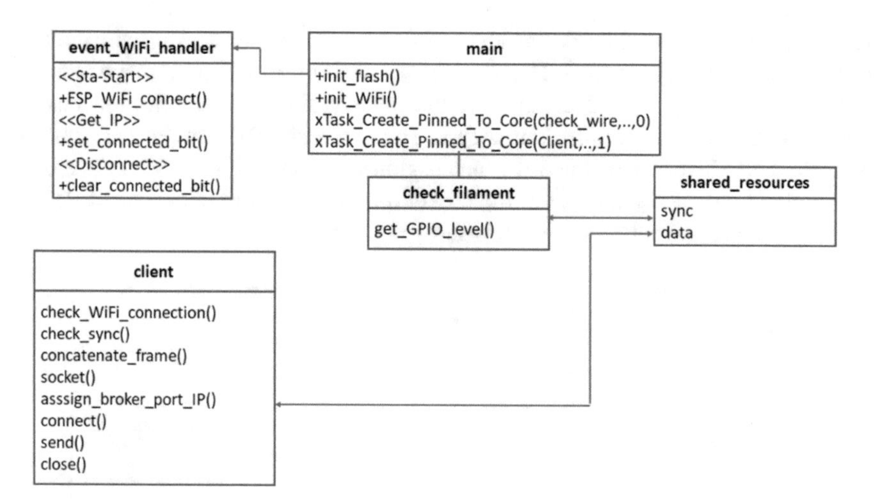

Fig. 3. Software architecture of filament breakage/runout detection sensor node

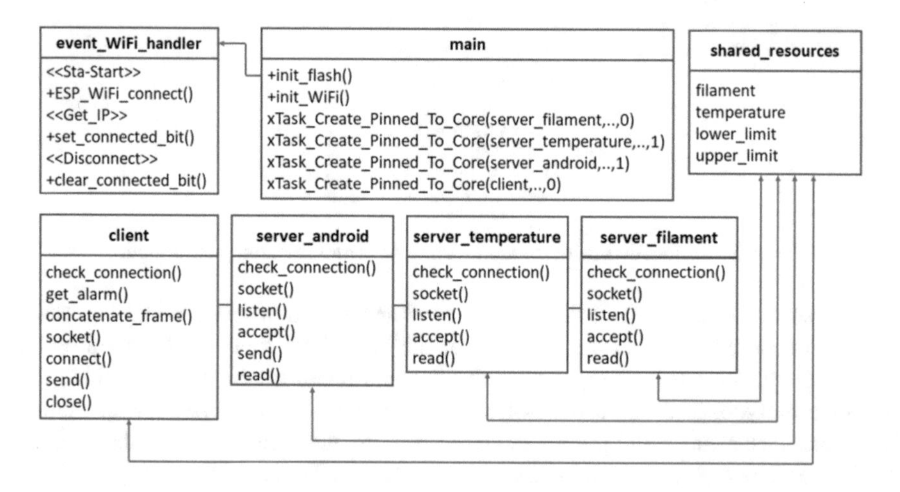

Fig. 4. Software architecture of broker node

The server mirror node has been designed by ESP32 with a SD card interface via a SPI bus. The power technique is battery-supported backup to acquire, save and upload the data whether the electricity on the machine is on or off. The embedded software design of the server mirror node checks the DHCP (Dynamic Host Configuration Protocol) connection of the internet. As long as there is a functioning internet connection, all fabrication process data will be sent to the cloud otherwise during a disconnection situation, the data will be stored locally on the SD card as an offline log. When the connection is reestablished, the server mirror node will upload the offline log data to the cloud. Figure 5 presents the finite state machine of the server mirror task.

The Android-based application is designed to set the limits of the nozzle temperature and to detect the filament breakage, as shown in Fig. 6.

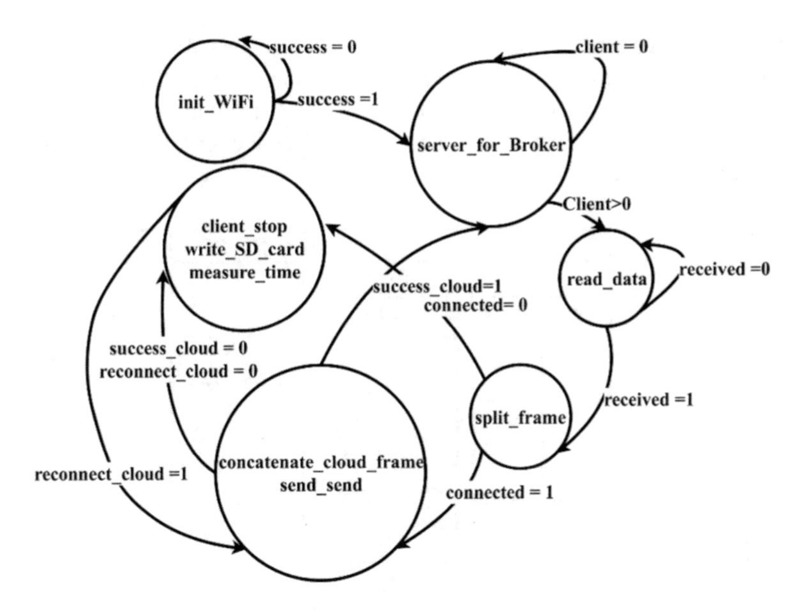

Fig. 5. Finite state machine of the server mirror node

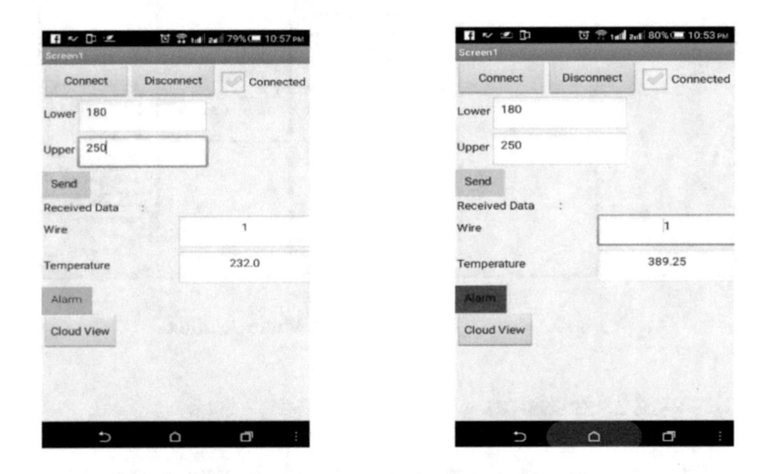

Fig. 6. GUI of Android-based application

The proposed dynamic structure of the nodes' embedded software presents as an activity diagram which shown in Fig. 7. The working scenario is performed as follows: the broker is defined as the main controller; it is concatenating the sensing data from the sensor nodes into the frame and sends it to the cloud via the server mirror node. In parallel, the data is also sent the android-based application to acquire thermal limit values. The inspector (user) will set the level of temperature that is suitable for the selected AM process via the Android-based application. The broker checks the sensed temperature and compares it to the original setting to identify potential deviations. In case of significant deviations, the broker will add an alarm signal to the concatenated frame of data. Likewise, an alarm signal will be added if the filament breakage sensor reports a signal read. In case of disconnection to the internet, the uploading of the collected data is impaired and the node will decide upon the storage place for the data (specific local memory storage) and send the data to the cloud once the connection is reestablished. The broker has been empowered by battery backup as the server mirror node.

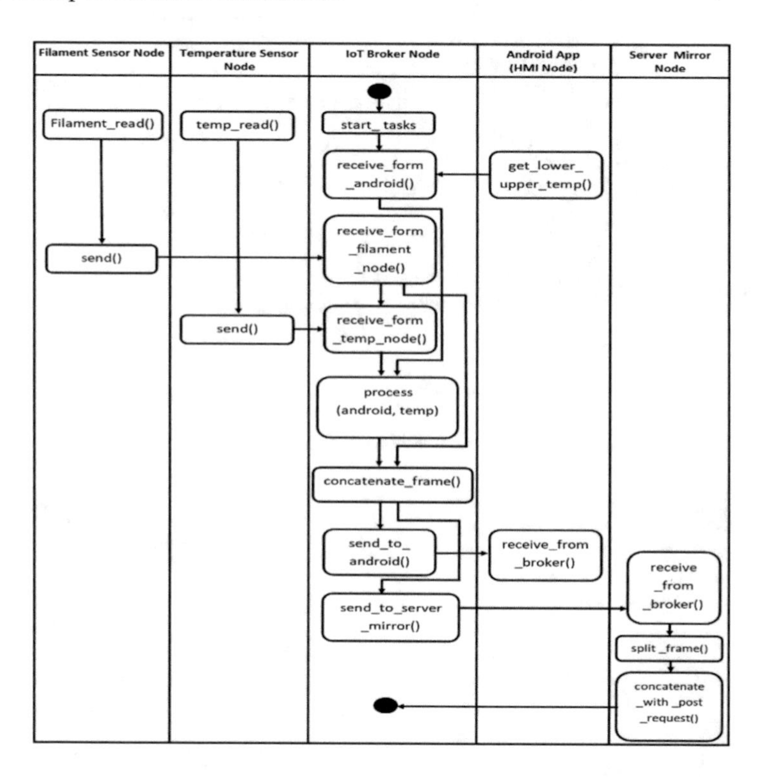

Fig. 7. Activity diagram of the whole solution

4 Results and Discussion

Figure 8(a–d) presents the data visualization on the cloud, where the results are plotted verses time with a time interval of 2 s to check the nozzle temperature (Fig. 8a) and the condition of the filament (Fig. 8b), where the alarm chart (Fig. 8c) presents the response in case of issues detected in the monitored processing conditions. If there is a breakage in the filament or the temperature reading is not in the range displayed on the android-based application, this will be considered an issue. For instance, in our example the temperature limits were set between 180–250 °C. Figure 8a shows that the temperature varied between 115 and 128°, which is below the acceptable range. This was recognized as an issue therefore triggering the alarm as displayed in Fig. 8c. Wire breakage was also detected (Fig. 8b), which indicates the feasibility of the monitoring system to accurately detect the conditions of the fabricating process. Figure 8d shows the data transfer throughput clearly presents the ability of the system to implement auto-scaling based on the uploaded information, which helps to optimize the bandwidth.

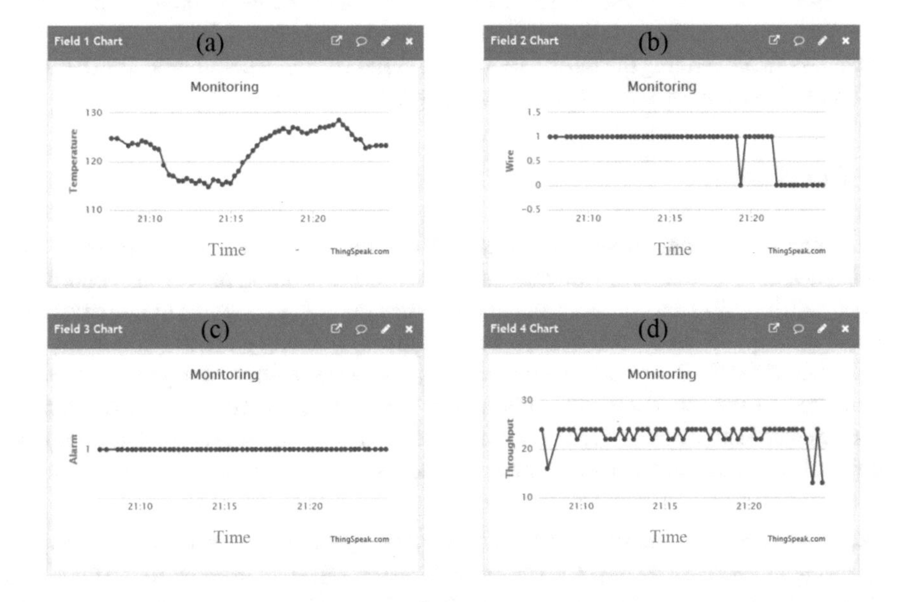

Fig. 8. Data visualization on the cloud

To demonstrate the ability of the software solution to handle unexpected behavior that could result in a communication issue between the nodes, the Android application was disabled on purpose as shown in Fig. 9. The results verify the stability of the system, as it continuously performed its functions and once the Android-based application was activated again, a new communication session was established to exchange information with the Android-based application nodes.

```
Received Data  From Wire=W,1  size: 3

Received Data  From Temperature=T,208.75  size: 8
Connected
□[0;32mI (162402) Main_Controller: ... connected□[0m
Error Send
□[0;31mE (162402) Main_Controller: ... socket send failed□[0m

Received Data  From Wire=W,1  size: 3

Received Data  From Temperature=T,207.0  size: 7
lower limit180,upper limit250in client task temeprature=207
Send to Mirror=M,1,207.0,0,22
□[0;32mI (164502) Main_Controller: ... allocated socket
□[0m
```

Fig. 9. Self-correction error handling and information exchange by the broker

As previously discussed, the proposed activity scenario between the Android-based application, the other nodes and the IoT broker was detected via the serial monitoring of the broker, as shown in Fig. 8. The execution time of the entire working solution from the sensing phase until the data delivery to the cloud was 2000 ms, 700 ms for the server mirror task while sending the information to the cloud and other communication between the nodes, data acquisition and data processing took 1300 ms.

5 Conclusion

In this research, a concurrent multi-task IIoT communication solution was developed aiming at real time monitoring of various additive manufacturing processing conditions, in our case the nozzle temperature and filament breakage. The network structure includes several sensing nodes, a HMI node, a server mirror node and a control unit node with the design of the IIoT network being enriched by an auto-scaling throughput of the data transferred to the cloud thereby minimizing the bandwidth. However, a convenient communication algorithm based on a wireless approach is utilized to eliminate communication latency and enables the system to handle real time communication issues.

Acknowledgment. The authors would like to thank the European Commission for funding the PAM^2 project under H2020-MSCA-ITN-2016 Program, Grant Agreement No 721383. The authors gratefully acknowledge the technical help provided by Eng. Saeed Mohsen, PhD student at ASU, Egypt.

References

1. Yeo, N.C.Y., et al.: Revolutionizing technology adoption for the remanufacturing industry. Procedia CIRP **61**, 17–21 (2017)
2. Uhlemann, T.H.-J., et al.: The Digital Twin: demonstrating the potential of real time data acquisition in production systems. Procedia Manuf. **9**, 113–120 (2017)
3. Lee, J., et al.: A Cyber-Physical Systems architecture for Industry 4.0-based manufacturing systems. Manuf. Lett. **3**, 18–23 (2015)
4. Dilberoglu, U.M., et al.: The role of additive manufacturing in the era of Industry 4.0. In: 27th FAIM, Procedia Manufacturing, vol. 11, pp. 545–554, 27–30 June 2017
5. Kang, H.S., et al.: Smart manufacturing: past research, present findings, and future directions. Int. J. Precis. Eng. Manuf. Green Tech. **3**, 111–128 (2016)
6. Wu, Z., et al.: IoT-Based techniques for online M2M-Interactive itemized data registration and offline information traceability in a digital manufacturing system. IEEE Trans. Ind. Inf. **13**(5), 2397–2405 (2017)
7. Xu, L.D., et al.: Internet of Things in industries: a survey. IEEE Trans. Industr. Inf. **10**(4), 2233–2243 (2014)
8. Atzori, L., et al.: The Internet of Things: a survey. Comput. Netw. **54**(15), 2787–2805 (2010)
9. Miorandi, D., et al.: Internet of things: vision, applications and research challenges. Ad Hoc Netw. **10**(7), 1497–1516 (2012)

10. Elkaseer, A., et al.: Approaches to a practical implementation of Industry 4.0. In: IARIA, ACHI, pp. 141–146 (2018)
11. MathWorks Cloud. https://thingspeak.com/. Accessed 25 May 2018
12. Silberschatz, A., et al: Operating System Concepts. Wiley (2006). ISBN 0470088486

IoT-Based Surveillance for Instant Marketing in Real Stores

Tomasz Chojnacki(✉) and Jarogniew Rykowski

Department of Information Technology, Poznań University of Economics and Business, Niepodległości 10, 61-875 Poznań, Poland
{chojnacki,rykowski}@kti.ue.poznan.pl

Abstract. The recent boom of e-shops and network-based advertising tools clearly showed the power of new technologies and solutions in the area of trade and marketing. However, so far little attention has been paid to apply e-* systems to real stores and marketplaces. This situation was mainly caused by the lack of efficient tracking tools, equivalent to network-based log analysis, server- and client-side cookies, timing registration and examination, etc. Thus, an efficient implementation of, e.g. a recommendation system or instant advertisement in a real store was quite problematic.

In this paper, we propose to apply Internet of Things devices and services to fill the gap and to get additional information about the behavior and the activities of the customers in a real store. Such information would be used as a basic input for different marketing tools, including above-mentioned recommendation systems, instant advertisement and promotion, JIT discounts, etc. We assume two-level architecture of the system: local networks of IoT devices and necessary hubs, and a shared marketing server, to implement several marketing utilities and strategies. The paper is devoted to the organization of the first level, namely a local multiplexer and virtualizer of IoT devices, including not only simple HVAC sensors, but also RFID/NFC/QR/BLE localizers, surveillance cameras with intelligent image-processing extensions, as well as a protocol to share the information obtained from the local sensors with the cloud.

Keywords: Internet of Things · Internet of services · Instant marketing
Device virtualization

1 Introduction

Along with the growing popularity of e-shopping, we observe a boom of new marketing tools, to mention recommendation systems, instant shopping after clicking an advertisement, automatic discounts and price reductions, and many others. These new utilities opened several new possibilities and marketing opportunities, important not only from the point of view of the store operators but also simply very convenient and useful for the customers. Boring and time-consuming classical advertisement was successfully replaced by the shopping assistance, "intelligent" enough to advice the products and services we are looking for at the moment. Similarly, human-based market research was completed by the automatic analysis of customer activity while searching Web pages

L. Borzemski et al. (Eds.): ISAT 2018, AISC 852, pp. 366–375, 2019.
https://doi.org/10.1007/978-3-319-99981-4_34

and using Web services – to mention here the server-log analysis, the cookies, timing inspection, shopping cart review, and many more.

However, so far almost none of the above-presented solutions was successfully applied to the real shops and marketplaces. There are several reasons for this situation. First, it is pretty hard to track a customer in a real shop efficiently. While for an e-shop it is enough to notify the links/pages visited, in a reality one has to monitor the position and the activities of a given person. On the contrary, it is not sufficient just to detect a person at given location, as the customers may pass by this place, and all the items in the neighborhood are of no interest for them. Instead, we should take into consideration such features as line-of-sight of the customer, timings for the observation and while making a choice, temporary and final decisions – all such information is readily detected from a server log in the virtual world, however hardly collected with a physical person and place. Second, efficient tracking of customer activity at a real site needs substantial efforts and resources, such as surveillance cameras with "intelligent" image analysis, radio-based localizers and identifiers (such as BLE beacons to mark the places, and RFID tags to identify the shelves and the items). So far such complex systems were not massively used in the real stores. And last but not least, the interaction between the customer and the store is limited, as traditionally realized by the human personnel. The productivity of a single staff member cannot be compared with the efficiency of a Web server, using multimedia for communication and automatically adjusting to the needs and skills of the customer. Even if human-based interaction is still preferred by some customers, in general, the computer-based contact is much more efficient, especially for the new generation of the customers relying on massive usage of new technologies such as a smartphone.

In this paper, we propose to improve the situation by applying a set of Internet-of-Things (IoT) devices and services at a real store to automatically monitor customer activity. Although IoT is considered very promising for many areas, surprisingly little of research is devoted for its application in marketing [1]. More precisely, the paper is devoted to a discussion which devices and services should be used to achieve a functionality similar to the one we observe for e-shopping. The main goal is to propose an efficient architecture of a distributed IoT system for (1) collecting the information about the customers and their activities from the sensors of different nature and purpose, (2) standardization of the information to be maintained in a centralized manner, and (3) efficient at-the-place, real-time interaction with the customers. To facilitate the process of analysis of the information collected by the IoT devices, we propose to use so-called virtual devices and a local hub capable of virtualization of real, both at-the-place and remote sensors. The set of virtual devices composes an ontology, which together with a protocol for the information exchange, makes it possible to export the information locally gathered in a store to the cloud, to compute the marketing recommendations.

The remainder of the paper is organized as follows. In Sect. 2 we describe basic system functionality and architecture, comparing our proposal with the solutions already proposed in the literature. In Sect. 3, we provide a justification of using virtual sensors, and we describe a basic set of such sensors to be applied for instant marketing purposes. In Sect. 4, a typical scenario of using the system is described. The paper is concluded with Sect. 5, containing also some indications for the future work.

2 Basic System Functionality and Architecture

After an inspection of a typical e-shopping system, we may mention certain essential functions, these are briefly enumerated and discussed below, comparing them with their counterparts from the real life.

Server Log Including Timings. This information, collected as a result of navigation through the set of server pages/links, is stored for each user activity, usually in the form of URL address and detailed system time. At the first view, an information from the server log for a single visit of given user is equivalent to the monitoring data about a person in a real store – URLs are counterparts of the shelves and other store locations, while server timings – the moments of bypassing the given shelf/place. Note, however, while analyzing this equivalence in more details, one detects a problem of precise identification of a shelf/place. As for the server, all the Web pages are precisely identified by their URLs. Nevertheless, as most of the stores is organized in such way there are some passages among the shelves, a single location usually points to at least two shelves. Moreover, the shelves are usually organized in a vertical manner, with the items of a different sort at a different height. So, the problem of identification of the real shelf is much more complicated in comparison with its e-counterpart. Observing the customers, we came to the conclusion that what is important in the real store is the observation of the stops rather than passages of the customers by the shelves. Thus, we should be able to detect the situation a customer is standing by a shelf for some time (at least counting in several seconds), and possibly interacting with the items – getting them from the shelf, unpacking, testing, etc.

Shopping Cart. The items chosen by a customer are usually located in a shopping cart – a real one, such as in a supermarket, and a virtual one for e-shops. As for the latter, it is trivial to count and assess the items. However, the real shopping carts are not equipped with an "intelligence" to enumerate the items. Anyway, such functionality would be very desired in order to track the customer choice. There were some proposals to include a display at the top of a cart to enumerate all the items and to count the total price for the contents of the cart [2]. Such carts were tested by several market leaders, however, never gaining significant attention, for several reasons. First, due to technology restrictions, it is hard to detect the contents of a cart automatically – either a customer is forced to scan a barcode/RFID tag manually, or some tags are omitted because of screening of radio signals or line-of-sight (LOS) restriction [3]. Second, as the customer was aware of the price at any moment, frequently they decided to remove too expensive items from the cart. Also, the accumulators for the displays to operate were heavy and needed frequent recharging. As a consequence, in general, the "intelligent" shopping carts are not massively used.

Cookies. The little files called cookies are frequently used to mark the users, their activities, choices, preferences, etc. Although this mechanism is usually treated as a potential violation of privacy, it is massively used for e-shops and other Web pages. There is no direct counterpart of cookies in real life – except maybe for such activities as pieces of paper with handwritten notes (frequently used "lists to buy/do"). Anyway, such functionality would also be very desired, especially from the customer point of view, guiding them from one shelf to another to collect the items.

Recommendation Systems and Shared Opinions. A customer of an e-shop, while choosing or hesitating to select an item, is frequently informed that "people who liked this item also bought X, Y, and Z". Such recommendations are based on statistical analysis of the history of shopping of the previous customers [4]. Again, a recommendation system, quite trivial as looking for its implementation is an e-store, is very difficult to achieve in the reality, for several reasons. First, as already mentioned, the store has no information about customer's choice at the moment. Second, the statistical data is also hardly available – usually, the only possibility is to analyze the final transactions of the customers. However, only partial information about their choice is available, in particular, the moments of deciding on buying a given item cannot be unfolded. One may expect that such recommendations would be of much worst quality. So, it would be desirable for the real store to detect the moments of taking the choice rather than paying for everything.

Note also that the opinions and comments of the previous customers, commonly used for e-shopping, are hardly applied at real-time for real stores, for the following reasons. Firstly, a customer cannot easily identify the just-inspected item – to this goal, they should either scan the barcode (which is restricted by law) or describe the item by its name/code etc., to get some shared information. Second, this activity needs a manual interaction by means of the personal mobile device – connecting to the network, playing with keyboard/touch screen/camera etc., and navigating through the list of opinions. What would be much better from the point of view of a customer would be an instant opinion "worth buying"/"give it up", automatically achieved by means of the personal device – as a voice message, color (red/green/orange), etc. Such notification could be also used for other purposes, such as warnings about potential medical problems related to this customer and this item.

Instant Marketing and Promotion. Human personnel, observing the clients in the real shop, may come and offer "special conditions", such as a discount, promotion, etc. In an e-shop, this functionality is frequently applied in the form of coupons of different sort and origin. However, even partial automatization of this process for the real place seems to be a hard task. First, the personnel is usually restricted in number and timings, thus only a limited number of the customers may profit from such assistance. Second, the decisions for offering, e.g. a discount are based on intuition or needs a long-term inspection of the behavior of the customer. It would be desirable for the store to provide such assistance by a computer system, based on real rather than imagined information.

As it may be drawn from the above-depicted description of system functionality, the automatization of at least a part of these functions in a real store requires new hardware and software. On the contrary to the real-store technologies, e-shopping techniques are based on online analysis and instant recommendations. However, they are built on the information that is not available for automatic treatment at a real place, such as timings and presence of given person at given place and time, careful inspection of the items before shopping, returning to the place/item after some time, choosing a product from a group of similar ones, hesitating before final decision, etc. The question is if we really cannot get such information in real time and place, to shift the profits of e-shopping analysis to the area of real stores and marketplaces. We think that gathering such information is now possible with the recent advances in the field of Internet of Things and Services [5]. Below a discussion is provided which IoT devices and services may be applied to achieve the desired system functionality.

We assume that we extensively apply the BYOD ("bring your own device") idea, with an application on a personal mobile device, playing the role of individual guide to the store, recommendation & opinion source, and messaging center. We count on the fact the smartphones are now massively used, thus treating this device as a primary interaction broker between a customer and a store would be natural for most of the people. We also think that the smartphone would play a role of an advisor rather than the surveillance inspector, providing real value to its owner. Below, we enumerate the smartphone/application functionality as well as a set of IoT devices and services to implement essential system functions.

Activity Log and Timings. User activity and location may be tracked in several ways. Obviously, however, the most expensive solution is to use several cameras and "intelligent" analysis of the image towards a number of people in the room, their locations, movement tracking, line-of-sight and eye-tracking, differentiating tall (possibly adults) and smaller (possible children) person, etc. Recently, efficient extensions to popular OpenCV library have been published capable of sophisticated image analysis and the tasks enumerated above. As the price of 3K and 4K IP cameras is now reasonable, camera-based inspection becomes an interesting option even for small stores.

Alternative technologies for detecting the presence and movements of people at given locations are mainly based on Bluetooth Low Energy (BLE) beacons. A BLE beacon is a small battery-powered device capable of radio-broadcasting its identifier. If one is in the radio-range of such device, it is possible, based on the received identifier and a map/database, to identify the site with the approximation of tens of centimeters. Applying a smartphone as a reader of BLE beacons, one may thus know the detailed place inside a store. Via registering on the local server the location each time a new beacon is observed, one may representant all the customer activities. BLE beacons may be also supplemented by traditional identification by means of RFID/NFC tags and QR codes, scanned at-the-place by the application and further broadcasted to the server.

Virtual Shopping Cart. To facilitate shopping, a customer may be asked to manually scan the barcodes printed on the items. Semi-automated detection of RFID/NFC tags may also be applied here. All the scanned identifiers may compose an individual, virtual shopping cart. Note that this functionality may only be partially automatized – even if it is possible to detect the radio tags of all the items carried by a customer for a longer time, most of the items in the stores are not marked with such tags, thus scanning the barcode is the only possibility to get some information about an item. However, code scanning and registering may be connected with inspecting users' opinions, obtaining a discount (equivalent to the "loyalty card" distributed by supermarkets), early-bird and on-line payments, etc. It may be expected that even manual scanning would be accepted by the clients.

Personal Recommendations and Requirements. Personal information, being a cookie counterpart, may contain individual data about the smartphone owner, such as clothes size, shoe number, favorite color, etc. This information may be used to search for items or to formulate queries to the personnel.

Local Context. Usually not taken into consideration, the local context, according to the first level of the Maslov's pyramid of needs, is very important for everyday life, including the visits in the stores. Several IoT devices may be depicted here dealing with the measuring of HVAC (heating, ventilation, air-conditioning) parameters. All these devices compose a local context, at least potentially influencing the shopping process. Complemented by noise, humidity, lightness and similar measurables, HVAC sensors provide important information about the background of the shopping.

From the implementation point of view, the system is composed of several layers (Fig. 1). Local sensors and cameras are connected to a multiplexer/hub. The hub hardware is based on Raspberry Pi (version 3), and physical connections among the devices and the hub depend on the device requirements. Basically, any of the commonly used connections and standards apply here, including 1-wire, I2C, SPI, and more. The connection details are not visible outside the hub. Instead, all the information coming from the sensors is standardized according to the specification of an ontology of virtual devices, and as such sent to the cloud to be processed at server-side.

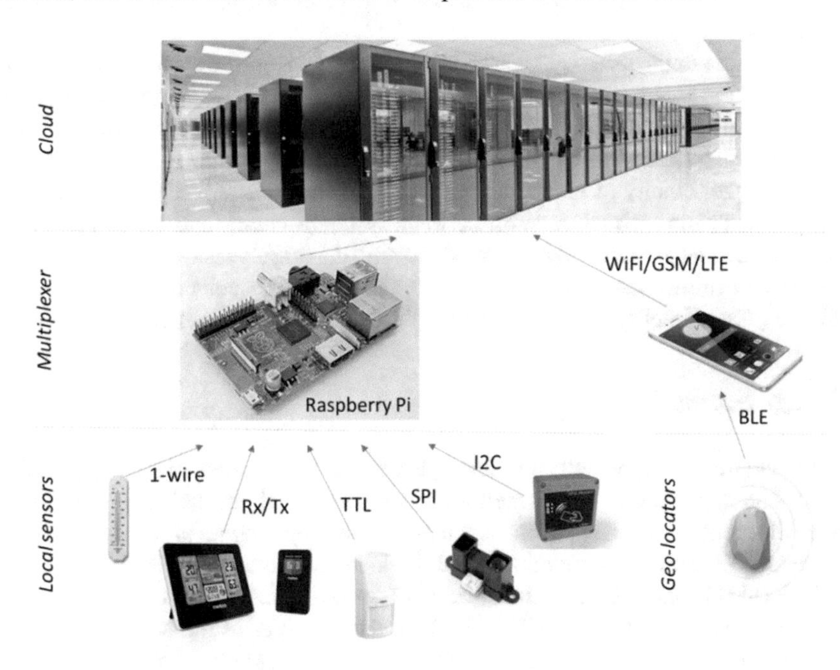

Fig. 1. Layered system architecture.

A virtual device is a specification of certain functionality, with well-defined both data and meta-data. As for the latter, the most important information covers geo-location of the sensor (fixed by the operator or dynamically read from the device), and sensor accuracy, formulated as a value 0..100% depicting the probability the sensor measurement is correct and up-to-date. The accuracy parameter may be used for the automatic choice of the sensors based on the location of the customer. For example, there may be a thermometer reporting the temperature close to a shelf – the accuracy is then represented as 100% maximum value. If, however, the store is equipped with only a single global thermometer, such centrally measured temperature is accompanied by the accuracy limited to 70%. Further, if the store has no own thermometer, but there is a global temperature sensor for the whole shopping mall, the information from such global sensor is reported with the accuracy at the level of 50%. Moreover, the accuracy may also be used to share some information among several locations. For example, it may be assumed that the temperature in each store in the same shopping mall is similar – then one spot may get the data from the neighbor, with the accuracy at the level of, e.g., 60%. Anyway, the information from the virtual sensor is always available, differing only in the accuracy and trust. This idea is applied to all the virtual sensors, shared by the stores if needed and always reported in a standardized way to the remote server.

We based the implementation of the local hub on OpenSensorHub project, and Raspberry Pi as the main host. Additional software is written in Java and C (depending on the microcontroller used), and image processing is based on OpenCV library and C++. Thus, the total cost of the local part of the system is reasonably small, while the cost of the server is shared by several stores, minimizing the end-user price for each store.

Note that, due to the usage of standardized virtual devices, the system functionality is fixed for all the stores. However, the quality of the marketing support (and the level of acceptance of this support by the customers) depends on the number and the accuracy of the set of local sensors. More investments in local hardware and software ensure better results and higher quality of the marketing. A final choice is reserved for the owner of a store, to choose the optimal trade-off between system accuracy and costs. Virtualization hides the details for the local hardware/software, making it possible to use the system with minimum initial costs, such as buying a set of BLE beacons or RFID tags, and finishing on, e.g., "intelligent" surveillance by means of Aurora cameras.

3 Usage Scenario

A user is entering a store with a mobile phone and a dedicated loyalty application installed. The application is awakened after approaching given shelf and detecting a beacon. The identifier of the beacon, as fetched by the application, is sent to the cloud. There, detailed location is determined, and the local multiplexer is asked for the data from the local sensors. Based on the data, local context is fixed. The context, together with the personal requirements sent along with the identifier of the beacon is used to compute the advertisement message. Finally, the message is sent, in almost real-time, back to the mobile phone to be displayed/played to the customer.

The above-mentioned basic schema may be adjusted to the current situation of a customer and local marketing strategy. For example, the server may decide to keep silent until a special-case situation is detected, such as a customer is standing near the same shelf (or regarding the same item, taken from the shelf) for more than a minute. Then, either someone from the personnel of the store is contacted to help, or a discount offer is sent in a form "if you buy this item within 5 min, you will obtain 5% discount of its price". Note that even the latter choice, completely automated, is to be accepted by the customer as an individual help, and as such is very welcome by most of the customers.

Regarding local context, the following additional information may be determined to better adjust the marketing decision:

- customer movement (fast-moving customers are probably not a target for instant discounts as depicted above) – to be detected by simple movement sensors (PIR and/or microwave sensors),
- staying in front of a shelf for a longer time (detected by means of beacons or proximity sensors – again quite cheap hardware),
- staying in a group, e.g., a mother with a child (information deducted after an analysis of camera image),
- differentiating in height (an adult or a child – again, by means of camera-image analysis),
- taking and regarding given item (detected by means of RFID tags and shelf-located RFID reader),
- talking – voice detection (if someone is speaking near a shelf, probably talks about the near items – maybe an instant help is needed?), etc.

As it may be seen, the above-mentioned system functionality may be achieved at minimum cost (as for the hardware) and quite quickly (as for the software). Note also that we extensively use BYOD idea – mobile phones are used not only for direct interaction with a customer but also as localizers and mini-databases storing private data.

The detailed analysis of the "intelligence" of the marketing server is out-of-the-scope of this paper. This is a very complicated problem. Also, the choice for the "intelligence engine" (e.g., neural networks, extensions to expert systems, rule-based choice, etc.) is an open question. As for the moment, a rule-based system is applied with fixed rules (the simplified solution). Due to the lack of space, we also did not mention here the details for the protocol among the local sensor multiplexers and the cloud, taking into account some meta-data such as geo-locations and accuracy of the sensors.

4 Comparison with Similar Work

There are several proposals for managing sensor networks created for different purposes and application areas, from home automation to SCADA systems. Usually they are focused on providing a convenient user interface for people controlling those networks, to efficiently manage the set of devices. One of such platforms is OpenSensorHub (OSH) [6], which main task is to provide distributed sensor hubs supporting a various range of sensors by means of unified interfaces. OSH incorporates Open Geospatial Consortium

(OGC) [7] Sensor Web Enablement (SWE) [8] standard as well as OGC SensorThings API. It is worth mentioning that OSH takes particular attention on geo-localization of sensors as well as easy integration with external geospatial data. One of the implemented standards is Sensor Observation Service [9], which enables to group a subset of available sensors and provides unified access to their data. There is no direct equivalent for a virtual sensor in OSH, but there are no restrictions to use one sensor repeatedly for different groups. The description of the sensor is based on SWE ontology, which has a sensitivity property telling about the resolution of the measurement, but it lacks accuracy defined as proposed earlier in this paper. As it may be seen, OSH (and similar proposals) are mainly focused on physical sensors, rather than their functionality and possible usage. On the contrary, in our proposal, even if we applied OSH as the implementation base, we tried to hide as many technical and organizational details for the sensors as possible, due to necessary virtualization of the devices. As such, we observe a set of functions rather a set of sensors. Due to the virtualization, in our case it is also much easier to create and use a non-standard sensor, such as a camera counting customers in a store, a movement detector that is able to measure and distinguish the speed of certain move, a sensor capable of selecting a mother and accompanying child, etc.

Regarding the camera-image analysis, there are some commercial systems for analyzing data from sensors located in the shopping area, e.g., RetailNext [10]. This system, based on dedicated cameras and other complex surveillance devices (such as Aurora sensor [11]) and online image analysis, opens the possibility to track people at a certain location, to count the customers, to identify the points-of-interest (a point people frequently stop and stay for some time), etc. However, the data collected from RetailNext system are mainly used for statistical purposes, e.g., to better arrange the store, to determine the place for an advertisement, to optimize the layout for the shelves, etc. There are no proposals to use camera-surveillance analysis at (almost) real-time for JIT marketing [12] and discount offers. It is also hard to imagine that the at-the-place generated information is shared by several stores to be analyzed as a whole and to apply a common marketing strategy – surveillance systems are by default standalone systems, for obvious security reasons. Thus, there is limited number of tools for RetailNext for data sharing, which is a crucial feature of our proposal. Also, sensor sharing and possible replacement are not possible. RetailNext shows also no possibility for sensor virtualization, instead, quite fixed system functionality is observed, hardcoded in the camera hardware. Moreover, there is a limited support for BYOD and the personal use of mobile devices by the customers, which is crucial for JIT marketing and instant discounts.

5 Conclusions and Future Work

In this paper, we propose to apply a set of Internet of Things devices and services at a real store to automatically monitor customer activity and to generate JIT marketing at (almost) real time. We analyze several IoT devices towards obtaining a functionality observed for e-shopping and used for instant marketing and customer-behavior analysis. To facilitate the process of analysis of the information collected by the IoT devices, we propose to use so-called virtual devices and a local hub capable of virtualization of real, both at-the-place and remote sensors. The set of virtual devices composes an ontology, which, together with a protocol for the information exchange, makes it possible to export the information locally gathered in a store to the cloud, to compute the marketing recommendations.

In the proposal we concentrated on the aspect of sharing the information from the local, distributed sets of sensors of different nature and purpose, and using them by a cloud in the standardized and centralized way. To this goal, we proposed to apply so-called virtualization of the sensors and a shared ontology of virtual sensors. The main goal for the virtualization is twofold. First, a virtual sensor normalizes an access to the information – as such, this information may be processed in the cloud in the same way regardless the place of origin, time, hardware and software used, etc. Second, the information from a non-existing or currently unavailable sensor may be easily replaced by similar information from a different place. To this goal, we applied the accuracy feature – this parameter is used to determine the trust for the current sensor value.

We clearly see that this is the beginning of the proposal to be really used in a real store. We are now working on determining a set of devices to be applied in practice for a store of given type, and on defining preliminary rules for instant marketing and discounts, to be further unfolded and adjusted to the specificity of a store/situation.

References

1. Nguyen, B., Simkin, L.: The Internet of Things (IoT) and marketing: the state of play, future trends and the implications for marketing. J. Mark. Manag. **33**(1–2), 1–6 (2017)
2. Mullin G.: Amazon launches supermarket with NO checkouts and uses cameras to track what shoppers remove from shelves (2018). https://www.thesun.co.uk/money/5394671/amazon-supermarket-no-checkouts-seattle-camera-technology/
3. Yewatkar, A., Inamdar, F., Singh, R., Bandal, A.: Smart Cart with Automatic Billing, Product Information, Product Recommendation Using RFID & Zigbee with Anti-Theft (2016). https://www.sciencedirect.com/science/article/pii/S1877050916002386
4. An Introduction to Recommender Systems (2018). https://www.springer.com/cda/content/document/cda_downloaddocument/9783319296579-c2.pdf
5. Intelligent Shelf Label Solution: Blueprint (2012). https://www.intel.pl/content/dam/www/public/us/en/documents/solution-briefs/intel-blueprint-intelligent-shelf-label-final-r.pdf
6. OpenSensorHub. Open Standards, open source platform to build sensor networks (2018). https://opensensorhub.org/
7. Open Geospatial Consortium. An international not for profit organization making quality open standards for geospatial community (2018). http://www.opengeospatial.org/
8. Botts, M., Percivall, G., Reed, C., Davidson, J.: OGC® sensor web enablement: overview and high level architecture. In: Nittel, S., Labrinidis, A., Stefanidis, A. (eds) GeoSensor Networks. Lecture Notes in Computer Science, vol. 4540. Springer, Heidelberg (2008)
9. Sensor Observation Service standard (2018). http://www.opengeospatial.org/standards/sos
10. RetailNext. The most advanced in-store analytics solution available, https://retailnext.net/en/how-it-works/ (2018)
11. Aurora – the next-generation sensor for shopper measurement (2018). https://retailnext.net/en/aurora/
12. Just In Time – JIT (2018). https://www.investopedia.com/terms/j/jit.asp

Author Index

L. Borzemski et al. (Eds.): ISAT 2018, AISC 852, pp. 377–378, 2019.
https://doi.org/10.1007/978-3-319-99981-4

Printed in the United States
By Bookmasters